Dear M[illegible]
Thank you for [illegible] birthing within me a love of medieval art & culture. I never could have guessed where this journey would take me, but this book is stop #1.
Thank you!
Eric

A Saving Science

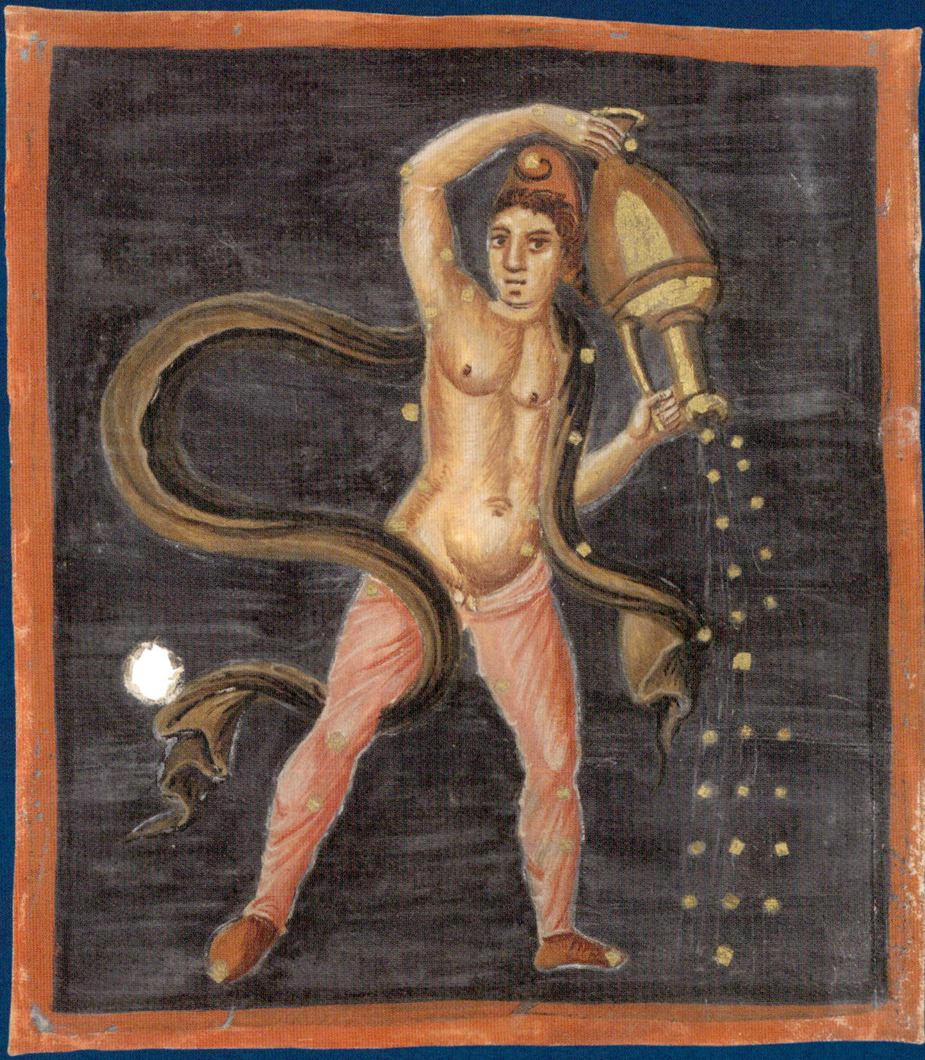

THE PENNSYLVANIA STATE UNIVERSITY PRESS
UNIVERSITY PARK, PENNSYLVANIA

A SAVING SCIENCE

Capturing the Heavens in Carolingian Manuscripts

ERIC M. RAMÍREZ-WEAVER

MM

Publication of this book has been aided by a grant from the Millard Meiss Publication Fund of the College Art Association.

Publication of this book has been supported by a Samuel H. Kress Foundation Research Grant administered by the International Center of Medieval Art.

Library of Congress Cataloging-in-Publication Data

Names: Ramírez-Weaver, Eric M., 1971– , author.
Title: A saving science : capturing the heavens in Carolingian manuscripts / Eric M. Ramírez-Weaver.
Description: University Park, Pennsylvania : The Pennsylvania State University Press, [2017] | Includes bibliographical references and index.
Summary: "Focusing on the Handbook of 809, explores how the liberal arts, and in particular astronomy, experienced a revival in the ninth-century court of Charlemagne. Documents the utility of the constellations for prelates who needed to fix the floating feast of Easter and reckon time"—Provided by publisher.
Identifiers: LCCN 2016009956 | ISBN 9780271071268 (cloth : alk. paper)
Subjects: LCSH: Biblioteca Nacional (Spain). Manuscript. 3307—Illustrations. | Illumination of books and manuscripts, Medieval—France—Metz. | Illumination of books and manuscripts, Carolingian. | Astronomy, Medieval. | Astronomy in art. | Biblioteca Nacional (Spain).
Classification: LCC ND3399.L53 R35 2016 | DDC 745.6/709443853—dc23
LC record available at https://lccn.loc.gov/2016009956

Printed in China
Published by The Pennsylvania State University Press, University Park, PA 16802–1003

The Pennsylvania State University Press is a member of the Association of American University Presses.

It is the policy of The Pennsylvania State University Press to use acid-free paper. Publications on uncoated stock satisfy the minimum requirements of American National Standard for Information Sciences—Permanence of Paper for Printed Library Material, ANSI Z39.48–1992.

Additional credits: page ii, detail of Aquarius, Leiden *Aratea*, folio 48v (fig. 50); page vi, "The Apsides of the Planetary Orbits Within the Zodiac," Madrid 3307, folio 65v (fig. 5); page 22, detail of Ursa Major, Ursa Minor, and Draco, Leiden *Aratea*, folio 3v (fig. 42); page 142, Hercules, Leiden *Aratea*, folio 6v (fig. 51).

MIX
Paper from responsible sources
FSC® C013314

FOR

OLIVIA

Make the art in your heart.

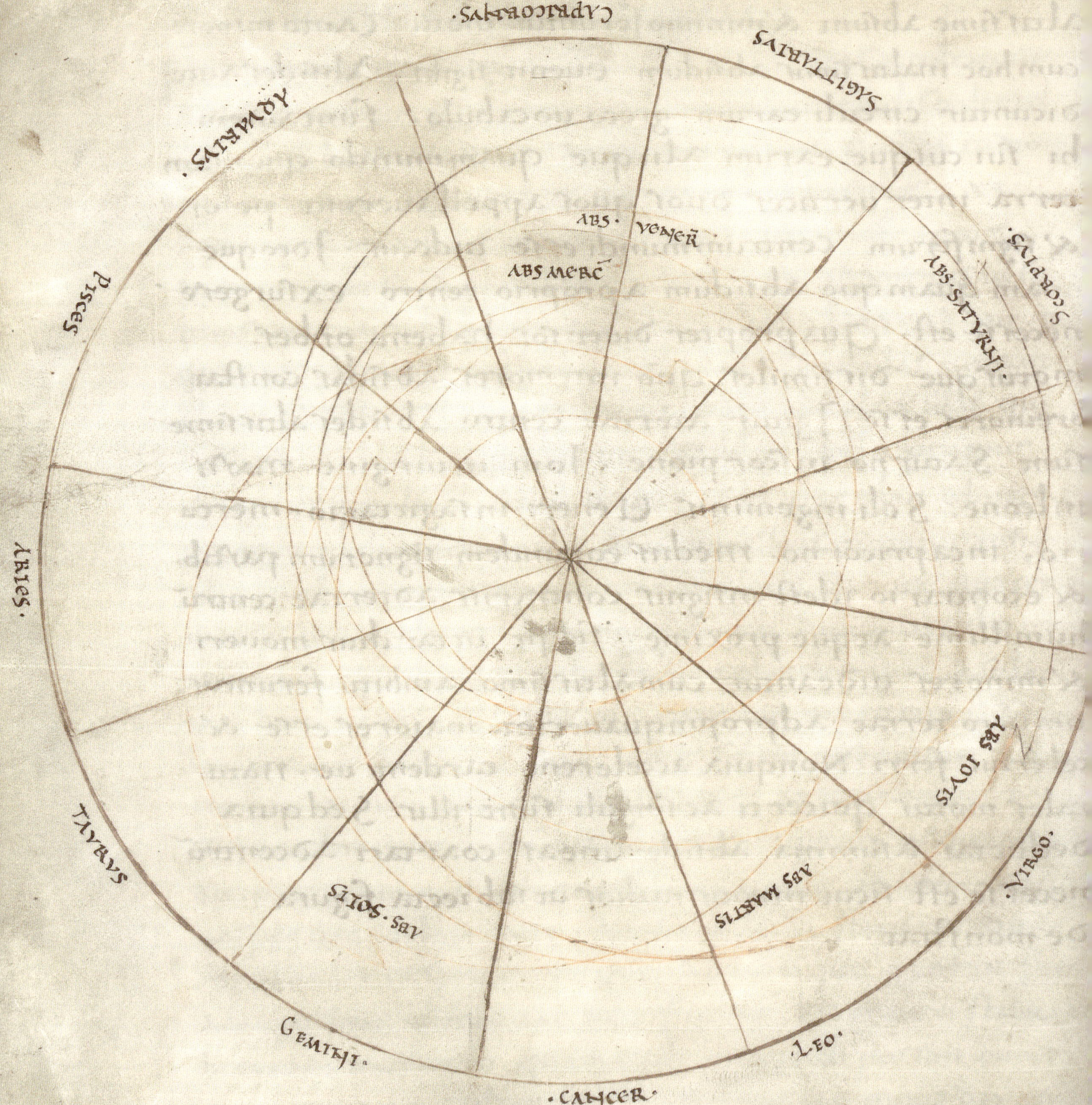

VI DE CURSU EARUM PER ZODIACŪ CIRCUL

Cur autem magnitudines suas & colores mutent & eaeder

CONTENTS

ILLUSTRATIONS

PLATES

FIGURES

ACKNOWLEDGMENTS

This book has benefited from the benevolence of multiple institutions and individuals. Generous financial support has been supplied by the following institutions: the Millard Meiss Publication Fund of the College Art Association; a Samuel H. Kress Foundation Research Grant administered by the International Center of Medieval Art; and the University of Virginia, through the McIntire Department of Art and the Carl H. and Martha S. Lindner Center for Art History, multiple summer stipend awards, and two grants for support of research in the arts, humanities, and social sciences. Composition and completion of this project also benefited from research funding supplied by a Sesquicentennial Associateship at the University of Virginia and one of the Mellon Fellowships for Assistant Professors within the School of Historical Studies at the Institute for Advanced Study, Princeton.

This book would not have been possible without the generosity of many libraries and cultural institutions worldwide that conferred gracious opportunities to work with their rich treasures or resources through diverse combinations of funding, access, and the inestimable guidance supplied by their personnel. In particular, I wish to thank the Bayerische Staatsbibliothek in Munich, the Biblioteca Apostolica Vaticana of Vatican City, the Biblioteca Capitolare of Monza and the Fondazione Gaiani, the Biblioteca Nacional de España in Madrid, the Bibliothèque nationale de France in Paris, the Erzbischöfliche Diözesan- und Dombibliothek of Cologne, the Institute of Fine Arts of New York University, the Metropolitan Museum of Art in New York, the Morgan Library and Museum in New York, the Österreichische Nationalbibliothek in Vienna, the Römisch-Germanisches Zentralmuseum in Mainz, the Staatsbibliothek zu Berlin of the Preußischer Kulturbesitz, the Stiftsbibliothek Sankt Gallen, the Universitätsbibliothek in Basel, and the Zentralinstitut für Kunstgeschichte in Munich.

Too many extraordinary individuals deserve mention for them all to be named. And yet the indefatigable model of scholarly excellence supplied by my dear *Doktorvater*, Jonathan J. G. Alexander, is ever before me as I daily set to work with renewed vigor. Lucy Sandler (also from NYU) stands alongside Meredith Lillich of Syracuse University in my mind as matriarchs of medieval art history who have played a formative role, shaping not only my research results but also the mental habits that led to them. This book is significantly indebted to the research day spent exploring manuscripts from Metz at the Bibliothèque nationale de France. Several of the manuscripts treated in this book are tantamount to national treasures, and I am humbled by the opportunities I had to research such books on site in Madrid, Monza, Munich, the Vatican, and Vienna. In this context, I reminisce with forlorn fondness, as well as the utmost gratitude, for the *mündliche Prüfung* and aid I received over the years from Florentine Mütherich, so recently departed. The original book manuscript also benefited greatly from the suggestions of its anonymous reviewers and the editorial acumen of Eleanor Goodman and Victoria Scott.

All of my colleagues have been sources of encouragement and guidance, honing my ideas through thoughtful conversation, conference interaction, and email exchange. In particular, special thanks are due Ahmed al-Rahim, Paul Barolsky, Gregory Hays, Danielle Joyner, Paul Kershaw, Herbert Kessler, Aden Kumler, Christopher Lakey, Susan L'Engle, Richard Leson, Lawrence Nees, Karl Shuve, Tyler Jo Smith, Joshua White, Dorothy Wong, and my friends at the Brown Residential College. The ongoing and faithful support of each respective chairperson under whom I have served within the Department of Art at the University of Virginia has been a mainstay: Francesca Fiorani, Lawrence Goedde, and Howard Singerman. Superior students have also challenged my views and helped me perfect my arguments. In particular, I wish to thank current and former students: Adrienne Albright, Tracy Cosgriff, Jennifer Grayburn, Catherine Hundley, Taylor McCall, and Victoria Valdes.

In closing, my wife, Elizabeth, remains the solar source of every joy, radiating steadfast kindness in the children's hospital and at home. With her came Olivia, to whom this book is dedicated. Keep dancing, my dear, upon the moonbeams that summon you, and may your beautiful brown eyes forever be filled with the light of heaven.

INTRODUCTION

Drogo and the *Handbook of 809*

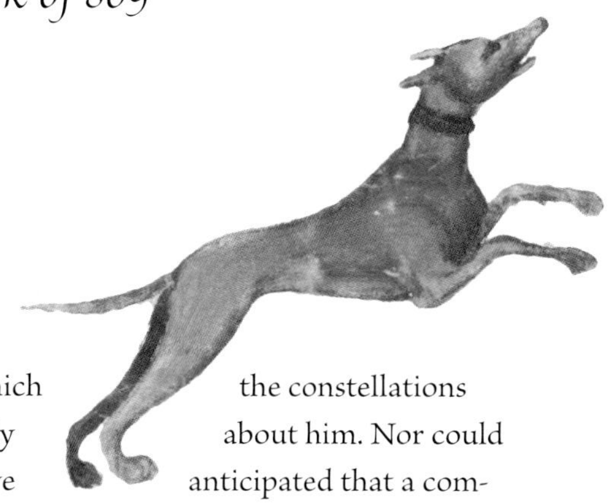

In the year 809, a seven- or eight-year-old illegitimate son of Charlemagne by his concubine Regina, a boy dubbed Drogo, was naturally a physical reminder of his father's royal body. It was believed at the time that Drogo—like everyone else—represented in human form a harmonious reflection of the cosmos (fig. 1). This early medieval conceit, commonly identified as the microcosmic-macrocosmic symmetry, is perfectly portrayed within the interlocking arcs and concentric circles of the fifth wheel diagram, or *rota,* in Isidore of Seville's highly influential *De natura rerum* (On the nature of things; also known as the *Liber rotarum*) of 612–13.[1] The ordered, intelligible, self-evident structure exemplified by such a diagram conveyed to a reader the ways in which the individual person, Creation, and the cosmos shared in an interconnected ongoing revelation of being in time. There is no way of knowing what young Drogo saw or wanted to see as he looked upward and outward from Aachen toward the celestial sphere of the fixed stars on which the constellations turned daily about him. Nor could anyone have anticipated that a comprehensive astronomical-computistical compilation in seven books, sometimes referred to as an encyclopedia but better labeled the *Handbook of 809,* would result from Charlemagne's synod convened that year.[2]

The synod of 809 was a royal event, involving some of the most important Frankish prelates who were adept at *computus.* They convened at Aachen, with Adalhard of Corbie (d. 826), Charlemagne's cousin, playing a vital role. Apparently, a smaller cadre of computistically skilled churchmen profited from wintering in Aachen and withdrew from their theological discussions of the *filioque* clause (the Frankish Christian creed's endorsement of the Holy Ghost's grace conferred by Father *and* Son) to probe the state of Frankish computus. They offered a record of their proceedings, referred to by the late Arno Borst as *Das Aachener Verhör von 809.*[3] Under the probable direction of Adalhard, they took stock of Carolingian computistics, the serious study of the calendar with astronomical implications and mathematically verifiable results.

As a child, Drogo (d. 855) would have been separated from all the synodal proceedings, but he would have been reared in the intellectually rich fervor that followed the interrogation of such questions as when Jesus was crucified (reckoned to be 25 March), how old Christ was at the time of his Passion (apparently 33 and a half), and when the vernal equinox transpired (assigned at the synod as 21 March). And most important, of course, this group at the Aachen synod of 809 convened to review practical calendrical information required to reckon suitable dates for the floating feast of Easter.[4] While their confreres at Aachen were debating dogma and defining doctrinal boundaries against Byzantine orthodoxy, these learned men were determining how well the Franks could put their contested faith into practice.

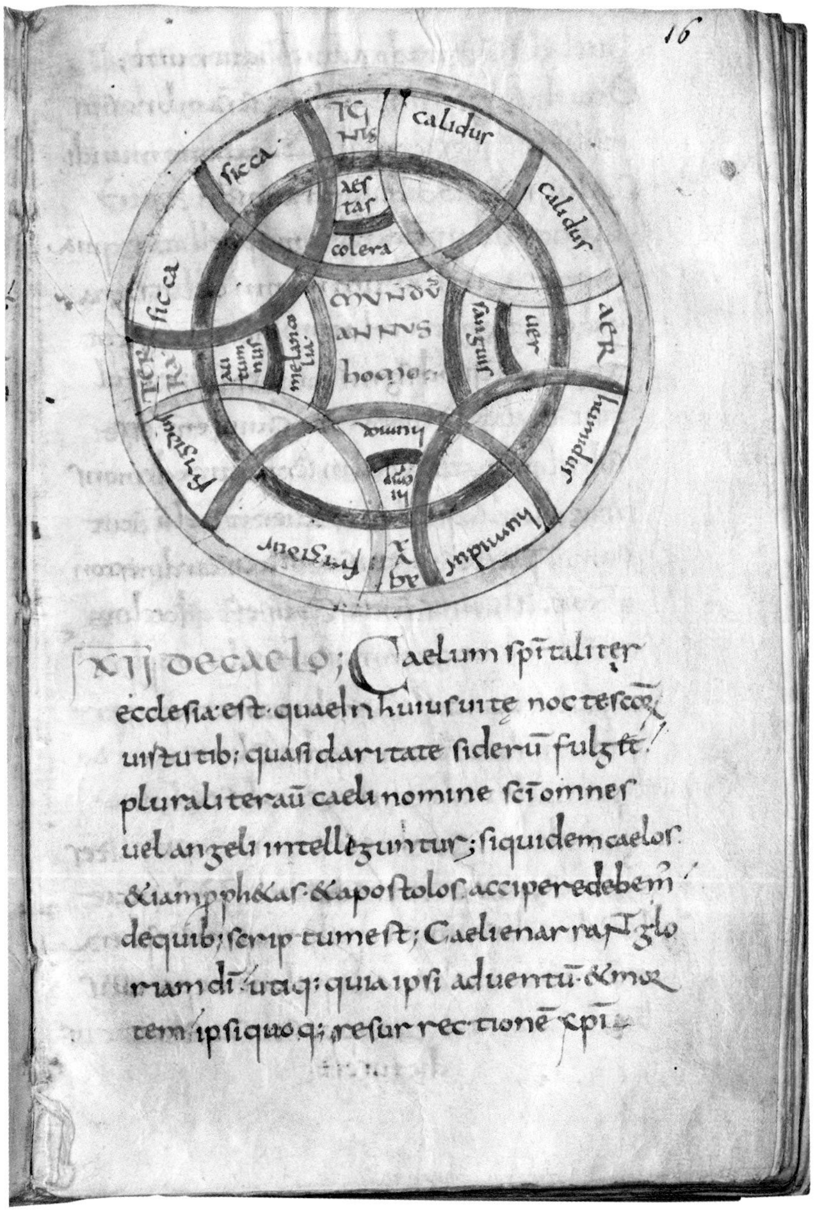

FIGURE 1

Mundus-annus-homo diagram. Isidore of Seville, *De natura rerum*, Munich, Bayerische Staatsbibliothek, Clm 16128, fol. 16r. Photo: Bayerische Staatsbibliothek München.

Creating a Carolingian View of Astronomy

In fact, this complete interpenetration of spiritually sanctioned liturgical practice since Nicaea, Frankish liturgical and calendar reform, and pragmatic applications of the calendar for successful governance required that the early medieval Franks turn to the liberal art of astronomy in their study of computistics.[5] As Brigitte Englisch has so aptly stated, "Every year thus appears like a mirror of heavenly activity, through the high feasts systematically structured by means of the holy feast days of the saints in accord with historical elements of time."[6] But that does not reduce the study of the constellations or fixed stars exclusively to computistical or liturgical concerns.

On the contrary, the sphere of the fixed stars was an object of inquiry in its own right, and a Roman Latin literature derived from Greek precedents—such as the *Phaenomena* of Aratus (third century B.C.E.) and the portions of Plato's *Timaeus* rendered accessible to Latin readers through Calcidius's fourth-century translation with commentary—were available to sustain meaningful ninth-century pedagogical dialogue.[7]

Fundamental to this Latin discourse is a set of passages from books II and XVIII of Pliny's first-century *Naturalis historia* (*Natural History*). These excerpts were originally believed to be Anglo-Saxon in origin, yet Vernon King and Bruce Eastwood have found that the astronomical passages instead testify to a creative Carolingian engagement with the literature of antiquity, reinforcing the Christian celebration of the liturgy in the present and laying the foundation for new exegetical interpretative maneuvers in Frankish astronomical treatises in both text and illustration. These excerpts, routinely copied as a set, became a composite treated as a textual unity and found their way into the *Handbook of 809*, accompanied by a set of Carolingian diagrams created about 809–12, after the synod in Aachen.[8]

Individual study of the liberal art of astronomy through both word and image was fundamental to the spiritual reform program of a young ecclesiast such as Drogo himself, who eventually acceded to the see of Metz in 823. In what follows, it will become clear that early medieval astronomy and computistics were a liberal art and an applied sister science. In tandem, they enabled the supervision of the Frankish kingdom and the proper conduct of an early medieval Christian church. The ecclesiastical hierarchy was linked to Rome, but for the *religio christiana* throughout the Carolingian empire, the work of the church was also synecdochically associated with the royal chapel, and eventually with Odo of Metz's Cappella Palatina. Drogo must have attended Mass in many of his father's chapels, and he would have listened to the prescribed readings from the Godescalc Evangelistary (fig. 2) as a boy.[9] It was certainly in Aachen, however, that Drogo and his father heard their royal Frankish *scola cantorum* melodiously give voice to the indigenous doctrinal preference for inclusion of the *filioque* clause in the creed that had drawn the clerical elite to the synod of 809 in the first place.[10] Two important debates intertwined with the *quadrivium* (i.e., the advanced medieval liberal arts curriculum of arithmetic, geometry, music, and astronomy) were advanced at Aachen in 809, then: first, Carolingian liturgical music defiantly adapted the Niceno-Constantinopolitan Creed, and second, computistical misunderstanding gave rise to astronomical clarification through word and image.[11]

A thorough cultural history of the Carolingian appropriation of the classical legacy for the study of computistics and astronomy cannot ignore the significance of star pictures in the development of early medieval intellectual traditions. Paul Dutton has argued that in one literary exchange (letter 121, ca. 797), Abbot Alcuin of Tours (d. 804) describes to Charlemagne (d. 814) what Dutton labels a "pedagogy of the stars."[12] Alcuin united an inquiry into the relative placement of the constellations on a celestial globe or on the concave surface of a domical interior, "as if painted on high in some great man's house," with study of the liberal art of grammar and a celebration of the "aged wine of ancient studies."[13] Indeed, the liberal art of astronomy was a joyous curiosity, accomplishing a sought-after spiritual transformation that renewed the intellect and restored the mind with the assistance of scientifically apposite diagrams, schematics, and Frankish images of the

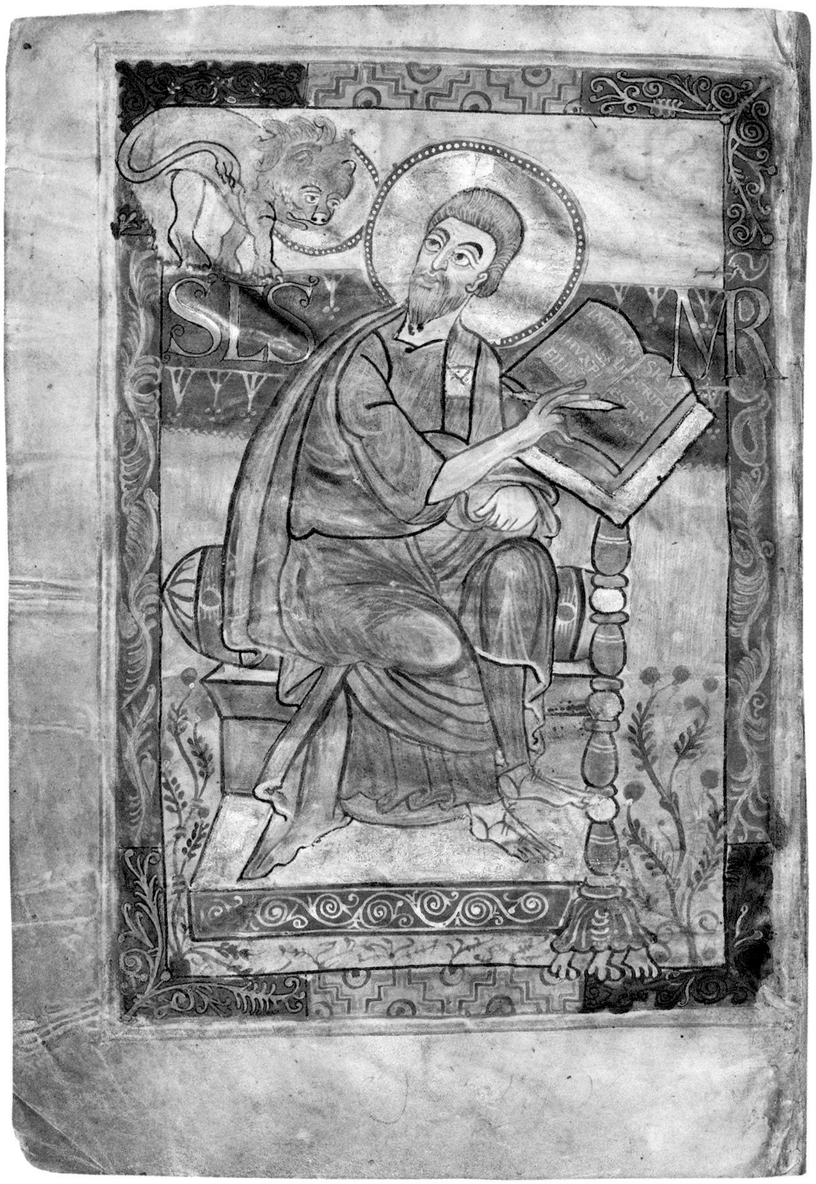

FIGURE 2

Gospel-author portrait of Mark. Godescalc Evangelistary, Paris, Bibliothèque nationale de France, Nouv. acq. lat. 1203, fol. 1v. Photo: BnF.

constellations that benefited from semiotic strategies of *exegetical emendation*.[14]

Arno Borst's texts remain widely influential resources to consult on Carolingian calendrical study, even if his arguments meet with resistance in the details. On the issue of the constellation imagery in the *Handbook of 809*, he had this fairly standard disclaimer: "Foremost, the imagery on the rotating heavenly sphere of the fixed stars and their interdependence according to the ancient myths about the stars were described, with colorful pictures that are to a greater extent evidence of pictorial fantasy than astronomical precision."[15] This position reflects a common assessment of the significance and import of the images of the constellations, which with rare exception are considered to populate the ancillary portions of standard copies of the *Handbook of 809* in book V (of seven books total). For this reason, interest in the *Handbook of 809* typically divides between historians of science in the majority (who study the computistical encyclopedia contained in books I through IV plus the diagrams accompanying book V, sections 3 through 6),[16] and art historians or students of mnemotechnics (who have tended to examine the classical iconographic properties or purely astronomical aspects of the star pictures in book V, section 2, and their pedagogical utility for Frankish audiences).[17]

Although Drogo's copy of the *Handbook of 809* has long been heralded as an example of Carolingian astronomical illustration, alongside the Leiden *Aratea* of ca. 816 (Leiden, Bibliotheek der Rijksuniversiteit, MS Voss. lat. Q79; hereafter Leiden *Aratea*),[18] Drogo's book has never benefited from a rigorous art-historical appraisal of its contents or its significance for the development of Carolingian painting. The present volume thus undertakes both tasks.[19] In so doing, it will become evident that outmoded art-historical appraisals of early medieval manuscript illumination, driven by traditional assignments of certain illustrated books to particular schools, must shift in favor of a view that emphasizes the professional skills and mobility of individual artists.

Borst determined that the copy of the *Handbook of 809* made for Drogo ca. 830 (Madrid, Biblioteca Nacional, MS 3307; hereafter Madrid 3307)—though now missing portions of its original seven books[20]—once supplied to early medieval readers the definitive record of the intentions of Adalhard and his team of compilers who created this computistical and astronomical compendium.[21] An overemphasis on either the computistical components or the astronomical aspects of the *Handbook of 809* permits a reductionist focus on computus or on the illustrations in book V. The very nature of the medieval compilation permits such selective reading, since any *compilatio* was created from diverse parts that could but did not have to be read in meaningful semiotic dialogue with one another. Whether the hermeneutic significance of these juxtapositions among texts in a compilation was intentional or was only discerned after the fact by modern researchers is inconsequential. Anyone investigating the kinds of medieval books available in libraries for consultation during the eighth and ninth centuries needs to examine the treatises and texts that were collected and compiled as unified sets, such as those contained within the *Handbook of 809* itself.[22]

This implicit dialogue among the various sections of an anthology, or set of books bound together, could derive from a desire for knowledge by accretion, from agreement between texts, or even from deliberate polemic within an anthology. In keeping with the general strategy of accuracy through accretion, the complete text of the *De natura rerum* composed by the Venerable Bede (d. 735) was appended to the end of the *Handbook of 809* as book VII. Borst considers this a meaningful maneuver, one by which an orthodox Christian cosmology could counteract any mythological or pagan sources that had found their way into the preceding texts. In other words, such problematic content would be polemically juxtaposed with a more standard view derived from Bede and thereby neutralized as Anglo-Saxon discussions of time and cosmology laid a Christian groundwork for computus and normative Frankish praxis.[23]

The culled, composed, and assembled texts (computistical or otherwise), together with the imagery in Drogo's copy of the *Handbook of 809*, provide multiple primary sources documenting a particularly Frankish view of the heavens. This vision of the stars is built on the earlier teaching of Isidore of Seville (d. 636), who proclaimed at the end of book III of his *Etymologiarum* (hereafter *Etymologies*) that all study of the stars and constellations achieves the early medieval goal of sacred study of the liberal arts, namely, to "draw minds tangled in secular wisdom away from earthly matters and set them in contemplation of what is above."[24]

The present volume endeavors to be the first serious appraisal of Drogo's enigmatic copy of the *Handbook of 809*. Toward this end, his handbook is treated as the primary record of the collaborative participation of multiple creative Carolingian communities—scribes who penned Drogo's copy of the handbook, four painters in Metz who filled it with lavish miniatures of the constellations and with diagrams, and intellectual teams under prelates such as Adalhard who formed a specific Christian Carolingian perspective on the liberal art of astronomy.

Dynastic kinship ties gave rise to official copies for Carolingian princes in clerical service, such as Drogo, who served as bishop of Metz from 823 to 855. These Carolingian ecclesiasts used their books to advance pedagogy and to foster an intellectual brand of spiritual renewal through the study of astronomy, relying on the texts and miniatures in their handbooks. Ultimately, this is the story of Drogo and a cultural history of his copy of the *Handbook of 809*. The present volume addresses the *scientia* that was saved in the handbook. It also reveals the *scientific soteriology* that filled Drogo's handbook with strategies for reckoning the liturgical calendar through computus and also for reforming the minds of individual readers.

The star pictures are more than elaborate decorative schemes intended to heighten the cultural impact of the calendrical portions of an encyclopedic compilation such as the *Handbook of 809*, or to reconstitute a vision of antiquity before Frankish eyes. Rather, the images of the constellations in discrete copies of the *Handbook of 809*, such as Drogo's personal book (Madrid 3307), are vital records of Frankish engagement with the heavens and of individual interventions of an overtly scientific, philological, and explanatory nature, as when Carolingian painters in places such as Metz made creative formal and iconographic decisions about the earthly appearances that their illustrations of the celestial bodies should take. The actual star pictures in Madrid 3307 (fols. 54v–62v; plates 1–17) filled book V.2, the *De ordine ac positione stellarum in signis* (On the order and placement of the stars in the constellations; hereafter *De ordine*), with an ordered manifestation of classical astronomy specifically prepared for a Christian audience.

On each parchment folio, one of the four painters working on Drogo's behalf in Metz made selections from or adapted classical precursors and image-types remaining from antiquity, which often harked back to the forms in illustrated books from Corbie[25] or to images on astronomical devices such as celestial globes, now lost due to the perishable materials often employed in their manufacture.[26] The choices these illuminators made document a sophisticated process of looking at the extant samples remaining from the past, which were relics of classical astronomical science, and rereading their significance for eighth- and ninth-century Frankish Christian communities. The present volume emphasizes astronomical manuscript illumination precisely because the painted folios of astronomical and cosmological books supply the best record of the formerly diverse varieties of celestial forms. In particular, chapter 2 offers a complete historiographic review of Aratean recensions and the early medieval star pictures that illustrate them.

Shifting Signs and Exegetical Emendation

The innovative compilation of image cycles in Carolingian astronomical compendia was neither a rote process nor a practical exigency resulting from lacunae in disparate Frankish models. On the contrary, the careful examination of certain star pictures, such as Hercules (plate 2) and Chiron the Centaur (plate 16), and their corresponding texts in Drogo's personal copy of the *De ordine* from the *Handbook of 809* reveal ongoing intellectual engagement with the visual models for the star pictures that had survived the ravages of late antiquity, and that were then realized in the lavish but diverse cycles of early medieval illuminated astronomical manuscripts.

It is important to pause here to underscore a significant methodological point. Each created copy of a medieval manuscript is an opportunity for reconsideration, inflection, and exegesis, as Richard Corradini argues with respect to computistical treatises, including Drogo's copy of the *Handbook of 809*. In his essay "The Rhetoric of Crisis," Corradini emphasizes the significant ways in which annals, calendars, and computistical texts worked together to fashion a Christian Carolingian identity: "*computus*, as a framework for memorial thought, was more than a means of reckoning time; it was a reckoning *with* and *on* time, and formed the basis for experience-based history in both the past and the future."[27] Carolingian artists were participants in their society, and their mores, fears, doctrines, and aspirations all informed their iconographic decisions about the forms their star pictures could take in a projection of the individual onto even official records of period science. Such a strong endorsement of the creative capabilities of early medieval authors and artists has not always been supported by researchers investigating premodern or early modern star pictures. Eastwood's studies of Carolingian diagrams, created and then modified for readability to accompany the aforementioned Plinian excerpts included in book V of the *Handbook of 809*, are but one more important exception (figs. 3–6).[28]

Scholars have treated the computistical texts or the astronomical pictures in the *Handbook of 809* without suitable integration. Previous beneficial work has focused on the computistical portions of the compilation and its diagrams. The present volume instead assesses the classical traditions and early medieval Christian interventions that account for the astronomical illustrations in Drogo's copy of the handbook. An endorsement of Carolingian creativity has recently received additional support from a collective effort by Dieter Blume, Mechthild Haffner, and Wolfgang Metzger, who, in *Sternbilder des Mittelalters*, advocate for the kind of creative astronomical intervention during the Frankish period called for here: "But, one does not only augment the pictorial effect through illusionistic painting, rather at the same time one turns the picture into a bearer of scientific information, which the accompanying text does not contain."[29]

One important result of this approach is the need to support the intellectually refined observational and artistic practices employed by early medieval artists. This attentive aspect of Frankish *poiesis* informed the processes of selection that omitted the traditional Tree of the Hesperides (visible, for example, in figs. 7–8) from the image of Hercules, as explained in chapter 2, and rendered the Centaur as Chiron, the teacher of Achilles, as explained in chapter 4. These techniques for interpreting imagery through the artistic transformation of pictorial details modify the meaning of classical precedents. Through this process, termed here *exegetical emendation*, Frankish painters worked in conjunction with the scribes and compilers who assembled an anthology such as the *Handbook of 809*, creating through textual and visual strategies of meaning a lasting record of their astronomical science. New forms did not necessarily become hegemonic, precluding further reinterpretation and reinvigoration. In fact, some forms did persist and recur across copies of the *Handbook of 809*, indicating their widespread and celebrated acceptability for Carolingian audiences. But every new codex was an opportunity for the creative decision-making process to renew itself, and in each copy Carolingian painters determined, for their own reasons, which forms to emend.

Artistic expression must, however, be considered alongside other rhetorical and hermeneutic strategies of interpretation in a compilation such as the

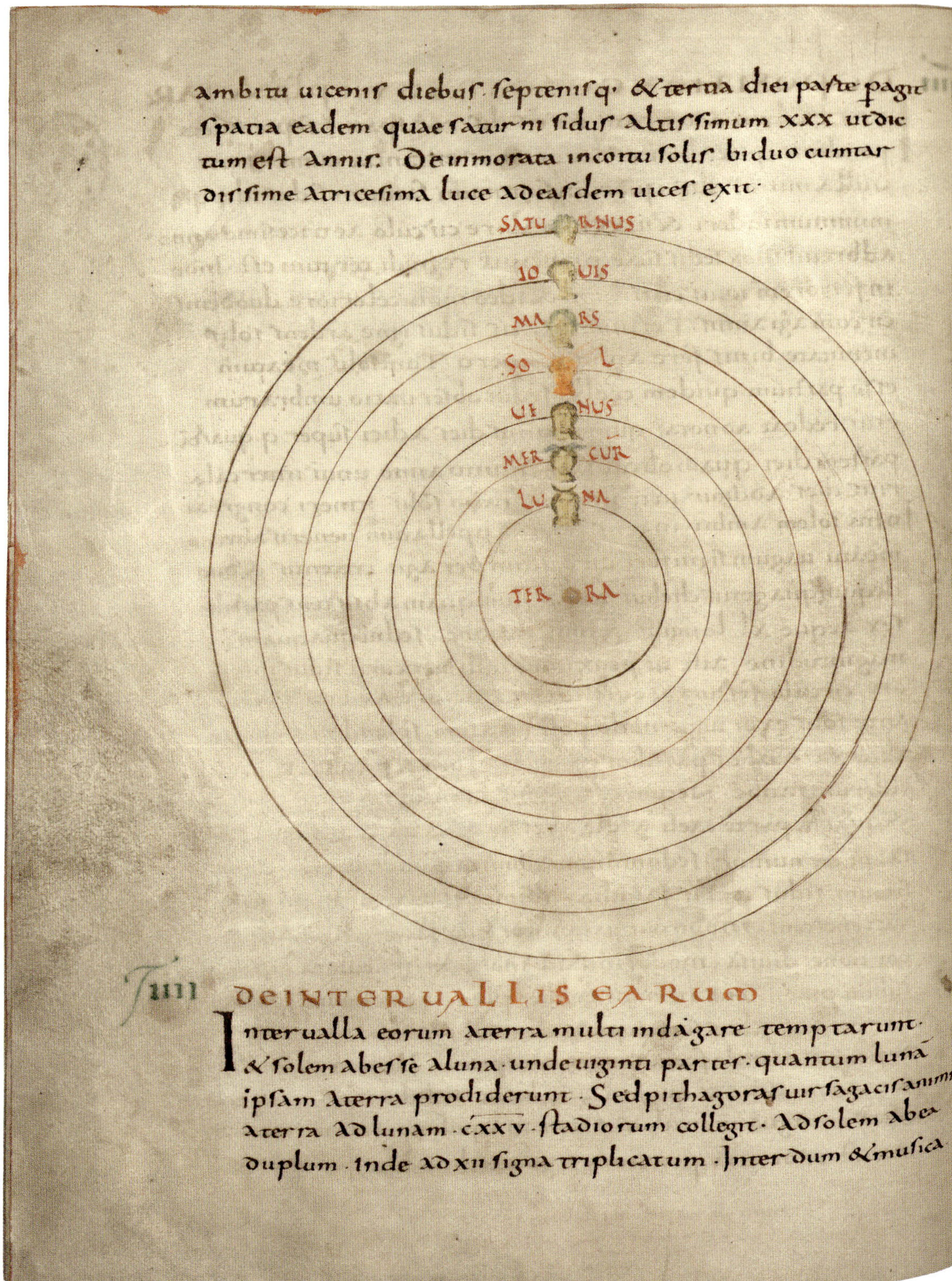
ambitu uicenis diebus septenisq· & terna diei parte pagit
spatia eadem quae saturni sidus altissimum XXX ut dic
tum est annis. De inmorata in coitu solis biduo cum tar
dissime a tricesima luce ad easdem uices exit·

IIII DE INTERUALLIS EARUM
Interualla eorum a terra multi indagare temptarunt·
& solem abesse a luna undeuiginti partes· quantum lunam
ipsam a terra prodiderunt· Sed pithagoras uir sagacis animi
a terra ad lunam ·CXXV· stadiorum collegit· Ad solem ab ea
duplum· inde ad XII signa triplicatum· Inter dum & musica

FIGURE 3
"The Order of the Planets" diagram. Madrid, Biblioteca Nacional de España, MS 3307, fol. 63v. Photo © Biblioteca Nacional de España.

Handbook of 809. Since the visual components of the compilation were built on the nonmathematical applications of computistics and the liberal art of astronomy, the present volume delves deeply into the philological and visual traditions that contributed to the creation of Drogo's copy of the handbook. It is this material alone that has not been thoroughly contextualized for cultural historians in the extant literature. A novel focus on the compilation from the standpoint of the art historian, looking outward from the astronomical pictorial program at the remaining contents of the anthology, sheds new light on the pedagogical utility and purpose of the *Handbook of 809* for various audiences within medieval studies.

To take full advantage of this perspectival inversion, certain limitations must be imposed. Rather than revisit the established literature pertaining to computus (by Arno Borst, Bruce Eastwood, Brigitte Englisch, Dáibhí Ó Cróinín, Kerstin Springsfeld, Faith Wallis, and Immo Warntjes, *inter alia*) for its own sake, only concepts relevant to the pictorial program in the *Handbook of 809* and its legacy are addressed. Similarly, this is not a book about the textual transmission of classical mythology to Renaissance audiences. Aspects of this important philological work have been undertaken already, with vital contributions by Jane Chance and Hubert Le Bourdellès, reforming and building upon aspects of the foundational work supplied by Ernst Maass and Jean Martin. The Aratean textual traditions are, however, compared critically in chapter 2.[30] The iconographic analysis of the Aratean image cycle in chapter 2 necessarily draws upon the work of Mechthild Haffner and Florentine Mütherich, paying particular attention to the relative degrees of continuity and creative intervention, or exegetical emendation, to be found throughout Drogo's pictorial program.[31]

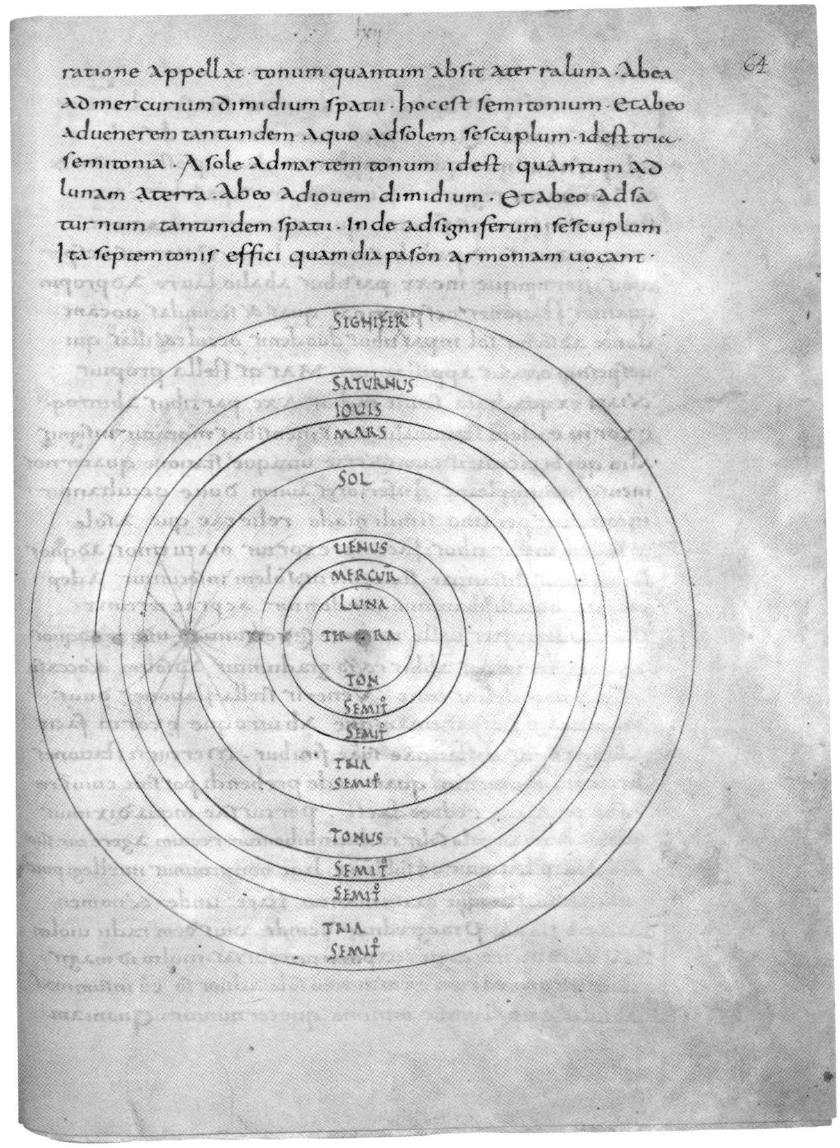

FIGURE 4

"The Pythagorean Theory of the Harmonic Intervals" diagram. Madrid, Biblioteca Nacional de España, MS 3307, fol. 64r. Photo © Biblioteca Nacional de España.

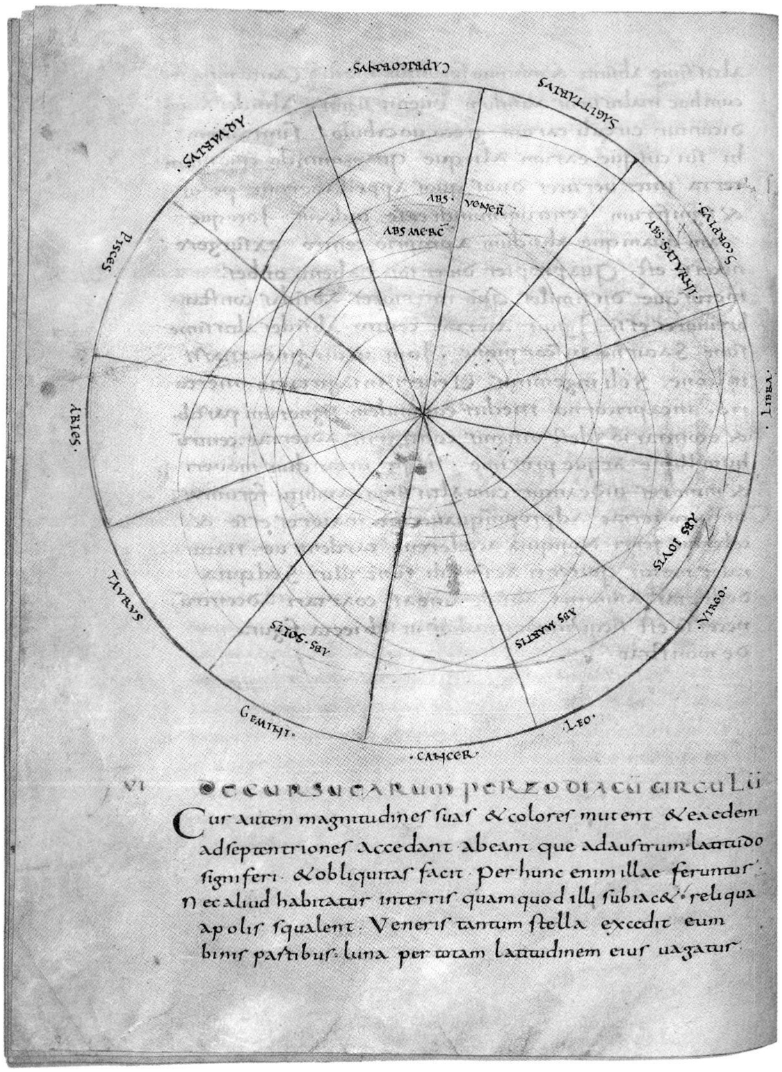

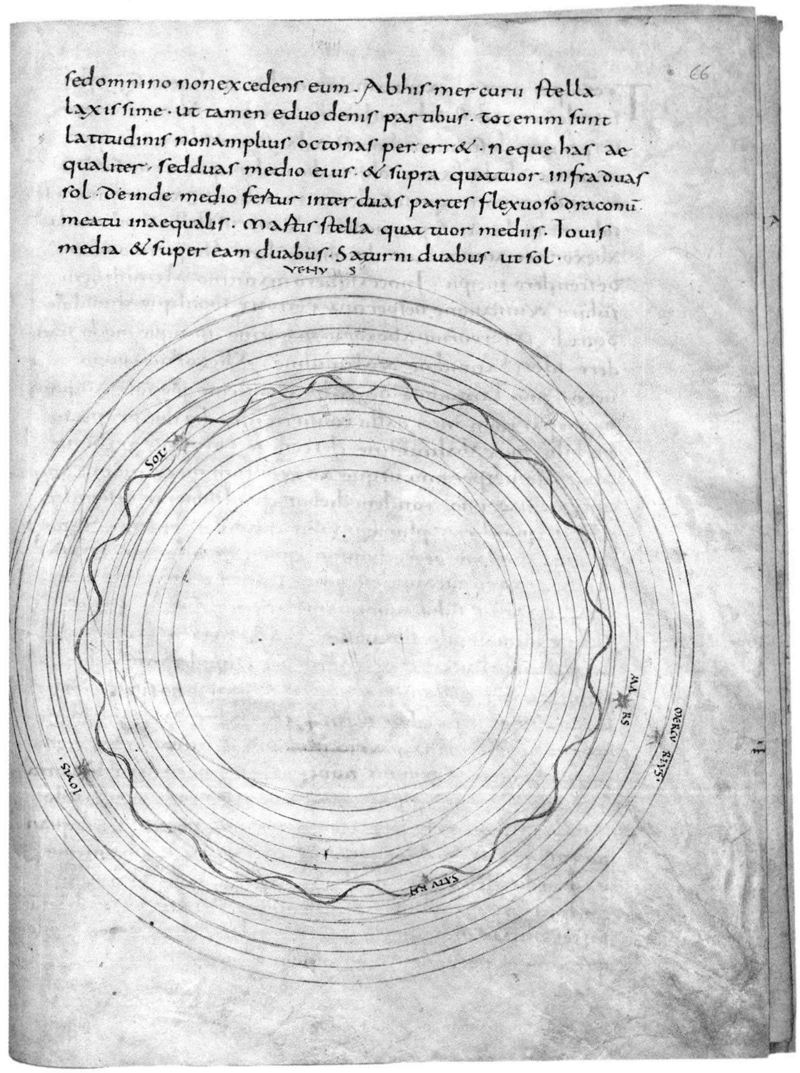

FIGURE 5

"The Apsides of the Planetary Orbits Within the Zodiac" diagram. Madrid, Biblioteca Nacional de España, MS 3307, fol. 65v. Photo © Biblioteca Nacional de España.

FIGURE 6

"The Latitudes of the Planets Through the Zodiac" diagram. Madrid, Biblioteca Nacional de España, MS 3307, fol. 66r. Photo © Biblioteca Nacional de España.

The *Handbook of 809* was truly a princely book treating astronomy and computistics with a particularly Frankish agenda. It was created to encourage both spiritual renewal of the reader's mind and implementation of the clerical calendar, which would ensure that the reader attained eternal salvation at the End of Days. In order to understand this, it is useful to review the conceptual framework in which earlier historiographic discussions of the star pictures took place and the institutions that fostered such activity.

Study of the Heavens: The Warburg Institute, Saxl, and Panofsky

Aby Warburg (1866–1929) and his Kulturwissenschaftliche Bibliothek Warburg (or KBW, relocated to London in 1933) have played a seminal role in our understanding of medieval and Renaissance astronomy and astrology, and the institute that bears his name remains at the forefront of new scholarly approaches to the study of such material.[32] In some ways, it is the *Nachleben* of Warburg's devotion to the intertwining effects of fate and feeling upon discrete historical contexts that supplies his most important legacy to the present project. As Matthew Rampley has rightly noted, "It is perhaps not too bold to suggest that [Warburg's] interpretation of Northern realism can be extended to form a more general theory of pre-Renaissance representation, a theory following which the medieval veneration of images, for example, would testify to the power of symbolic magical-associative representation."[33] Yet Warburg's Renaissance was a fecund era imbued with dissonance, according to Rampley. An opposition arises between the medieval fetishization of the sacred symbol and a modernizing tendency toward allegory,

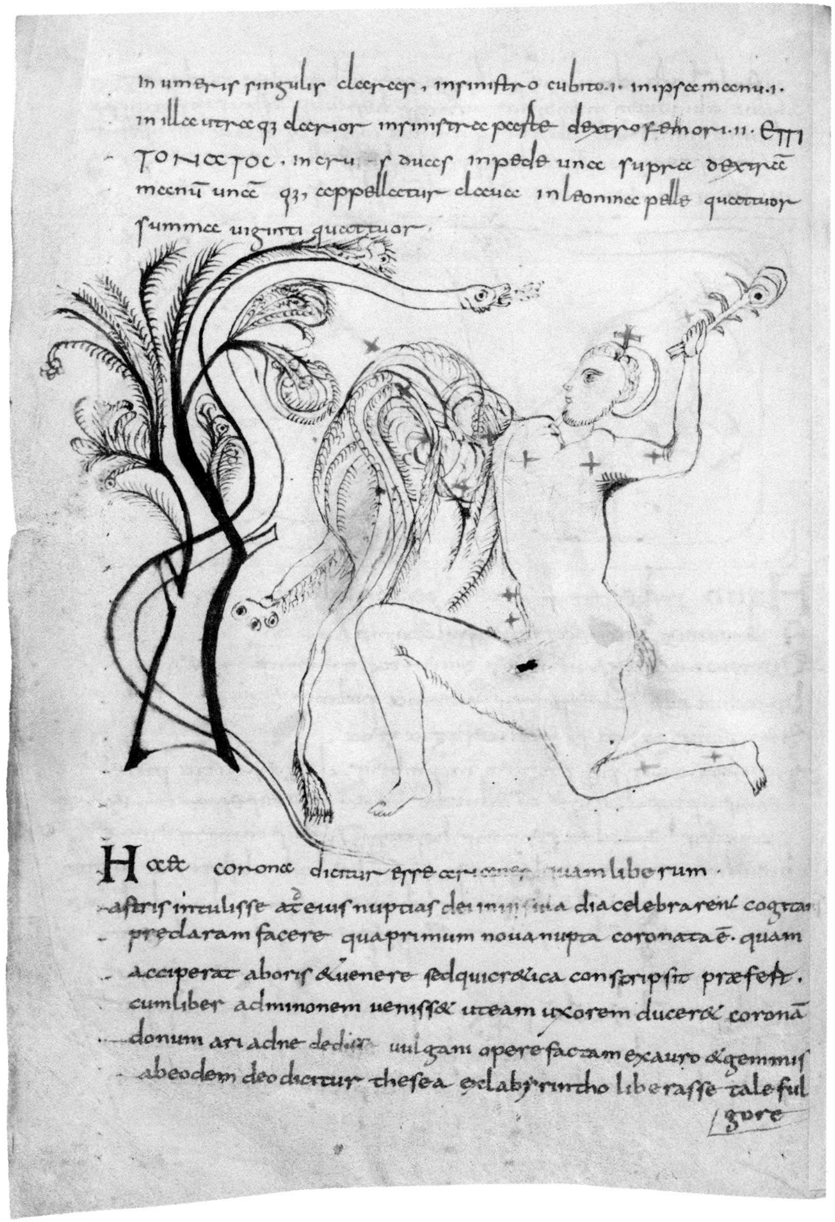

FIGURE 7

Hercules attacking the serpent in the Tree of the Hesperides. Basel, Universitätsbibliothek Basel, MS An. IV 18, fol. 14v. Photo courtesy of Universitätsbibliothek Basel.

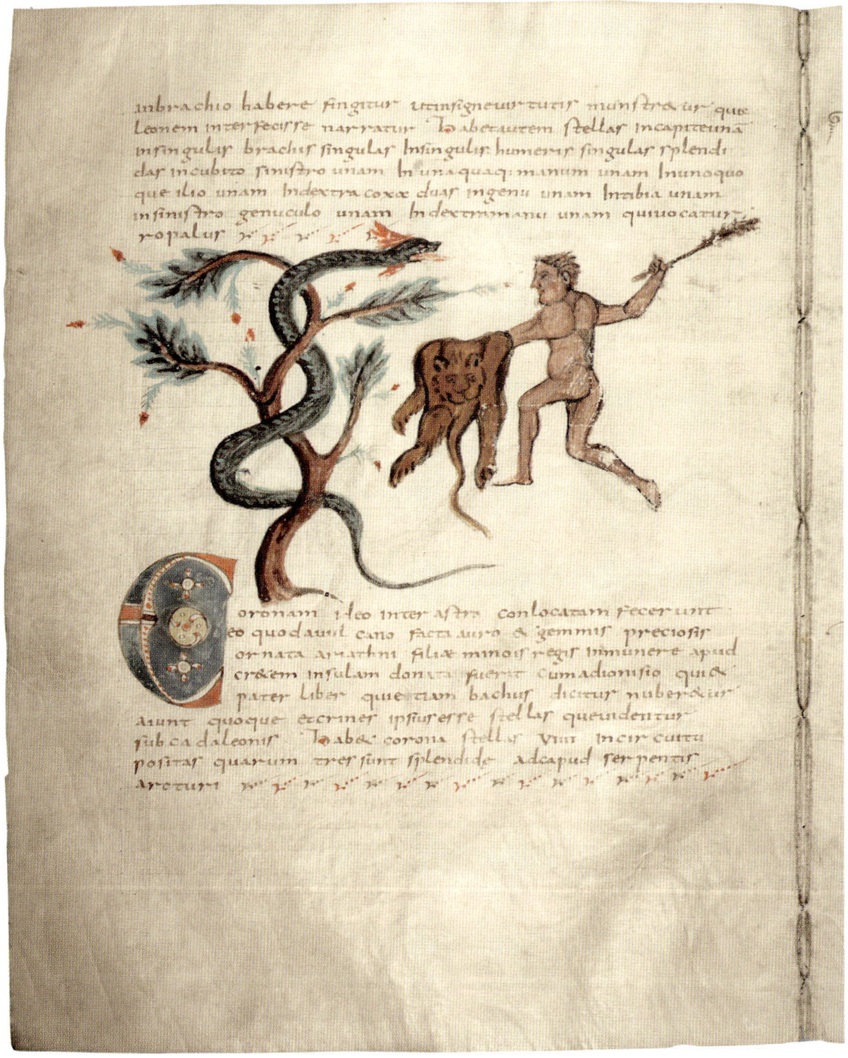

FIGURE 8

Hercules attacking the serpent in the Tree of the Hesperides. Cologne, Erzbisch. Diözesan- und Dombibliothek, Cod. 83-II, fol. 156v. Photo courtesy of Erzbisch. Diözesan- und Dombibliothek.

resulting in a dissociative rupture between the sign and its referent, overcome solely through empathic connection to the object of mimetic representation. For Rampley's Warburg, this outpouring of empathy infused Renaissance Florentine artists and their patrons with an alleged ability to record, through art, a realist expression of their objects of representation and to identify straightforwardly with both the thing depicted and the depiction of the thing. This emotive mutability fostered the fertile environment requisite for the efflorescence of Florentine expression in the Quattrocento.[34]

On the one hand, it is important to underscore that Warburg would never have applied his considerations to Carolingian star pictures. On the other, the possibilities for redefinition that emerge in seemingly incongruous artistic citations of the classical Roman past permit the chief proponents of the iconological method, such as Erwin Panofsky in his *Renaissance and Renascences in Western Art* (1960), to formulate a straightforward critique of the Carolingian Renaissance.[35] Ironically, Warburg's presentation of the Florentine Renaissance (beginning with his dissertation on Sandro Botticelli, published in 1893) as an expressive age of transition between the medieval and early modern periods allows his self-critical appraisal of cultural appropriation troubled by empathy. And by extension, Warburg's view therefore possibly enables an empathic resonance to be felt with premodern cultural periods of transition as well. This arguably makes Warburg's perspective more useful for scholars of Carolingian cultural history than the perspective of those like Panofsky, who nonetheless believed that he had built his discussion of successive periods of renewal on a solid Warburgian foundation.[36]

One of the most important names in any historiographic review of the development of medieval and Renaissance presentations of the heavens is Fritz Saxl (1890–1948), who arranged for the relocation of the KBW from Hamburg to London in 1933.[37] Even though Saxl originally considered the idiosyncratic acquisition and shelving practices of the library "baffling," he became one of the KBW's greatest benefactors, beginning with his efforts as Warburg's assistant in 1913.[38] After Franz Boll (1867–1924), who drafted the seminal text *Sphaera* in 1903, Saxl's catalogues of astronomical and astrological manuscripts—the *Verzeichnis astrologischer und mythologischer illustrierter Handschriften des lateinischen Mittelalters* (Catalogue of astrological and mythological illustrated manuscripts of the Latin Middle Ages)—have enabled all subsequent study.[39] In 1915, Saxl devoted the first installment of the four-part series to Roman libraries, followed by a volume on the Viennese celestial codices in 1927.[40] The third catalogue attests to the relocation of the Warburg Institute to English soil, since the volume appearing in 1953 bears its imprimatur and Saxl inventoried the holdings of English libraries in conjunction with Hans Meier.[41] Finally, the fourth catalogue documents the passing of the mantle to Patrick McGurk, who compiled the catalogue devoted to libraries in Italy outside Rome.[42]

In 1933, Saxl joined with Erwin Panofsky to identify salient historical and cross-cultural contextual considerations that permitted the diachronic evolution of medieval and Renaissance star pictures. In "Classical Mythology in Mediaeval Art," Panofsky and Saxl argued for the inability of early Christian or Carolingian artists and compilers to celebrate the venerable alterity of antiquity with an apposite respect for the normative codes of life that informed Greco-Roman poetry and painting. Building expressly upon Warburg's foundation, they attempted to trace "the way in which classical thought continued through the postclassical era."[43] For Panofsky and Saxl, the Carolingian artists of the eighth and ninth centuries were preservationists who ensured the survival of Horace, Ovid, Pliny, Vitruvius, and Aratus. In their words, "[i]n the same spirit the Carolingian illuminators endeavored to copy the illustrations in the ancient astronomical picture books. . . . They conscientiously, and sometimes most successfully, imitated their prototypes in style and technique as well as in mythological subject matter."[44] More will be said about the Carolingian negotiation with classical form and content in chapter 2. For now, it suffices to underscore that according to Panofsky and Saxl, "during the Middle Ages in the western European countries it was inconceivable that a classical mythological subject should be represented within the limits of the classical style."[45]

Panofsky developed this thesis further in his highly important book, *Renaissance and Renascences in Western Art*. The Italian Renaissance owed an inexhaustible debt to the scribes and illuminators of the Carolingian renewal, who preserved models of classical pictorial forms, such as the images of the constellations in Drogo's copy of the *Handbook of 809*, for later generations. The legacy to Western cultural historians of those Carolingian painters, according to Panofsky, was precisely their lack of creativity: "And the classical values—artistic as well as literary—were salvaged but not 'reactivated' (as we have seen, no effort was made either to reinterpret classical images or to illustrate classical texts *de novo*)."[46]

One chief thesis and methodological bias of the present volume stands in opposition to Panofsky's argument. It shifts the focus of analysis away from the computistical import of the *Handbook of 809* and onto the creative aspects of the annals, the development of certain astronomical concepts related to lunar-solar motion, and the interpretation of the star pictures found in book V. Indeed, the images of the constellations are a point of departure for this intellectual repositioning. The Carolingian compilers of the *Handbook of 809* performed a task similar to that of the painters who examined alternative visual examples of a constellation such as Hercules (plate 2, figs. 7–8). In both cases, the compilers under the supervision of Adalhard and the painters in Drogo's Metz made deliberate decisions about the form their anthology or visual representation would take.

Moments of interpretative intervention in any act of creation permit exegetical emendation. Images of the constellations allowed the reinvestment of classically inherited forms with current Carolingian meaning precisely because the constellation pictures operated simultaneously as mnemotechnic tools, conceits requisite for the liberal art of astronomy, and visual summaries of the iconographic traditions that gave rise to them. Just as the texts compiled into the *Handbook of 809* expanded the semiotic potential of the anthology through juxtaposition and recombination, so the various pictorial elements assembled into Frankish images of the constellations document a skilled, and not haphazard, engagement with the classical past. This made possible the creation of the specific eighth- and ninth-century Christian perspective examined in the book. The images of the constellations, the diagrams, and the remainder of the compilation all participate in the creation of meaning within the *Handbook of 809* through complex hermeneutic strategies that defy a reductionist reading. Only through an integrated reevaluation of the coeval, complementary status of the texts in the *Handbook of 809* and their corresponding images can we achieve a fuller appreciation of the import of Drogo's copy.

The *Handbook of 809* (the *Aachener Enzyklopädie*)

In order to examine the Carolingian significance of the *Handbook of 809*, it is important to index the contents of a standard copy. It was Saxl, again in the *Verzeichnis*, who established the normative cycle of texts included within the table of contents of the seven-book "Mammutwerk," as it has been labeled by Borst.[47] Borst dubbed this nameless anthology the *Libri computi* and the *Aachener Enzyklopädie von 809*, but confessed to dissatisfaction with his own titles, since the compilation defies facile classification into existing genres.[48] As Borst rightly notes, the *Aachener Enzyklopädie* does not encompass the same range of topics covered, for example, in Isidore's *Etymologies*. Nor is it exclusively a pedagogically oriented computistical treatise like the *De temporum ratione* (On the reckoning of time) of the Venerable Bede.[49]

Previous scholarly work has tended to gloss over the cycle of imagery in the *Libri computi*. Eastwood has supplied the most important corrective, evaluating the creative ways in which Frankish artists appended new diagrams to the Plinian excerpts belonging to book V of the compilation. Eastwood terms the *Libri computi* the "Seven Book Computus," emphasizing the aspects of the compilation that are significant to the history of science,[50] following Springsfeld.[51] Springsfeld, however, also points the way to a compromise, considering the compendium from Aachen (that is, the *Libri computi*) to contain enough information to supply a computistical handbook for its owner.[52] This perspective has the advantage of focusing on the normative Christian presentation of a cosmology recombined with an applied computistical framework that constituted the fundamental structure of the *Libri computi*. After all, the compilation was never exclusively a computus.

Akin to a classical handbook in its concise, authoritative presentation, the *Libri computi* was intended to encourage pedagogical instruction while quelling dissent.[53]

As Englisch has masterfully argued, in Cassiodorus's engagement with classical arithmetic, he sought to reclaim for a Christian audience the logical parameters within which a meaningful discussion of ciphers and operations could take place.[54] In the same way, the vast array of topics covered in the *Libri computi* demonstrates a conscious effort to create, in text and image, an authoritative ninth-century compilation that would set the agenda for further discussion of natural history, cosmology, and computus. But the organization of the compiled texts and images in the *Libri computi* is even more revealing.

To better understand just how meaningful the structure of the compilation happens to be, it is helpful to review a further point raised by Englisch vis-à-vis the use of finger reckoning (*de loquela digitorum*) in Bede's *De temporum ratione*. By combining one traditional means for counting according to long-standing schemes of calculation (linked to the arithmetical liberal arts of the *quadrivium*) with the praxis-oriented problem-solving aspects of computus, Bede achieved a new intellectual hybrid. In other words, the early medieval predilection for authoritative ideas originating in antiquity was marshaled in service of the celebration of the Christian liturgy. Temporal reckoning in a Christian world thereby drew upon ancient practice, creating a ratiocinative problem-solving method aimed ultimately at achieving verifiable computistical results.[55]

In the present volume, I contend that Adalhard of Corbie and the subsequent copyists, such as those who manufactured Drogo's *Handbook of 809*, performed a similar hermeneutic maneuver. By introducing the innovative pictorial program of the constellations and four original diagrams into their definitive statement of computus and the liberal art of astronomy, the compilers and copyists of the *Libri computi* created a novel holistic presentation of the cosmos for early medieval Christian eyes. Visual precursors had remained in Italy and throughout Rhenish Frankish villages since Roman times, but as a text and image composite, the *Libri computi* offered an integrated official presentation of the classical and computistical traditions from which Frankish astronomy emerged. This authoritative textual and visual praxis-oriented hybrid known as the *Libri computi* is better labeled as an up-to-date Carolingian version of a handbook, and is therefore called here the *Handbook of 809* on these methodological and cultural historical grounds.[56]

Before turning to a discussion of the significance of the *Handbook of 809*, a brief guide to its scope and size follows. Any such reconstruction benefits from Saxl's early efforts. He identified the original seven books of the *Handbook of 809* because he edited the complete table of contents included within a second copy of the work, presently in the Biblioteca Apostolica Vaticana, Reg. lat. 309 (fols. 4v–6r), and an incomplete version of the table of contents in a third copy of the *Handbook of 809*, likewise in the Vatican, Vat. lat. 645 (fols. 1r–2v).[57]

Book I oriented all that followed relative to the annual cycle of liturgical feasts, with a *rota* and a martyrology followed by the nineteen-year computistical tables—explained for art historians, with an introductory user's manual, in chapter 1 below. Students of computus will also find there additional references signaling places where more advanced information can be gathered.

In book II, computistical references to the year 4761 *ab initio mundi* recurred, indicating the date for the commencement of work on the *Handbook*, namely, 809. Book II treats essential computistical concepts such as epacts, indiction numbers, and *concurrentes*, relying upon them for calendrical calculations of note (all of which are explained in the present volume in chapter 1).[58] Englisch's point mentioned above, highlighting the intersection of computistical acumen and learned appreciation of a liberal mathematical art such as astronomy, is also relevant here.[59] Fundamental cosmological truths such as the age of the world emerged from these rudimentary computistical exercises. Such arithmetical operations permitted the Carolingian reader further access into the hidden biblical history of a world in need of perfection.

Book III addressed the annual course of the sun, while book IV moved on to lunar movement and the period defined by a lunar month (or lunation). This prepared the reader to examine the pivotal calendrical issue of correlating lunar and solar orbits in order to reckon accurately the floating feast of Easter, according to a geocentric model of the solar system. In other words, the preliminary portions of the *Handbook of 809* underscored the importance of the liturgical calendar and Paschal season. In keeping with standard Christian tradition, these were the antidote for the historic spiritual problems facing Creation.

Book V introduced the pictorial program into this dialogue. Both the pictures of constellations and the diagrams for Plinian excerpts were added to the *Handbook of 809* as complementary data sets informing the reader about their classical legacy in novel ways. In addition, the entire pictorial program participated in the pragmatic soteriological agenda established by the preceding four books. Computus and crafted visualizations of the heavens were equally essential but performed alternate modes of psychical labor within the *Handbook of 809*. Computus enabled the appropriately trained cleric to reckon the feasts of the liturgical year in community with the fellowship of saints, present and past. The image cycle continued that labor within the individual in a more specialized personal or educational context. The immediacy of the presented constellations and diagrams enhanced the effective pedagogical and devotional import of the *Handbook of 809*, enabling the reader to connect with the celestial realm and to benefit from the spiritual renewal such study promised, as explained in chapter 4 below.

Book VI of the *Handbook of 809* discussed pragmatic concerns such as weights and measures, and the handbook concluded with Bede's *De natura rerum* in book VII.[60]

Communities of Faith and Knowledge

Derived from the Aachen synod of 809, the full impact and import of the pictorial program in the *Handbook of 809* can only be assessed relative to the intersection of temporal concerns and the administration of power throughout the Frankish kingdoms. One aspect of this multifaceted issue was the call for educational reform contained within Charlemagne's *Admonitio generalis* (General admonition), compiled and promulgated on 23 March 789.[61] In the relevant section of this capitulary, Charlemagne focused on the intertwining spiritual and social responsibilities of pedagogical reform that confronted the generation of ecclesiasts such as Adalhard, who rose above his kindred *sacerdotes*. The king appealed to high-ranking clerics such as bishops, given the use of the term *vestram almitatem* (your benignity) in the opening line of the passage.[62] The clergy were exhorted to lead by example, which meant that the education of children required first an erudite assembly of elders. Leading by example also required the serious supervision of schools, intended for the practical aims of reading and correcting lacunae (*emendatio*). This is the cultural and scholarly approach to the Christian transformation of classical forms that likewise influenced Carolingian painterly practices of exegetical emendation:

> And let schools be established in which boys[63] may learn to read. Correct carefully the Psalms, the signs in writing [*notas*], the songs,[64] the calendar [*compotum*], the grammar, in each monastery or bishopric, and the catholic books; because often some desire to pray to God properly, but they pray badly because of the incorrect books. And do not permit your boys to corrupt [the books] in [the process of] reading or writing them. If there is need of writing the Gospel, Psalter, and Missal, let men of mature age do the writing with all diligence.[65]

This Carolingian stipulation made education accessible to the servile and free classes[66] and emphasized the universal applicability of the educational reform initiatives. All of the boys needed to learn to read as the gateway to further study in the liberal arts. Reading was also the basic skill that would enable novices to enter meaningfully into the monastic life.[67]

I have already suggested that the interpenetration of spiritual goals with training in the liberal arts established the basis for a scientific soteriology that involved the study of astronomy. In fact, as John Contreni has expertly argued, "When Carolingians wrote about schooling they invariably described its goal in spiritual terms. It was not only that the study of the arts would provide some technical assistance in comprehending the sacred texts, but also that the arts themselves had value as conduits to the knowledge of God. . . . But philosophy and theology were not the only beneficiaries of the Carolingian reform programme. The process of governing increasingly buttressed traditional sources of power with its own literary forms."[68] Reforming the study of the liberal art of astronomy and the related applied science of computistics serves as but one example of the intersection of political, spiritual, and astronomical concerns in Charlemagne's Frankish realms. Control of the spiritual and administrative aspects of the calendar was another, discussed at length by Arno Borst throughout his distinguished career.[69] Borst has emphasized in his numerous publications that Charlemagne's interest in educational reform and calendrical study (or computus) transcended the pragmatic clerical and administrative requirements of dutiful service and became a strategy for the imperial consolidation of authority.[70]

The quintessential product of authoritarian calendrical reform, according to Borst, was therefore the Lorsch calendar of 789 that recombined celestial observation and sacral veneration of the liturgical feasts throughout the Frankish kingdom with computistical erudition as a creative compilation effort in its own right, or at the very least including preestablished components into a cohesive suitable format. This calendar supplied the basis for the initial wheel diagram in book I.1 of the *Handbook of 809*, called the *Rota quae continet natalicia sanctorum*, but was not the model for the martyrology that followed in book

I.3.[71] This highlights the nonhegemonic or at least nonexclusive status of the prototypical Lorsch calendar. It was but one important early medieval record of calendrical reckoning among an expanding set of divergent and site-specific alternatives aimed at solving real computistical problems, at least since the synod of Soissons in 744, as argued perspicaciously by Englisch.[72] This kind of "creativity" *qua* computistical intervention from the mid-eighth century has also been documented by James Palmer as evidence of a prescient, politically engaged effort aimed at perfecting Easter reckoning (and anticipating a scenario parallel to the one in 809 under discussion here).[73]

The creative confrontation of Carolingian artists and scholars with computus and with the classical legacy of the constellations gave rise to a series of astronomical-computistical-pedagogical handbooks, discussed in chapter 1 of the present volume. There, the art-historical significance of Drogo's copy of the *Handbook of 809* is thoroughly situated within the modern literature, resolving long-standing questions concerning the creation of lavish copies of the *Handbook of 809*, the origin of Drogo's book in Metz, and that work's subsequent history. Chapter 2 offers an iconographic history of celestial iconography as it pertains to Drogo's copy of the *Handbook of 809* as a resource for art historians. Chapter 3 reinterprets the relationship between painters working in Metz and the formation of Drogo's copy of the *Handbook of 809*, presenting evidence in support of itinerant teams of traveling painters working during the early medieval period. Finally, chapter 4 confronts two important philosophical issues: first, its discussion of the Carolingian image debate includes a proposal for a theory of Carolingian aesthetics; second, the ways in which the Carolingians conceived of the spiritual benefits awaiting students of the *Handbook of 809* and other such celestial compilations reveal the importance of the handbook's texts, computistical underpinnings, and cycle of illustration.

In other words, the present volume moves from a comprehensive history of Drogo's copy of the *Handbook of 809* to a discussion of its imagery. It then considers the significance of the handbook's paintings for the history of art and early medieval Metz, followed by an appraisal of the significance of these images of the constellations relative to Carolingian ideological justifications for imagery in the service of individual spiritual reforms.

The society of trained and skilled participants in the Frankish debates about the liberal arts contributed to the creation of a culture and not just an empire through study of the heavens, computus, and the liturgical calendar. This company included those close to Charlemagne's shifting band of itinerant courtiers eventually affiliated to varying degrees with Aachen (after 794), such as Alcuin, Einhard (beginning 791–92), Paul the Deacon (until 786), Paulinus of Aquileia (departing before 787), and Peter of Pisa (leaving before 791), as well as an extended network at a great remove from this coterie, such as Adalhard (Corbie), Alcuin again (posted to Tours in 796), Arn (Saint-Amand and Salzburg), Dungal (Paris), Hrabanus Maurus (Fulda and Mainz), and Theodulf (Fleury and Orléans).[74] Granted, these luminaries had weaknesses and at times possessed incompatible perspectives, but these differences encouraged the creative problem-solving spirit throughout the Carolingian era that fostered an environment appropriate for the creation of Drogo's copy of the *Handbook of 809*.

Although Charlemagne was geographically mobile until late in life, his intellectual community established a network of administrative centers for the relocation and distribution of texts and the dissemination of ideas. Period philosophical treatises and the evidence from illustrated astronomical manuscripts indicate that pedagogy and study of the liberal art of astronomy participated in Carolingian strategies for the individual restoration of the soul in keeping with Christian doctrine.[75] (This theme is treated in chapter 4.) In so doing, the Carolingian teachers created within the schools sites of pedagogy yielding a form of doctrinal conformity, as education worked in tandem with ideals of sanctification for the edification of the empire. In other words, study of astronomy was a pathway to individual salvation, and throughout the Carolingian empire, astronomy truly became a saving science.

Astronomy was a tool for individual renewal, one educated member of the empire at a time; hence the textual and visual traditions belonging to Carolingian study of the liberal art of astronomy were slated for preservation. The creation of a doctrinally informed, intellectually renewed, fortified populace was itself a form of political fashioning through the cultivation of knowledge. As each person mastered himself, he became prepared to train others and to participate in a well-formed community of scholars or prelates. Johannes Fried has labeled this aspect of the Carolingian empire a *Wissensgesellschaft*, or a community based on its shared *scientia*, a politico-economic realm with a technology of Christian knowledge formation at its core.[76] As Rosamond McKitterick has emphasized concerning the Frankish administration of the realm and its knowledge(s), "the *Admonitio generalis* needs to be seen in the context of . . . the integration of the Christian faith within the institutional and political framework of the Frankish realm," and "the concept of *correctio*, the acquisition of knowledge and the exercise of power were yoked together."[77]

In this fertile and protean context,[78] it is unsurprising to learn that artists, too, moved about and drew upon diverse models in discrete locations, varying their finished products in accord with the general expectations that unified the widespread Carolingian milieu, but with individual inflections for specific patrons and audiences. As the painters moved about, so did their forms, and thus the varied presentations of the constellations benefited from this permissible mutability. Although hardly an isolated early medieval phenomenon,[79] it was precisely this creative diversity of forms that allowed a novel Christian presentation of the star pictures throughout the Frankish realms, drawing upon classical Roman and autochthonous iconographic traditions in a process of exegetical emendation.

Centers and Wandering Painters: Four Men in Metz

No one has laid the foundation for this reinterpretation of the star pictures within Drogo's copy of the *Handbook of 809* more than Lawrence Nees. In his groundbreaking 2001 essay "On Carolingian Book Painters: The Ottoboni Gospels and Its Transfiguration Master," Nees demonstrates with careful description, buttressed by thick art-historical contextualization of mid-ninth-century painterly styles and codicological considerations pertaining to Biblioteca Apostolica Vaticana, Cod. Ottobonianus, lat. 79, that "Carolingian patrons may have . . . [used] local or in-house talent for some jobs . . . seeking distinguished outside assistance when available, or for special purposes."[80] In particular, Nees adopts an artist-centered perspective that reconsiders the impact of specific painters on regional styles. The so-called Transfiguration Master thus comes to be understood as a renowned and sought-after painter who plied his trade as an itinerant illuminator in the same regions under discussion throughout this book. Historic connections between the Frankish cities of Metz and Reims fostered the communication of pictorial information along traditional lines of cultural exchange that can also be traced using art-historical methods. Working in a conventional Franco-Saxon style during the ninth century, the Transfiguration Master made his way through the regional centers of Neustria and Austrasia, learning and adapting his craft as he proceeded from one commission to the next.[81] Nees's model is of vital significance for the present book. He demonstrates the rich potential for constant exchange, as an itinerant painter traveled or was reassigned to new sites after successfully completing his work. An early medieval itinerant painter continued to acquire new iconographic ideas—and to master new painterly approaches to the resolution of pictorial problems—while fulfilling a variety of commissions throughout a professional life.[82]

Nees's paradigm undermines the traditional approach to Carolingian painting, which sought to isolate Frankish production according to monolithic ateliers. Wilhelm Koehler and Florentine Mütherich's monumental and invaluable catalogues *Die karolingischen Miniaturen* taxonomically cluster illuminated manuscripts according to putative court schools and monastic or episcopal scriptoria.[83] The catalogues also, however, establish hegemonic categories of difference that cannot account for the creative modalities of exegetical emendation found in the books contained within or left out of those exhaustive catalogues. Nees's model supplies a suitable synchronic corrective, analyzing specific books with an attentive interest to the specific parameters that affected an individual commission at a particular time and in a specific place.

It therefore is also problematic to assume that all iconographic details copy an earlier source, to fabricate unnecessary lost models for the evidence that remains and, arguably, to prefer what is missing to the codices at hand.[84] In order to track the precise diachronic links that permitted exegetical emendations in the eighth and ninth centuries, a middle ground is required. In a case such as Drogo's copy of the *Handbook of 809*, this middle ground must take into account the classical Greco-Roman iconography of the star pictures, countervailing Carolingian presentations of the constellations or planets, and the subsequent reinterpretation of those celestial forms in Metz by a band of four traveling painters. Since the extensive astronomical material is in need of explanation, this volume has limited itself to a focused study of early medieval astronomical manuscript illumination. Fluid Frankish practices for forming image cycles, however, would never have precluded simultaneous consideration of stellar representations in plastic media, and for that reason the manuscript records of lost celestial globes are also assessed in what follows.

Mindful of these concerns, the present volume focuses on the aspects of Carolingian manuscript illumination that were of import for Drogo's copy of the *Handbook of 809*. Books traditionally believed to come from Reims and Metz influenced its manufacture or were influenced by it. Because the imagery in the *Handbook of 809* was of primary importance for the subsequent transformation of western European star pictures and cosmological Plinian diagrams, as discussed above, this art-historical study is overdue.

Two master painters and their assistants in early medieval Metz realized the expectations of their patron and created a glorious copy of the *Handbook of 809*. Drogo's book remains, however, but one copy, and the differences between its cycle of illustration and other princely examples help define the significance of its pictorial program. In closing, it is essential to state that Bishop Drogo would have valued his book first for enabling him to perform his spiritual service. Having mastered its contents, Drogo did more than reckon the feast of Easter with accuracy and participate in Carolingian clerical reform. Through the folios of his copy of the *Handbook of 809*, Drogo also passed through a personal intellectual portal revealed to him by his star pictures and diagrams and, in Isidore's words, set his mind "in contemplation of what is above."[85]

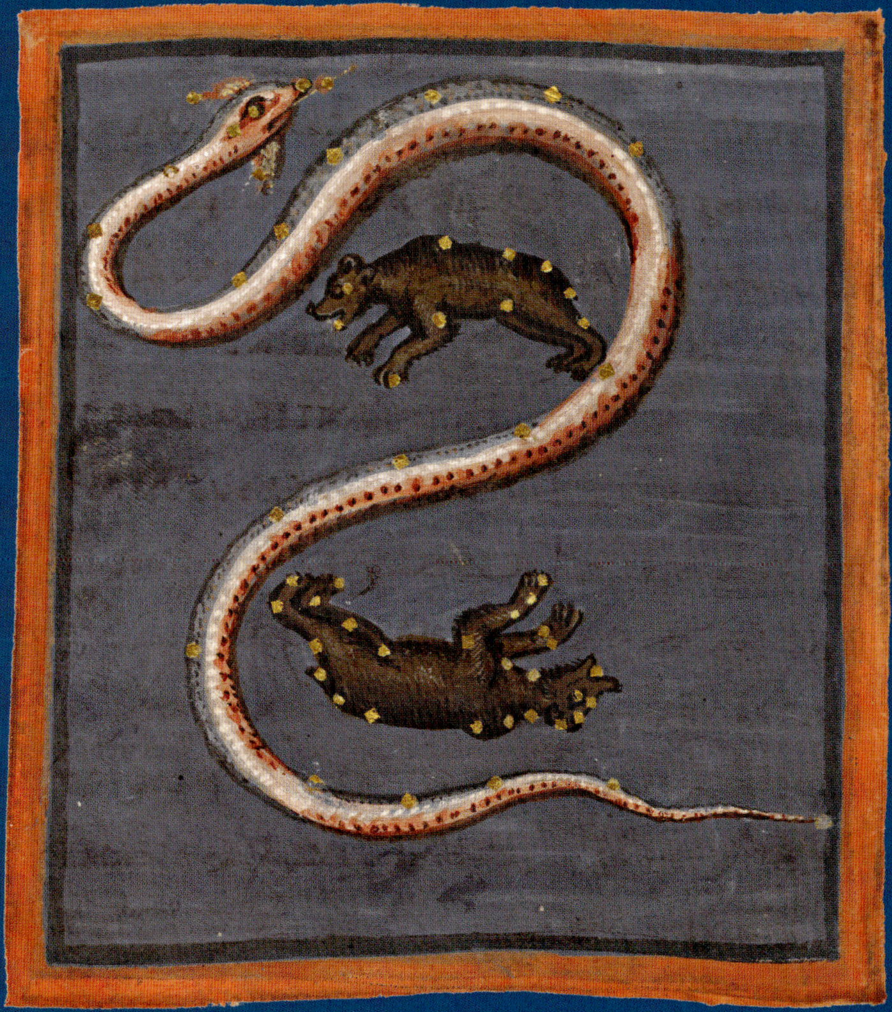

PART ONE

Unveiling the Heavens over Carolingian Skies

CHAPTER ONE

ILLUMINATING SCIENCE

Creating the *Handbook of 809*

When the ninth-century prelates Adalhard of Corbie (ca. 750–826) and Drogo of Metz (801–55) left their monastic chambers to study the night sky, they did not step into the confused *saeculum* of the mythopoetic Dark Ages. Instead, they walked into a *mundus* replete with illumination.[1] According to the Christian Creation narrative (Genesis 1:14–19), God had placed the sun, moon, and stars in the sky on the fourth day in order to illuminate day and night. These great luminaries in the heavens divided the medieval day, year, and lifetime into discernible periods of work and rest that came and went predictably. Human beings were then created on the sixth day (Genesis 1:24–31, 2:18–25), installed within the Garden of Eden, and charged with its care. From the beginning, the night sky supplied records of momentous events that had already taken place or that were still to come, and it enabled the reckoning of time for those who would watch and wait, permitting the universal *mundus* to reveal the unflagging reassurance of its ordered regularity.[2]

The stars in the heavens served as fixed reference points, moving along predictable pathways ordained by their Maker, locked into their place on the rotating celestial sphere.[3] They could be used by sailors for navigation. Their disappearance and return helped map the changes of the seasons for agricultural purposes. Early human ingenuity had long since grouped the flickering celestial fires into constellations, and the nascent taxonomic systems for identifying star clusters, which for the most part resemble those used to this day, had already emerged by 700 B.C.E.[4] Charting the passages of the constellations through the night sky aided ideological control, too, as the prelates and rulers of the early medieval world exerted their influence over Christendom by ordaining which activities were to occur and when. Carolingian stargazers appreciated the starry host as an array of emblems denoting temporal continuity—a temporal continuity that made change possible, thus

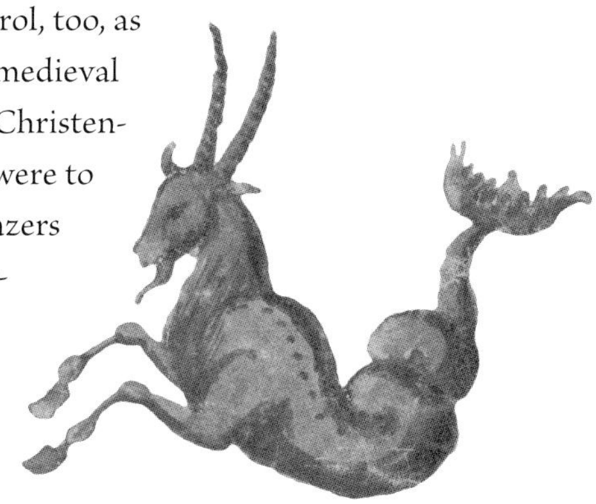

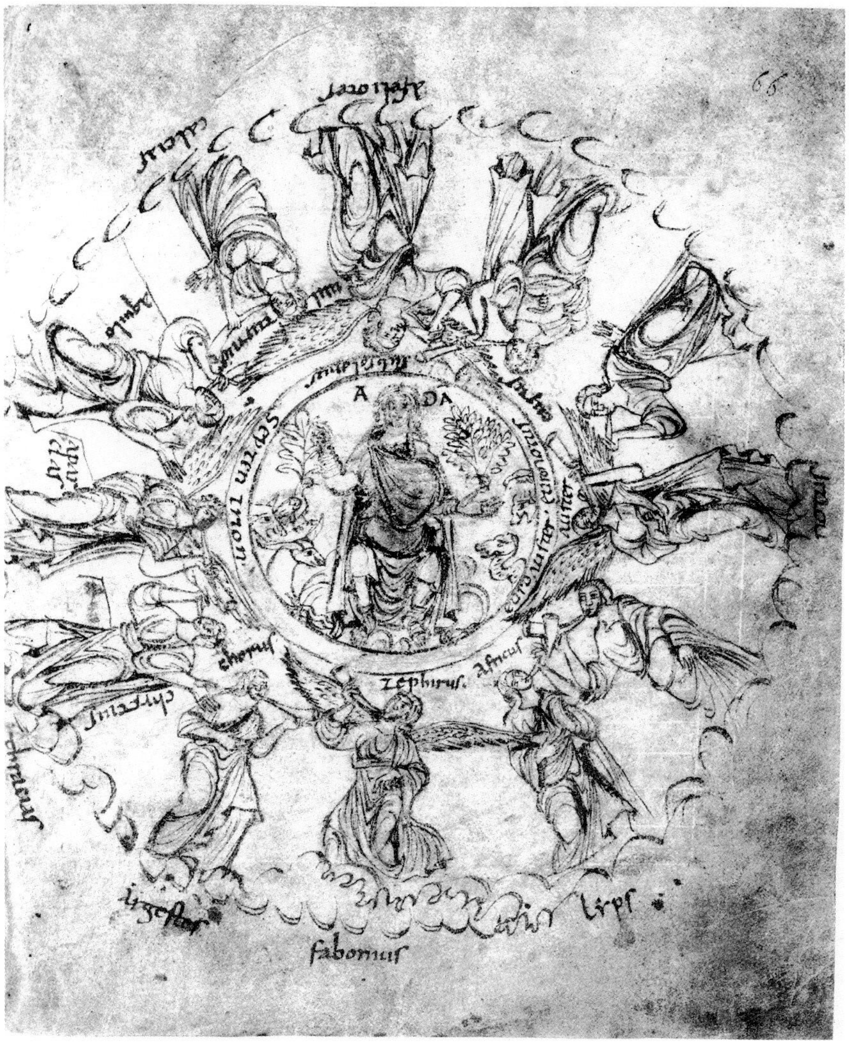

FIGURE 9

Wind rose. Vatican City, Biblioteca Apostolica Vaticana, Vat. lat. 645, fol. 66r. Photo © 2017 Biblioteca Apostolica Vaticana. Reproduced by permission of Biblioteca Apostolica Vaticana, with all rights reserved.

permitting widespread Christian renewal. As Augustine had explained in the *City of God*, "the world was not created *in* time but *with* time."[5]

Adherence to the Rule of Saint Benedict (d. 543) in the medieval monastery also required the careful telling of time.[6] As Paul Dutton has suggested, it is highly likely that, following Charlemagne's example, the refined study of astronomy became a fashionable *cause célèbre*, encouraging ninth-century courtiers to cultivate their mental acuity under the night skies. Reassessing the utility of celestial observation and the cultivation of the liberal art of astronomy throughout the Frankish schools, Carolingian renewal also involved the possibility of social change. Comets and other celestial events were routinely taken as omens of tragedy, and the Christian negotiation of these heavenly signs therefore required interpretative care.[7] Deviations in the natural order of things in the heavens were thought to reflect the transformative impact of the controversial Frankish efforts at renewal below. A vigilant and ambitious ruler such as Charlemagne recognized the need to foster the study of astronomy, since only through skillful observation of the heavens could the Franks best chart their reforms and ultimately ensure their success.[8]

Working within a geocentric model of the universe, the Carolingian stargazer also monitored the paths of the wandering planets. By day the course of the sun was observed, since it, too, was once considered a planet. The moon was likewise counted among the planets and observed when visible by day or by night. Arguably, the most important date for the Carolingian Franks to fix was the feast of Easter. This was no small feat, for it required a monastic stargazer to track the relative movements of the sun and moon. When early medieval Franks tracked the movements of the heavenly bodies, they did much more than observe their environment: they donned the Adamic mantle, taking up the conscientious care of the Creation. Adam named the animals and thereby established the possibility of taxonomy while he laid the groundwork for a systematic explanation of an ordered cosmos. The cosmological significance of one diagrammatic wind rose from a copy of the *Handbook of 809* in the Vatican is clear. Ratiocinative individuals were installed within the *mundus* in order to understand it and to give Christian meaning to its parts (fig. 9). In addition, according to the *Tractatus in Iohannis Evangelium*, Augustine made the name "Adam" a Hellenic acrostic for the four corners of the world, linking Greek and Latin biblical tradition with Christian cosmology: *anatol* (east), *dysis* (west), *arctos* (north), *mesembria* (south).[9] At the center of the wind rose, Adam sits enthroned; more than an everyman, he is a placeholder for a divinely appointed arbiter, centrally situated as a reflection of the *mundus-annus-homo* diagram discussed above (fig. 1).

Astronomy and computistics were intrinsically intertwined theoretical and practical ways to participate in the investigation of the cosmos. Their study became a component of an ideological and spiritual mission. Charlemagne—following the synod in 809 where calendrical, computistic, and astronomical issues had been discussed—personally encouraged his learned prelates to compile a creative record of their science. That document is the *Handbook of 809*, and in this chapter Drogo's personal copy is assessed in some detail.[10] In what follows, I will first review the historical considerations that inspired modern interest in Drogo's copy of the *Handbook of 809* before assessing its recent significance for understanding Carolingian art. At that point, I will discuss the princely recipients of *de luxe* copies of the *Handbook of 809* as well as the work's origins within a legacy of Frankish interest in computus. This requires a cursory primer introducing key computistical concepts, offered as a guide for those unfamiliar with computistical texts. (Advanced students of computus might wish to jump to the discussion of the *Annales Prumienses* [the Annals from Prüm] that follows and the ensuing review of the contents of Drogo's book from Metz, as well as its *Nachleben*.)

Drogo's Copy of the *Handbook of 809*

Although the original is evidently lost, as discussed in the introduction, the miniatures and texts of one copy of the *Handbook of 809*, made for Charlemagne's illegitimate son Drogo, reveal its transcendent achievement. The manuscript—hereafter referred to as Drogo's copy of the *Handbook of 809*, or Madrid 3307—treats astronomy and calendrical matters, and it includes forty-six finely executed diagrams and star pictures. In 1933 Jesús Domínguez Bordona made an inventory of the manuscripts in the collection of the Biblioteca Nacional, and he listed astronomical manuscript 3307 as cat. 568:

> *Treatise on computus and astrology*
> Its writing, Carolingian, appears for the most part from the ninth century. It offers plenty of interspersed paintings of Hellenistic origin. It is an important codex for its text and iconography, and so it seems, has not been explored by specialists.[11]

Wilhelm Neuß (1880–1965) published the first studies of the Madrid manuscript in two articles released in 1940 and 1941.[12] The Spanish dictator

Francisco Franco had flirted with an alliance between Spain and the Axis powers in 1939–40. Although Franco and Hitler never achieved a lasting accord, the diplomatic exchanges between Spain and Germany in the 1930s and early 1940s permitted German scholars access to the Iberian peninsula to conduct research in Spanish collections. Neuß was thus able to advance studies of medieval art in Spain.

Neuß was reared in Aachen, Charlemagne's former administrative center and eventual court city, a setting conducive to the cultivation of an interest in Carolingian art. While studying theology in anticipation of his career as a priest, he also studied art history. His 1911 dissertation, *Das Buch Ezechiel in Theologie und Kunst bis zum Ende des 12. Jahrhunderts, mit besonderer Berücksichtung der Gemälde in der Kirche zu Schwarzrheindorf* (The Book of Ezekiel in theology and art until the end of the twelfth century, with special consideration for the paintings in the Church of Schwarzrheindorf), expresses this dual interest, and his 1913 *Habilitation*, *Katalanische Bibelillustration um die Wende des ersten Jahrtausends* (Catalan Bible illustration at the turn of the first millennium), examines medieval Spanish material. In 1917 Neuß joined the faculty at Bonn University, where his courses focused on church history from the Middle Ages to the modern day and on the history of Christian art. Neuß was also a crusader against Nazi ideology, for he co-edited and participated in an anonymous anthology protesting the racial ideology of Alfred Rosenberg (1893–1946).[13] This took place while Neuß carried out his research on the Madrid manuscript, leading to the two seminal articles of 1940 and 1941, which first brought widespread attention to the glorious star pictures in Drogo's copy of the *Handbook of 809* (plates 1–17).[14]

A catalogue of Spanish astrological or astronomical manuscripts, which would require an onerous effort, has never been achieved. There is no Spanish volume, for example, within the important corpus volumes of astrological manuscripts under the general title *Verzeichnis astrologischer und mythologischer illustrierter Handschriften des lateinischen Mittelalters*.[15] As a result, the significance of Drogo's copy of the *Handbook of 809* now in Madrid has not been fully appreciated. Although the bias of many cultural historians against secular medieval manuscripts has contributed to the handbook's unfortunate obscurity, its seemingly impenetrable nineteen-year tables (explained momentarily) can easily deter even ardent students of medieval culture. The philosophical and theological import of the star pictures in the *Handbook of 809*, however, eclipses such earthly difficulties. The texts and images in the Madrid copy of the *Handbook of 809* are an integral component and vital record of the legacy of learning in medieval Metz and of Bishop Drogo's role in the care and cultivation of his flock. The *Handbook of 809* is also a unique cultural record of Charlemagne's ubiquitous politics of ideological control and soteriological economy of domination, accomplished by the literal renewal of the empire, one mind at a time.

Drogo's copy of the *Handbook of 809* is occasionally included in surveys as an example of Carolingian scientific illustration,[16] and the manuscript has only been exhibited three times in major museums: in the great Aachen show of 1965,[17] in a showcase for Carolingian drawing, *The Utrecht Psalter in Medieval Art* (1996),[18] and in an exploration of early medieval Catalan art, *Catalunya a l'època carolíngia* (1999).[19] A series of publications associated with the creation of *Códice de Metz: Tratado de cómputo y astronómia* (1993) has also made the manuscript better known.[20]

The *Handbook of 809* Manuscript Recension: A Princely Book

Wilhelm Neuß's 1940 and 1941 articles on Drogo's copy of the *Handbook of 809* address the literary and textual traditions that influenced the redaction of the text. They also explore the origins of the forty-two paintings of the constellations in the handbook. The 1941 article, "Eine karolingische Kopie," revises and expands upon the ideas presented to a more limited readership in the 1940 article.[21]

Although Drogo's copy of the *Handbook of 809* has clearly suffered losses, Neuß was the first to link Madrid 3307 to a closely related family of astronomical-computistical-pedagogical handbooks. Two sister manuscripts in the Vatican Library (Biblioteca Apostolica Vaticana, Reg. lat. 309 and Vat. lat. 645), which include tables of contents, or *capitula*, provide evidence for an original seven-book format that was once shared by the now fragmentary copies of a hypothetical lost original handbook, as explained in the introduction.[22] Wilhelm Koehler has noted a third copy of the handbook in Monza (Monza, Biblioteca Capitolare, Cod. f–9/176; hereafter Monza f–9/176) with a list of *capitula* that corroborates the complete table of contents found in the Vatican manuscript from the Reginenses collection (hereafter Reg. lat. 309).[23] Reg. lat. 309 is therefore an important copy of the handbook, but it also includes the chronicles, or annals, of Saint-Denis in its calendrical folios and is typically dated to 859.[24] The other Vatican handbook (hereafter Vat. lat. 645) came from Saint-Quentin, where historical notes were also inserted into the calendrical folios.[25] These annals were ordinarily added alongside the sequentially arranged dates (*ab incarnatione domini*) found within the nineteen-year tables belonging to book I.[26] (The *Annales Prumienses* included within Drogo's book are discussed in this context below.)

The nineteen-year tables constituted a vital component of any ninth-century scholar's reference library. Computistics, or computus, according to medieval authors, denoted the mathematical applied science of calendrical calculation and temporal reckoning used to determine the date for the feast of Easter.[27] Fixing a date for Easter had long been a problem for churchmen, precisely because of the partial reliance of the floating holiday's date upon a lunar calendar, namely, the Jewish calendar for annual life-cycle events and feasts. A learned prelate needed to understand basic operating principles of the computus in order to celebrate the feasts of the Christian church calendar. It was also necessary to establish the dates of the appropriate feasts in order to officiate the Mass, using the appropriate liturgy for any given day of the cyclical church year. During Charlemagne's reforms, efforts were made to establish a consistent rite for the Catholic *Ordo* of the Mass. A working knowledge of the computus was necessary for abbots and bishops, if not all ecclesiasts. This was made explicit in an official notice entitled *Quae a presbyteris discenda sint* (Things that ought to be taught by priests, ca. 805). In its list of fifteen skills clerics needed to know, the computus is number eight, just before the need for all churchmen (*omnes ecclesiasticos*) to learn the nightly hymns of the Roman Ordo, which is number nine.[28]

For all these reasons, a copy of the *Handbook of 809* was a meet and proper gift for a young prelate from the Frankish royal family. Two younger half-brothers of Louis the Pious occupied their new ecclesiastical posts roughly contemporaneously with the creation of their personal copies of the *Handbook of 809*. The copy now in Madrid was created to celebrate Drogo's elevation to the see of Metz in 823; shortly thereafter, Drogo's (also illegitimate) full brother Hugo (802/6–844) became abbot of the monastery of Saint-Quentin.[29] It is certain that Vat. lat. 645 was copied in Saint-Quentin, given the addition of the *Annales Sancti Quintini* to its nineteen-year tables.[30] The handbooks thus seem to have been luxury commissions that demonstrated the spiritual and political efficacy of a religious leader during the Carolingian era, or rewards conferred on members of the Carolingian royal family recently elevated to religious office. Drogo and Hugo's older half-brother Louis the Pious (778–840), co-emperor with his father, Charlemagne, from 813, had had these two illegitimate sons of Charlemagne's concubine Regina tonsured, educated in the liberal arts, and committed to a life of ecclesiastical service. Hugo was also elevated to the abbacies of Saint-Bertin, Saint-Denis, Lobbes, and perhaps Novalesa.[31]

The Monza copy of the *Handbook of 809* contains a list of eclipses through the year 812 on folio 74v, a feature shared with other copies of the *Handbook of 809*, such as Drogo's, where the list appears on folio 68v in book V.10, the *Quando solis eclypsis visa sit moderno tempore*.[32] In Monza f–9/176, entries for the death of Duke Hildulfus, who was interred within the monastery cemetery at Lobbes in 707, and of Abbot Hugo of Saint-Quentin and Lobbes (reported incorrectly in association with the year 846 rather than 844), confirm the manufacture of the Monza copy of the *Handbook of 809* in Lobbes.[33] Lobbes is located halfway between Aachen and Saint-Quentin. It is thus not surprising that Monza f–9/176 harks back to either the archetypal *Handbook of 809* in Aachen or the Vat. lat. 645 copy in Saint-Quentin. The close connection between the text and the linear pictorial programs in the *De ordine* are in fact shared by Vat. lat. 645 and Monza f–9/176. Vat. lat. 645 could arguably have provided the model for the Monza copy during the tenure of Abbot Hugo at Lobbes. This must remain a plausible scenario.

The copies of the *Handbook of 809* in Madrid (which is the focus of this book) and Vat. lat. 645 were most assuredly made for Drogo and Hugo, respectively. Rather than assuming that a second luxury copy of the handbook, the Monza copy, was offered to Hugo at Lobbes, it makes more sense to seek another honored recipient related to Carolingian royalty. Lothair II's brother-in-law, known as Hubert, became abbot of Lobbes in 864, and this may have supplied the catalyst for the creation of Monza f–9/176, in keeping with what was by that time a long-standing Carolingian tradition: when a Frankish prince ascended to his see, he received a book of star pictures, replete with information about astronomy and astrology.[34] Moreover, 864 fits better with the traditional dating of Monza f–9/176 to ca. 875,[35] although Patrick McGurk has suggested the mid-ninth century[36] and Wilhelm Koehler favors the year 846, after which he believes that additions were entered in the annals of the Monza copy.[37] Bernhard Bischoff, however, suspects that Monza f–9/176 was made in the third quarter of the ninth century, supporting the ca. 864 date posited here. Annalisa Belloni and Mirella Ferrari have accepted Bischoff's date for Monza f–9/176 in the catalogue of the holdings in the Biblioteca Capitolare di Monza.[38] Entries for annals in the nineteen-year tables typically refer to earlier events, so such dates are not always helpful for determinations of the dates of manufacture of specific manuscripts. No one, after all, wants to ascribe these handbooks to the eighth century just because Charlemagne's father, Pippin (d. 768), routinely appears in Carolingian annals.

The second copy of the *Handbook of 809* in the Vatican Library, Reg. lat. 309, has been linked to the monastery of Saint-Denis because it contains the *Chronicon Sancti Dionysii ad cyclos paschales*. This copy was made during the tenure of yet another royal abbot, Louis, who served from 840 to 867. He was the first abbot of Saint-Denis who belonged to the Carolingian bloodline descending from Saint Arnulf. Louis was the grandson of Charlemagne, and during his abbacy he maintained successful diplomatic relations with Charles the Bald and Lothair.[39] This fragmentary copy of the *Handbook of 809* contains the important table of contents that permits a complete reconstruction of the archetypal handbook. Reg. lat. 309 is a true compilation, including an interfoliated ninth-century copy of the *Handbook of 809* with additions on folios 122r–157v. On folios 122r–125v, the year of the Lord is listed several times as 885, indicating that these additions were made at that time. Still later eleventh-century additions follow on folios 126r–156v; these include an additional illustrated calendar cycle that derives in part from the Carolingian portion of the codex, dating to 859.[40] A list of eclipses similar to the list in Drogo's copy of the handbook is found on folio 102r, indicating that Reg. lat. 309 is an independent copy harking back to the archetypal *Handbook of 809* made in Aachen.[41]

The evidence from Reg. lat. 309 fully supports the hypothesis that luxury copies of the *Handbook of 809* are linked to prelates with kinship ties to the Carolingian royal house after their accession to ecclesiastical offices.[42] If this supposition is correct, then the copy in Monza was probably made for Abbot Hubert, and Reg. lat. 309 was made for Abbot Louis. A reasonable chronology therefore presents itself for copies of the *Handbook of 809*: Drogo's copy (Madrid 3307), ca. 830; Vat. lat. 645, ca. 827; Reg. lat. 309, ca. 859; Monza f–9/176, ca. 864.

Carolingian Astronomical-Computistical-Pedagogical Handbooks

Before discussing the reconstructed contents of the lost archetypal copy of the *Handbook of 809*, it will be helpful to review briefly the development of the first of three important astronomical-computistical texts during the Carolingian period. Arno Borst's analysis of three discrete versions of such early medieval handbooks and their contents is an important contribution to the joint histories of medieval science, literature, and art.[43]

Einhard (ca. 770–840) composed his *Life of Charlemagne* in 825–26.[44] In it, he describes Charlemagne's keen study of astronomy under the royal tutor Alcuin (ca. 732–804), "the most learned man in the entire world": "[Charlemagne] invested a great deal of time and effort studying rhetoric, dialectic, and particularly astronomy with him. He learned the art of calculation [arithmetic, *artem computandi*] and with deep purpose and great curiosity investigated the movement of the stars."[45] According to the traditional account, Charlemagne first met Alcuin at the court of Parma in March 781. He soon persuaded Alcuin to leave his post as leader of the cathedral school at York and relocate to Charlemagne's court at the very earliest in the spring of 782.[46] Rosamond McKitterick prefers to push this date back to at least 786, following the estimable lead of Donald Bullough.[47] There the polymath Alcuin joined other notable scholars Charlemagne had invited to his court—which was only later identified, tenuously, with Aachen[48]—including the grammarian Peter of

Pisa, who was assigned the task of teaching grammar to the king. Einhard affectionately (and perhaps somewhat sardonically) reports that the king would place writing tablets under his pillow at night in the hope of overcoming his calligraphic inadequacies.[49] Charlemagne's courtly cadre of intellectual elites included Paulinus (who eventually was elevated to the patriarchate of Aquileia). As Josef Fleckenstein has made clear, even after the influential contributions of the Langobard Paul the Deacon, the Irishman Dungal, and the Visigoth Theodulf to Charlemagne's court culture, Alcuin's position as preeminent instructor of the liberal arts can be tendentiously sustained.[50]

According to Arno Borst, Alcuin demonstrated his astronomical expertise by approving or composing a series of texts compiled in the years 782–97. Where and under what circumstances he drafted or collected them should remain open questions. Later, in 797–99, Alcuin included these texts within a series of letters that formed part of a correspondence with Charlemagne. They reveal that Alcuin was involved in identifying, clarifying, and resolving astronomical questions at or for the mobile Carolingian court during the 780s and 790s. His first text, *Ratio de luna*, discusses the length of time needed for the moon to pass through the signs of the zodiac. The difference between the velocity of the moon, according to Pliny, and the velocity of the sun is evaluated using the ancient twelve-part system of Roman fractions known as *uncia*. Alcuin's text demonstrates a preference for accurate natural science over spiritualized allegory.[51] This is a significant point, because in the decades leading up to the synod of 809, there was a widespread effort to survey the state of astronomical knowledge from late antiquity, and the expertise of Alcuin, a respected master of the liberal arts, was partly manifest in the expansive and catholic nature of his learning. He was conversant with both the Christian Bede and the classical pagan Pliny, which permitted him to take full advantage of the Roman natural historian's ideas with impunity.

In Alcuin's second putative astronomical text of 797, *De saltu lunae*, he informs Charlemagne about the need to reduce the lunar calendar by one day at the end of every nineteen-year cycle. This reduction is referred to in computistical literature as the *saltus lunae*. To appreciate the impetus for the "leap of the moon," it is necessary to understand that reckoning the year according to a concatenated series of solar months, such as in the standard Western calendrical system employed today, results in eleven more days than the same year reckoned according to a series of lunar months. This discrepancy over time is offset by inserted, or intercalated, lunar months otherwise known as "embolismic" months. Invariably, over the course of a normative nineteen-year cycle (such as those contained within the tables of Drogo's copy of the *Handbook of 809* on fols. 7r–20v), this results in an additional disparity of one day. In addition to the other standard procedures used over a nineteen-year cycle, that final thirtieth day of an intercalated embolismic month must be skipped over in order to return the system to coordinated calibration for Easter reckoning. The final day was deleted by changing the embolismic month from a "full" thirty-day lunar month into a day-short, twenty-nine-day, "hollow" version. Where this leaping took place within a nineteen-year cycle was a matter of controversy.[52]

The treatise examines the computistical effects of dating with a nineteen-year solar calendar, according to both the Latin followers of Victorius in Gaul and the Greek followers of the more accurate computistical tables prepared by Dionysius.[53] Authorship of the *De saltu lunae* has occasionally been attributed to Alcuin, but Kerstin Springsfeld has shown the attribution to be spurious, since *De saltu lunae* documents Alcuin's inability to influence the official Frankish procedure for resolving the date of the lunar leap. Rather than conforming to Roman practice, the Alexandrian "Egyptian boys" prevailed, commencing the official year on 1 September and placing the "leap" earlier, on 30 July.[54]

According to Borst, Alcuin claimed to have completed a work—the lost *Libellus annalis*—that hypothetically could have fulfilled Charlemagne's desire for a comprehensive astronomical and computistical handbook. Alcuin asserted that Charlemagne's appreciation of computistical matters negated the need for such a book (a protest that Borst rightly deems sycophantic). He probably did write some sort of astronomical treatise called the *Libellus annalis* (ca. 793) for Charlemagne, although its significance is difficult to ascertain.[55]

There was a medieval belief that Alcuin's efforts had in fact supplied Charlemagne with his desired handbook, but this is probably apocryphal. The confusion arose during the Middle Ages because of the similarity between the name Alcuin had adopted for his lost treatise, the *Libellus annalis*, and the name given the actual first official Carolingian compilation meriting the title of an astronomical-computistical-pedagogical handbook. This work, the *Annalis libellus*, confounded matters further because it was also composed in 793. Borst declares that there are no surviving manuscripts of the putative Alcuin text *Libellus annalis*, and no excerpts from the *Annalis libellus* bear firm attributions to Alcuin, either. According to Borst, it is better to consider a copy of the *Annalis libellus* originally from the vicinity of Verona (the best example is now Staatsbibliothek zu Berlin, Phillips 1831, fols. 116r–125v; hereafter the Verona compilation) to be a truly new synthesis of celestial learning prepared for educational use.[56] Verona was one of the faithful outposts of Carolingian culture and flourished from 796 to 799 under Bishop Egino.[57] In the Verona copy of *Annalis libellus*, excerpts derived from Isidore of Seville's *Etymologies*, Irish precursors, and pseudo-Bedan texts address computistical problems in a jejune and cursory way. There is no effective tabulation of data in the anonymous first version of the handbook (referred to as "type A" by Borst, and as the *Annalis libellus* in the scholarly literature).[58] This first version was a pedagogical compilation that offers preliminary insights for novices in the craft of computistics. It could not, however, supply an autodidact or neophyte with a comprehensive introduction to the liberal art of astronomy and its utility for ecclesiasts. The very convenience and scale of the *Annalis libellus* prevented it from being a definitive computistical curriculum or astronomical reference work going forward.

It is important to revisit Donald Bullough's challenges to the likelihood of Alcuin's residence in Frankish Gaul on an official basis before 786. On the one hand, Bullough rightly underscores the precarious political circumstances in which Alcuin's relocation to Charlemagne's court was supposed to have taken place. In the wake of the turmoil surrounding the ouster of the Northumbrian king Æthelred in 779, it is highly probable that Alcuin was simply seeking patronage and a new courtly context rather than a post in which he could exert maximal influence over the development of early medieval intellectual traditions. It is hard not to see the move to Charlemagne's court as an ideal opportunity for the sage-courtier to alter the state of liberal arts instruction on the Continent, yet the conservative nature of Alcuin's documented pedagogical approach once he was firmly entrenched in the Frankish milieu does suggest that pragmatic interests contributed to his decision to depart York.[59] Bullough's misgivings derive primarily from one synodal report concerning a council believed to have convened in Corbridge in 786, naming two "illustrious men, legates of the [Northumbrian] king and archbishop, namely the lectores Alcuin and Pyttel."[60] Careful exegesis of the text could suggest that Alcuin was already in Northumbria when he was called upon to assist Bishop George of Ostia and Amiens at the synod.

On the other hand, an alternative exegesis also presents itself with potential implications for Alcuin's period of influence. Bishop George had been an official papal legate in the service of Hadrian on an ecclesiastical mission to the north, with Bishop Theophylact of Todi and a priestly Abbot Wigbod, probably linked to Saint Maximin of Trier. Naturally, they passed through Frankish territory on their way to the Channel, and Alcuin's assistance, not to mention familiarity with Hadrian, would have conferred a boon on the papal envoy, benefiting the mission. Under Bullough's interpretation, Alcuin's arrival at Charlemagne's court would have been delayed until 786 at the earliest; in fact, Alcuin could have simply been enlisted en route, perhaps even at Hadrian's behest, as relayed to him by George and Theophylact.[61] If so, then Alcuin's involvement with the Frankish court, and Charlemagne in particular, following the original 781 invitation in Parma might have been more formalized than previously considered, even if it is impossible to identify the precise nature of the relationship. Indeed, courtly service in the 780s did not entail stationary commitments. On the contrary, the correspondence between Alcuin and Charlemagne overcame the distances that legatine and administrative commitments required, permitting the continuity of astronomical dialogue between Charlemagne and his personal tutor throughout the late 790s, well after the creation of the *Annalis libellus* in 793. As Springsfeld rightly notes, however, a correspondence between Alcuin and Paul the Deacon (who left the court by 786, as noted earlier) argues in favor of a familiar camaraderie that could only have developed if they had jointly served at Charlemagne's court at some point in the years 782–86.[62]

Given the mobile nature of Charlemagne's early court, it is perhaps prudent to reconsider many aspects of this debate. Peripatetic and itinerant ecclesiasts and courtly officials would routinely have come into overlapping spheres of influence, some of which can be documented through sedulous archival work. In many cases, however, understanding the interaction among early medieval intellectuals working within and around Frankish courtly circles requires the interpretative and exegetical inferences called for by Springsfeld.[63] In any case, the concerns voiced by Borst, Bullough, and McKitterick argue against Alcuin's direct involvement in the creation of a prototypical first handbook. Alcuin's letters drafted during the period preceding his appointment as abbot of the monastery of Saint-Martin at Tours in 796 nevertheless do remain the best record of his participation in the sorts of astronomical discussions that informed the eventual composition of the *Handbook of 809*.[64] The correspondence demonstrates Alcuin's engagement from afar and supports the opinion, endorsed here, that he was more integrally involved with courtly exchanges before 786.

Bullough also prefers to place the origin for Borst's first, type A version of the Frankish astronomical-computistical-pedagogical handbooks, known as the *Annalis libellus*, in Austrasia (or western Germany and northern France) rather than Verona.[65] An origin in Austrasia would insert the original Carolingian computistical manuals in the area not far from Metz, capital of Austrasia. This would suggest that there was a long-standing Austrasian interest in astronomical matters during the Carolingian period. Such enthusiasm could have contributed to a local desire for a lavishly painted astronomical-computistical-pedagogical work such as Drogo's copy of the *Handbook of 809* (Madrid 3307).

Borst sensibly situates the creation of the original copy of the first handbook (type A, or the *Annalis libellus*) in Verona, since that was the site of manufacture for the oldest surviving (ca. 800–810) complete copy, the Verona compilation presently located in Berlin.[66] However, this manuscript, the best record of the original *Annalis libellus*, has been appropriately critiqued for its lack of illustrations and tables.[67]

The complete lack of illustration in the *Annalis libellus* is significant. In the second version of an astronomical-computistical-pedagogical handbook, the *Handbook of 809*, the Carolingian scribes and artists recognized the importance of visual supplements, resulting in the images of the constellations and cosmological diagrams in Madrid 3307, Drogo's copy. In Drogo's book, the scientifically astute prelates who compiled the treatise recognized the significance and value of mediated presentations of the heavens in partial fulfillment of their clerical and civic duty to promulgate effective liberal arts instruction throughout the Frankish lands. The pictures and diagrams in the Madrid copy enrich the scientific texts. This synthesis of text and image is one of the innovative aspects of the revised, second version of the compilation, the *Handbook of 809*.

For these reasons, it is important to recall that the learned churchmen who offered such pagan pictures to their coreligionists situated their discussion and discernment of the utility of these illustrations within otherwise widespread debates concerning the suitable role of pictorial programs in ecclesiastical spaces and public arenas. The introduction of visual material into the *Handbook of 809* underscored the liminal nature of celestial scenery, built upon the classical *Latinitas* of Rome, derived from the traditions linked to the Three Kings from afar, and formalized by Greek poets and intellectuals from Aratus to Ptolemy. Unsurprisingly, this Greek Byzantine connection would cause a Carolingian apology for the star pictures to confront Babylonian and Macedonian theological and astronomical traditions, their Western classical reinterpretation, and current Frankish debates about Byzantine image theory.

There is no reason to think that the creators of the illustration-free Italian handbook from Verona had an appropriate model for illustration on hand, or that the monastics making editorial decisions about the Verona compilation discerned any need to include imagery in it. However, the charted information, diagrams, and pictorial representations of the constellations that appeared in the *Handbook of 809* were also desirable justifications for its compilation and hence a major impetus for the creation of Drogo's book.

Nineteen-Year Tables—a User's Guide: Contents of the *Handbook of 809*

As explained in the introduction, Drogo's copy of the *Handbook of 809* (belonging to Borst's second, or type B, version of an astronomical-computistical-pedagogical handbook) is regrettably incomplete. It supplies, however, the superior textual record of the lost original handbook,[68] even though the contents of all seven original books from the archetypal *Handbook of 809* are reassembled using a complete table of contents in Reg. lat. 309 (fols. 4v–6r) and the Vat. lat. 645 copy of the *Handbook of 809* (fols. 1r–2v), following Fritz Saxl.[69] A general introduction to the contents of the *Handbook of 809* best prepares a modern reader to appreciate the breadth of topics and lacunae in the compilation created under the probable direction of Adalhard of Corbie.

The compilers of the *Handbook of 809* placed a diagrammatic calendar of saintly feast days at the outset of book I, making it clear that the church calendar was the earthly liturgical pathway to Paradise. The diagram—originally presented in the form of a *rota quae continet natalicia sanctorum* (book I.1), that is, in the form of an Isidorean wheel diagram—was derived from one of the fundamental achievements of the Carolingian renewal, the Lorsch calendar cycle (789). Key collaborators in the calendrical reform movement throughout the Frankish lands included Hildebald of Cologne[70] and another important innovator who participated in the development of a third edition of the astronomical-computistical-pedagogical handbooks, Arn of Salzburg, discussed below.[71] Churchmen were the fundamental participants in these reform projects and the main beneficiaries of the tools they created. The primary motivation for the creation of the *Handbook of 809* was signaled by a parenthetical embrasure, opening and introducing the remaining contents of the anthology, which duly concluded with an equally suitable coda and closing parenthesis in book VII, reproducing in full a copy of Bede's *De natura rerum*. On display for information and delight in the *Handbook of 809* was a presentation of the soteriological possibilities of Frankish science. This was reinforced by the inclusion of a martyrologium (book I.3), contained in fragmentary form within Drogo's book on fols. 5r–6v, 79r–80v, and a restored slip of parchment formerly used to bind the book.[72]

A *sine qua non* for a normative and functional computistical handbook in the ninth century was a set of nineteen-year tables such as those contained in book I.4 of the *Handbook of 809*. Art historians could be tempted to turn the pages when confronted with these confusing tables, but to do so would be to miss a rich repository of vital information contained within manuscript copies of the *Handbook of 809*. The tabulated information, which identifies the dates for the feast of Easter and permits computistical operations, establishes at the outset of a copy of the *Handbook of 809* such as Drogo's (figs. 10–18) that the book was made to enable the work of the church.

This ecclesiastical bent pervades the formulation and consideration of the nineteen-year tables, which were handy reference guides, charting physical relationships between the lunar months and the solar months for liturgical reasons. As early as 432 B.C.E., a pair of Athenian astronomers, Meton and Euctemon, identified a cyclical relationship (also of interest to Babylonian stargazers) uniting the lunar and solar calendars, which jointly had to be considered when reckoning the date for the floating liturgical feast of Easter.[73] Without fail, the standard date for a Paschal new moon (called *luna I*) or Paschal full moon (typically referred to as *luna XIV*, that is, the fourteenth day of the Easter lunation, or monthly lunar cycle), will repeat every nineteen years. This can quickly be confirmed by scanning any of the nineteen-year tables illustrated here from Drogo's copy of the *Handbook of 809*: notice that the last line in every table ends with the date in column seven (proceeding left-to-right and top-to-bottom) of *xv kalendas maias*, or 17 April. In other words, every nineteen years, a new moon will arise on luna I, the spring equinox will pass, and day 14 of the lunation (luna XIV) for the (then-full) Paschal moon will fall again on 17 April.[74] This classically observed astronomical consistency was revealed by luni-solar nineteen-year cycles, otherwise referred to as decennovenal cycles (that is, the *cyclus decemnovenalis*). These tables were key tools for the student of computus, used to fix the date for the floating feast of Easter and perform arithmetical calendrical operations. The nineteen-year cycles resolved the disparity between a series of twelve solar months, lasting 365.25 days, and a concatenated series of twelve alternating hollow and full lunar months, lasting 354 days (that is, $[29 \times 6] + [30 \times 6] = 354$). Over a nineteen-year period, the roughly eleven-day annual difference in days between the lunar and solar calendars results in a series of seven embolismic intercalated months.[75]

Dionysius Exiguus (d. 556) compiled a set of nineteen-year tables, also known as Alexandrian tables, that ran from 532 until 626. Bede (d. 735) championed this approach in his *De temporum ratione*, in which he supplied a more comprehensive set of tables, complete through the year 1063. This, in fact, is the last year included within the tables of Drogo's copy of the *Handbook of 809* on folio 20v.[76] For Dionysius, the authority of the Council of Nicaea in 325 was to be thoroughly respected, and the stipulations for Easter reckoning declared at the synod informed his creation of nineteen-year tables in the Alexandrian (meaning Greek) Egyptian tradition. Dionysius hailed from modern Romania, and this Scythian extended by ninety-five years (or five cycles of nineteen years) the hitherto noteworthy nineteen-year tables, which a figure referred to as the Pseudo-Cyril had prepared for his Western brethren in Alexandria.[77]

The respect for Bede's authority among the Carolingian prelates at Aachen resulted in the use of the nineteen-year tables associated with Dionysius. These tables eventually supplanted the rival set promulgated throughout Gaul by Victorius of Aquitaine, who was active in the late 450s and had created his nineteen-year tables at the behest of Pope Leo the Great's (served 440–61) archdeacon Hilarius in 457.[78] James Palmer has rightly cautioned that the ultimate success of Dionysius's tables came late throughout the Frankish lands. Victorius's tables were still favored in some computistical treatises, including a text that could have originated in Austrasia as late as 764. As Palmer notes, this was of course one reason for the creation of the *Handbook of 809* in the first place.[79]

The tables reveal that in order for a specific day of the week and date for the feast of Easter to repeat, it would be necessary to await the repetition of all seven days of the week and take into account the intercalary day inserted into February every fourth year, during the leap year. This number has to be multiplied by the nineteen years required for the date of the Paschal moon to repeat, as mentioned already, and listed in the nineteen-year tables. This constituted a Great Paschal Cycle of 532 years ($7 \times 4 \times 19 = 532$).[80] These tables provided one vital computistical contribution of the *Handbook of 809* for use in schools or churches.

The nineteen-year tables found within the second version of the astronomical-computistical-pedagogical handbook (the *Handbook of 809*) tabulated all requisite information in ten columns, following Bede's example. A copy of his *De temporum ratione* was made in Cologne in 795, according to Anton von Euw (Cologne, Dombibliothek, MS 103, fols. 52v–184r).[81] Von Euw noted that Bede manuscripts lacked

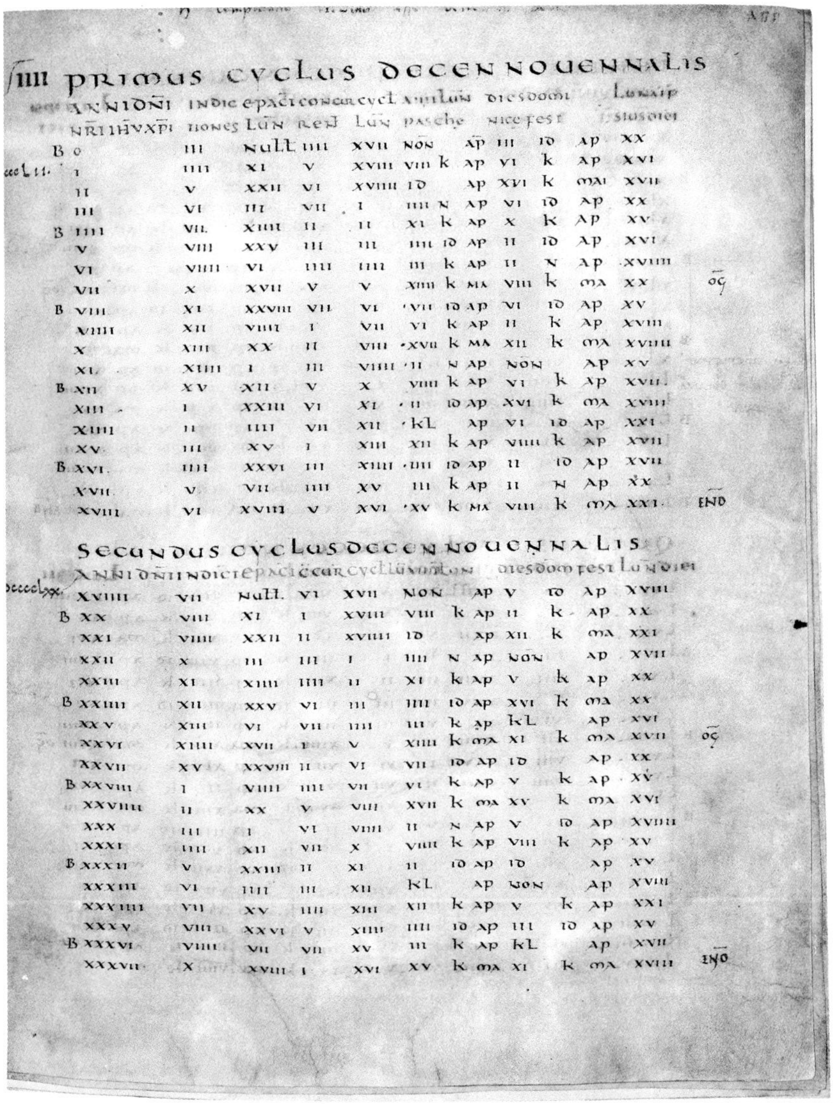

FIGURE 10

Nineteen-year tables 1 and 2. Madrid, Biblioteca Nacional de España, MS 3307, fol. 7r. Photo © Biblioteca Nacional de España.

illustrations until the Romanesque period, as the non-illustrated, first astronomical compilation from Verona had also lacked imagery. The earliest compilation of works by Bede with a pictorial program of any kind is apparently the complex but rich anthology at the Walters Art Museum (MS W. 73).[82]

A brief explanatory note is in order before working through the columns of the nineteen-year tables.[83] Any included piece of data constitutes a column, ranging in detail from a single letter, such as the letter *B* in column one (proceeding from left to right), to a complete date rendered in the old Roman system of dating. The letter *B* is listed alongside every fourth year in any given table, precisely because it signals the so-called *bissextiles anni*, the *bissextus*, or the bissextile day, designating those years requiring the intercalary day appended to every leap year in the solar Julian calendar. For example, on folio 11r of Drogo's copy of the *Handbook of 809*, the leap years are appropriately indicated for the years 304, 308, 312, 316, and 320 in the seventeenth nineteen-year table (fig. 12). The second column indicates the years since the birth of Jesus, the A.D./C.E. dates, or *anni domini*. Bede had followed this organizational procedure for his tables, so the Carolingian adoption of the practice is one of expediency as well as ideological significance. It comes as little surprise, on the one hand, that charts intended to identify the proper feast day for Easter would proceed with dates relative to Jesus's Nativity. The tabulated information provided a progressive record of Christian time. The tables thereby commemorated the life and death of the medieval Christian spiritual king, Jesus. This ideological framework was advocated by such theologians as the Venerable Bede, Alcuin, and Archbishop Hildebald of Cologne (d. 818), archchaplain of the court in Aachen after 791.[84] The chronological history of the Christian world dated from the birth of its divine ruler. This fixed point, Christ's birth, permitted the calculation of the day appropriate to celebrate his Passion. This submitted the progressive and repeated cycle of ecclesiastical time, and by extension every cyclical year, to the overarching dominion of the Christian Creator. Frankish time was placed in relation to the perpetual assurance of eschatological fulfillment.[85]

It is also true, however, that the failure to situate 1 B.C.E. as the first of the nineteen years in table 1, following the example of Dionysius, would have resulted in serious organizational difficulties for the system of nineteen-year tables. In fig. 10, the first entry for column two is 0. This corresponds to the year -1, or 1 B.C.E. To maintain the accuracy of the Easter feasts (which is paramount) and ensure that the leap years fall on years readily divisible by 4, it was incumbent upon Dionysius, and later Bede, to preserve this version of the system. The Carolingian compilers of Drogo's copy of the *Handbook of 809* followed their example.[86] In the end, the impetus for this decision was the celebration of Dionysius's preferred nineteen-year Alexandrian tables and not a conscious revision of the traditional reckoning of tables related to the reign of Diocletian.[87]

The Franks adopted the tables that a well-trained cleric adept at computus would be required to manipulate and read. In the service of the church and Carolingian schools, as called for in the *Admonitio generalis*, the nineteen-year tables participated in the intellectual spiritual reform of the Franks through erudition. In other words, Dionysius's formation of his tables was without ideology, but the Franks' inclusion of the nineteen-year tables within the *Handbook of 809* altered their significance.

The third column of the nineteen-year tables offers the indiction year. The *indictio* numbers run from one through fifteen. In late antiquity, the indiction number had marked the amount of cyclical time that had transpired within a fifteen-year periodic but flexible system of taxation beginning with 297 and recommencing fifteen years later, on 1 September 312.[88] In Drogo's copy, the indiction number for the year 297 is appropriately listed as year fifteen in the cycle and is adduced in table sixteen of the nineteen-year tables on the bottom of folio 10v (fig. 11). Year 313 is designated in table seventeen as year 1 (fig. 12).[89]

There follow in the next several columns straightforward pieces of information that are useful for solving certain computistical problems. In column four, the *epacts* are listed.[90] There are different ways to analyze the utility of epacts for computistical operations. The age of the moon on 22 March is a helpful indicator of the correct date to celebrate Easter, which cannot occur any earlier than this date. An epact is a varying number (ranging from 1 to 30 in theory, but 1 through 28 in nineteen-year tabular practice) of days remaining during any given calendar year upon the completion of a cycle of twelve lunar months. As noted above, this lunar year runs its course first, concluding eleven days sooner than a typical calendar (or solar) year.[91] It was an established convention in the early medieval world for the lunar months to progressively alternate between months with either twenty-nine days, called "hollow" months, or thirty days, called "full" months.[92] In general, given the annual eleven-day surfeit in the lunar year relative to the solar year, an intercalary month must be inserted into the cycle of months roughly every third year.[93] For this reason, still in table seventeen in Drogo's copy of the handbook (fig. 12), the epact for 305 is 11, the epact for 306 is 22, and the epact for 307 is 3: that is, the result of 33 excess

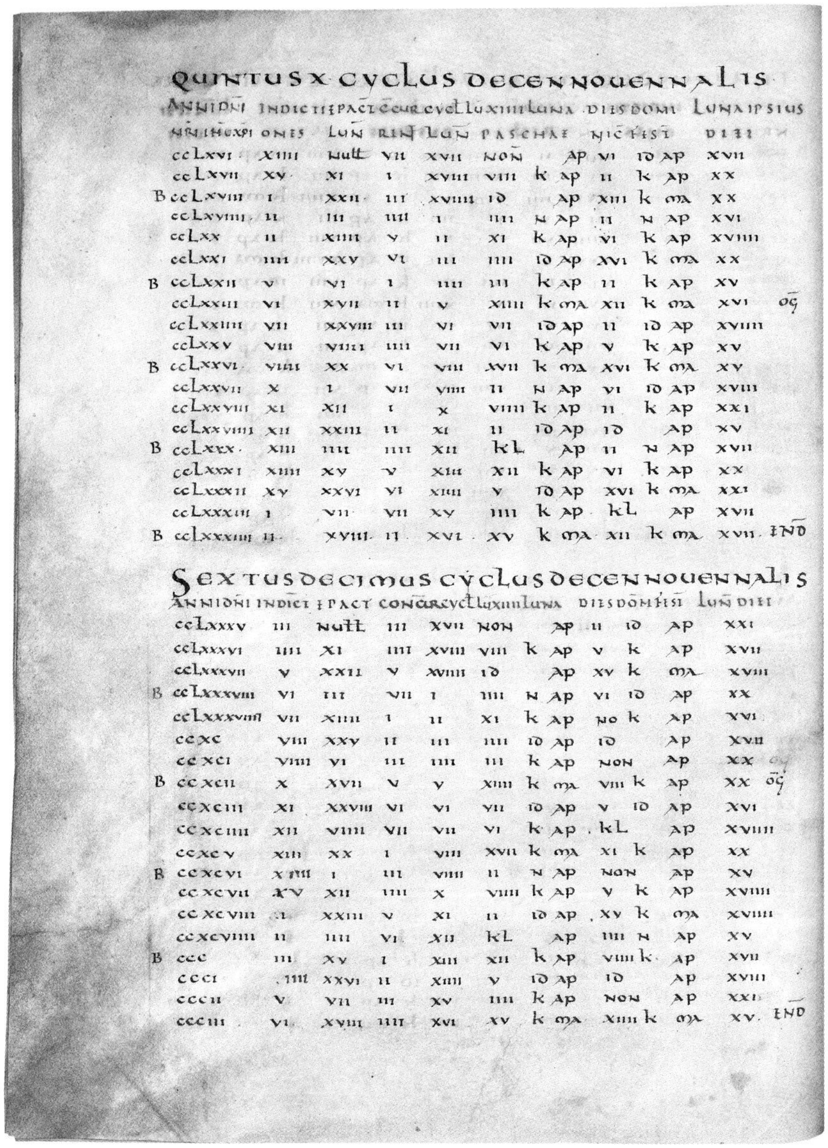

FIGURE 11

Nineteen-year tables 15 and 16. Madrid, Biblioteca Nacional de España, MS 3307, fol. 10v. Photo © Biblioteca Nacional de España.

lunar days from which a 30-day intercalary month has been subtracted.

An ingenious result of medieval computistics was the discovery that epact numbers added to a certain fixed value yielded the current daily age of the moon on the first day of the calendar month (known as the monthly *kalendae*) for any given month during the solar year.[94] Again, the most useful application of this information was in calculating the fourteenth day of the Paschal moon. It could be beneficial to estimate when a full moon is likely to fall relative to the beginning of any ordinary calendar month of the solar year, and these computations supplied timekeepers with readily adaptable principles and temporal guideposts. This was not complicated mathematics. The fixed values, known as the lunar *regulares*, were divided as follows: January–February (8), March–August (6), September (7), October (6), November (7), December (6).[95] For example, in chart seventeen, during the year 313, the epact is 9 (fig. 12). In November the lunar regular is always 7 for every year, so 9 + 7 = 16. The moon is already sixteen days old on the *kalends* of that month, 1 November. The previous full moon was two days earlier, on the fourteenth day of the moon, or 30 October. By this simple calculation, the age of the moon, and by extrapolation the date for a new moon, can be determined; in this case, the new moon beginning the cycle would fall on 16 October. These procedures made it possible to determine when the Paschal new moon would appear for any given year, using the tables and background knowledge about the lunar regulares.

In column five, the concurrentes offer the day of the week that coincides with 24 March. This invaluable information can assist with the reckoning of Easter, which, in accordance with the stipulations established at Nicaea in 325, follows the vernal equinox. In addition, since March follows February—which naturally fluctuates between 28 and 29 days in a month, depending on whether there is a leap year—March is a better gauge of the day of the week for discrete dates within a calendar year. According to this use for the concurrentes, a 1 designates a Sunday, Monday is a 2, and so forth through Saturday, which is a 7. In other words, in 305 the value of 7 assigned to the concurrentes indicated that the seventh day of the week, or Saturday, would fall on 24 March.[96]

By extrapolation, the concurrentes can therefore also supply the number of days separating the final Sunday celebrated before the end of one calendar year and 1 January, when a new solar twelve-month cycle commences. So these concurrentes would also inadvertently help someone figure out when the first Sunday of any given year would transpire. After all, if the year ends on a Sunday, then it is necessary to wait a week for the first Sunday of the following calendar year. The numbers are therefore listed 1 through 7 in order to accommodate all possible cases.[97] As just described, when 31 December is a Sunday, as it was in table seventeen during the year 304 leading up to the year 305, it takes the maximum number of days to reach the first Sunday of the new year, which fell on January 7 in 305 (fig. 12).

Column six, the *cyclus lunae*, listed the present position of the given year within the nineteen-year cycle, beginning with XVII and proceeding to XVIIII/XIX before returning to I and continuing through XVI at the completion of the cycle. These numbers derive from the Byzantine calendar that was out of synchronization with the Western calendar by a period of sixteen years.[98] This provided a somewhat useful index analogous to the more famous and simple "golden number." Both numbers run from one through

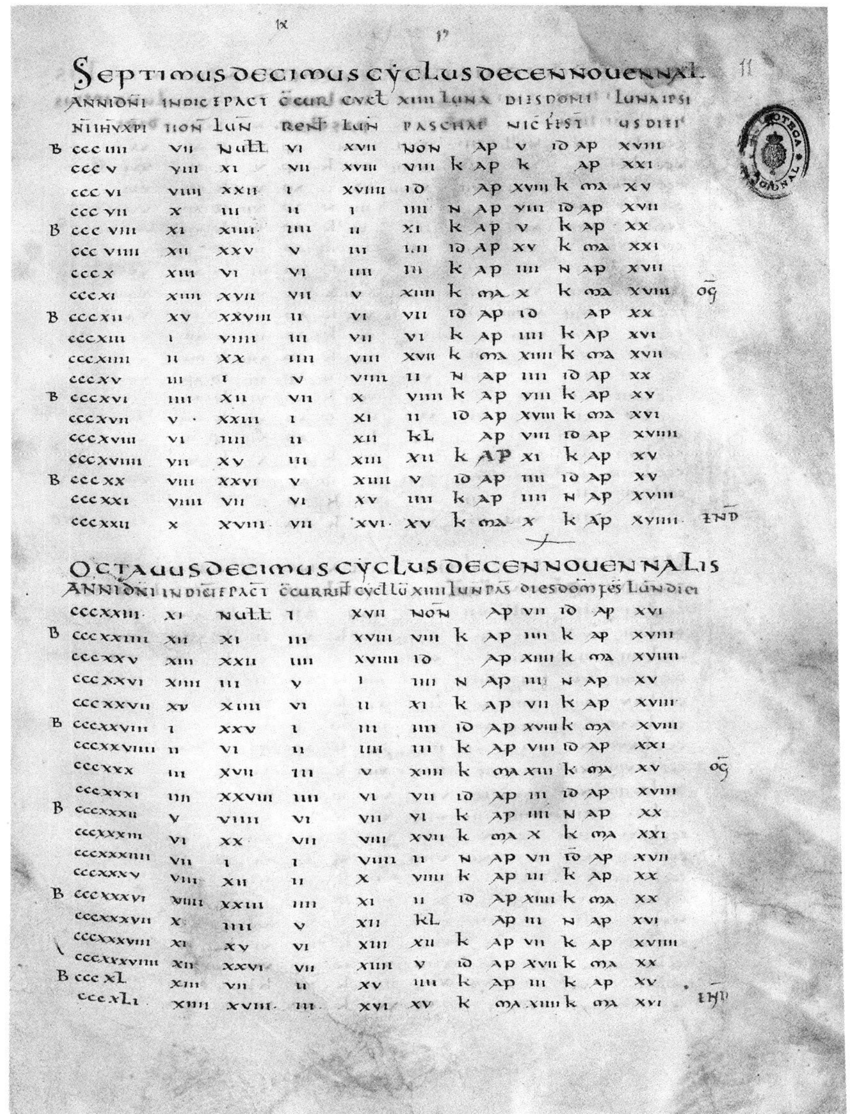

FIGURE 12

Nineteen-year tables 17 and 18. Madrid, Biblioteca Nacional de España, MS 3307, fol. 11r. Photo © Biblioteca Nacional de España.

nineteen and determine the place of a given year within the nineteen-year cycle.[99] This information would help determine dates for a full moon during any given year. Since the dates of full moons roughly repeat themselves in every nineteen-year cycle, by establishing the dates for full moons during one nineteen-year cycle, a prelate could fairly accurately predict the dates on which full moons would appear in future years. Any years that shared the same cycle year, or golden number, shared nearly identical dates for the appearance of the full moons during the calendar year. They ordinarily could be off by at most one day. This holds true for all the years in the different tables of the nineteen-year cycles with the same cycle year or golden number. The golden number is attained by taking the year, dividing by nineteen, and adding one to the value of the remainder. So, for example, in table seventeen, 305/19 = 16 + remainder 1. The golden number for 305 is then 1 + 1 = 2, which is sensible, as 305 is the second year in the cycle (fig. 12). Although the cyclical index number XVIII likewise identified the second year of the cycle in table seventeen, the golden number was a more intuitive and preferable index, used since at least the early tenth century.[100]

In column seven, the table reported the date for the first full moon to succeed the vernal equinox according to the classical Latin system of dating.[101] The equinox was believed to fall on 21 March during the Carolingian period, and it ushered in the spring. The label for this column is *Luna XIIII Paschae*, because the date for the full moon of Easter occurred fourteen days after the Paschal new moon, following the vernal equinox.[102] Nicaea had adopted 21 March as the official date, which was considered incontrovertible.[103]

In column eight, the dates for Easter Sunday complied with the dictates established at the Nicaean Council in 325 and could fall any time between 22 March at the earliest and 25 April at the latest. In general, Easter took place on the first non-coterminous Sunday after the full moon following the vernal equinox, which (as just stated) was assigned to 21 March. The label for the eighth column was *Dies dominica*. The easiest way to understand the range of dates is to recall that a hollow, twenty-nine day Easter lunar month could supply a full moon on the vernal equinox at the earliest, which could yield an Easter date of 21 March + 1 day to attain Sunday, or 22 March. In the event that the fourteenth day after the Paschal new moon occurred on a Sunday, Easter was postponed one week, according to the mandates established at Nicaea. So this leaves 22 March as the earliest possible date for Easter Sunday, but only when the full moon comes on Saturday, 21 March, in the same year. At the other end of the range of dates, in the event that a full moon had appeared on the day before the vernal equinox, then the next full moon would occur roughly 28 days later than 21 March—and 21 March plus the remaining 28 days of that (twenty-nine-day hollow) lunation would fall on 18 April. In the event the fourteenth day in the cycle of the Paschal moon fell on Sunday, 18 April, then according to the rules established at Nicaea, Easter occurred one week later, yielding the latest possible date for the holiday: 18 April + 7 = 25 April.[104]

In column nine, the *Luna ipsius diei* provided the age of the moon on Easter Sunday.[105] In 305 on table seventeen, for example, the moon was twenty-one days old on Easter Sunday (fig. 12).

In column ten there were two marginal additions to the text. After row eight, the compilers of Drogo's copy of the *Handbook of 809* abbreviated the term *ogdoas*, and then in the final row of every table of the nineteen-year cycle, the term *(h)endecas* was likewise indexed. These terms divided up the nineteen-year cycle into two subsets of years that enabled the tables to account for the need to include intercalary lunar months, as discussed above. In 307, the epact was 3 because of an intercalary month that had been added (fig. 12). In fact, the eight years of the *ogdoas* require the following numbers of lunar months per solar calendar year, again using table seventeen as an example: 304 (12), 305 (12), 306 (13), 307 (12), 308 (12), 309 (13), 310 (12), 311 (13). Then, the final series of eleven years, yielding the name *endecas*, divided the years accordingly: 312 (12), 313 (12), 314 (13), 315 (12), 316 (12), 317 (13), 318 (12), 319 (12), 320 (13), 321 (12), 322 (13), although for the final year of the nineteen-year cycle, the *saltus lunae* diminution of the lunar calendar by one day must also be taken into account.[106]

The *Annales Prumienses*

As was common practice in the Middle Ages, the nineteen-year tables provided convenient lists of dates to which monastics could add marginal notes. These lists recorded both key events in the Carolingian Empire and occurrences of merely local significance. The so-called *Little Royal Annals* (*Kleine Königsannalen*) were included in Drogo's copy of the *Handbook of 809*, for the *Little Royal Annals* had become linked at an early stage to the nineteen-year tables of the astronomical-computistical-pedagogical handbooks. Scribes also added historical notes important to the monks in the abbey of Prüm, and these, together with the *Little Royal Annals*, built up the *Annales Prumienses* by accretion. Scribes noted celestial phenomena and historical events until 1044, long after Drogo's copy of the handbook had found its way to Liège around 922 (as described below).[107]

Claudia Höhl has studied the manuscripts made at Prüm during the Middle Ages and the relevant history of the site. Prüm was an important monastery for the Carolingian ruling houses, located roughly halfway between Aachen and Metz, near the epicenter of the Austrasian heartland. Since 721 the abbey of Prüm had played an important role in the local dynastic celebration of Bertrada's aristocratic descendants and in the veneration of the cult of the Virgin. The Carolingians received the abbey from the important Irmina-Hugobert-Sippe family when Bertrada's granddaughter, who was also named Bertrada (d. 783), wed Pippin III, known as the Younger (d. 768), the father of Charlemagne (d. 814). As a result, Prüm was endowed with the protection of the Frankish king after 762. Rogue Carolingian relatives whose subterfuges or revolts had been thwarted were incarcerated there, but Prüm also served as a sanctuary for those fleeing internecine strife. Noteworthy examples include the usurper Pippin the Hunchback (d. 811) and Charles the Bald (d. 877), Charlemagne's grandson and the youngest son of Louis the Pious (d. 840). Emperor Lothar I eventually adopted the tonsure and died in 855 at the monastery, where he was subsequently laid to rest.[108]

The *Annales Prumienses* provide incontrovertible evidence that Drogo's copy of the *Handbook of 809* (Madrid 3307) came to Prüm. They also provide evidence for when that transfer took place. It does not appear likely that the manuscript was made in Prüm. Instead, the book came to the monastery as yet another example of the patronage and privilege the abbey enjoyed as a foundation in the service of the ancestors of the Carolingian dynasty belonging to the Arnulfing royal family. An analysis of the hands that annotated the nineteen-year tables in the Madrid manuscript yields no definitive results,

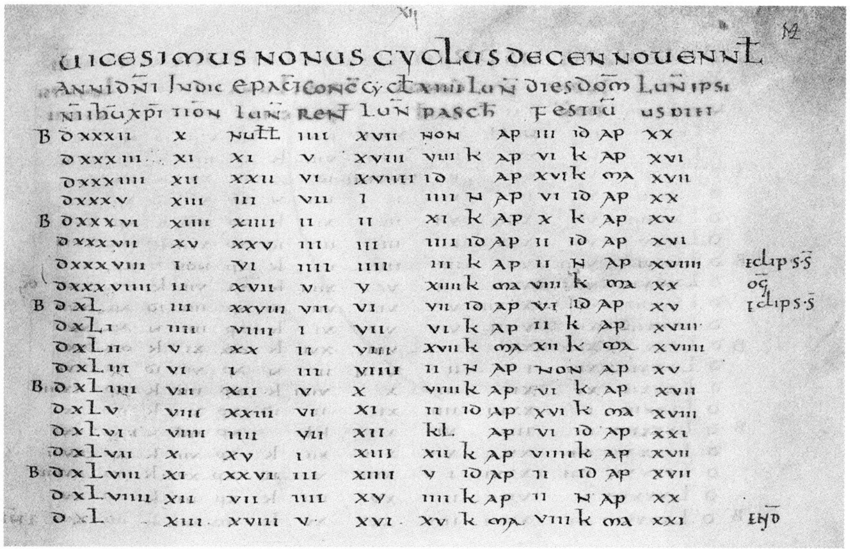

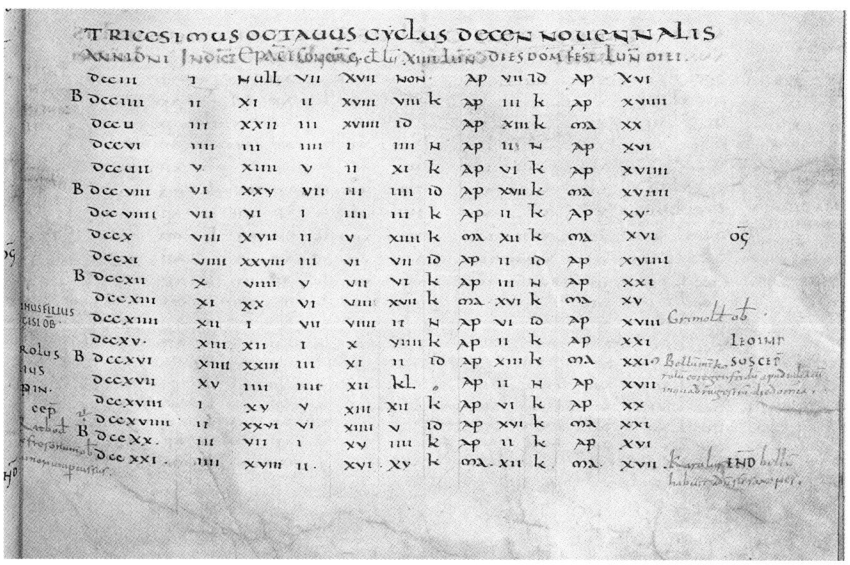

FIGURE 13

Nineteen-year table 29. Madrid, Biblioteca Nacional de España, MS 3307, fol. 14r, top (detail). Photo © Biblioteca Nacional de España.

FIGURE 14

Nineteen-year table 38. Madrid, Biblioteca Nacional de España, MS 3307, fol. 16r, bottom (detail). Photo © Biblioteca Nacional de España.

unfortunately. It is highly probable, however, that the book came to Prüm when Lothar retreated there to die in peace and seek burial within the cloister garden (29 September 855). Drogo himself died later that same year in a freakish drowning accident (8 December 855), perhaps linked to his support for his nephew Lothar.[109] As will be shown in what follows, there is also reason to believe that Drogo—made bishop of Metz in 823 as well as papal vicar north of the Alps in 844[110]—personally sent his copy of the *Handbook of 809* to Prüm along with Lothar.[111]

The scholar Lothar Boschen has investigated the paleographical evidence provided by the hands that worked on the entries of the *Annales Prumienses* in Drogo's copy of the *Handbook of 809*. Boschen argues that three hands participated in the creation of the various capital scripts and that no fewer than fourteen hands supplied annotations in the Caroline minuscule found in the *Annals*. Boschen holds that the three discrete hands that entered additions in capitals supplied the following entries (used here as examples of style for the sake of comparison): Hand A, folio 14r, which notes two solar eclipses that occurred in 538 and 540 (fig. 13); Hand B, folio 16r, which notes the death in 714 of Pippin of Heristal, son of Ansegisel and mayor of the palace (*Hausmeier*) of Austrasia, Neustria, and Burgundy (fig. 14); and Hand C, folio 18v, which notes the death of Louis III (d. 882), king of Saxony (fig. 15). However, the way the lower curve of the *u* bends in these last two entries—namely, slanting toward the right at the juncture with the right-hand vertical stroke arrested at the headline—is so close that it cannot sustain the hypothesis of two separate hands (B and C). Moreover, since Boschen believes that the entry he attributes to Hand C arose in isolation and displays an individual, one-time entry added by an anonymous scribe to Drogo's book, it becomes plausible to collapse the work of Hand B and the putative Hand C.

The first two hands presented by Boschen already divide fairly well into two sets of entries: the group executed by Hand A, which is roughly contemporaneous with the manuscript's production,[112] and the group executed by Hand B, which is of the utmost importance for the dating of Drogo's copy of the *Handbook of 809* (Madrid 3307) and for tracking its travels through various libraries during the early medieval period. If this second group is expanded to include the sole entry by Boschen's Hand C, then the dates encompassed by Hand B/C range from Pippin of Heristal's death in 714 through the death of Louis III in 882. These additions to the annals by Hand B/C would have been added after the manuscript had already found its way to Prüm, whereas the person designated Hand A would have worked in the original scriptorium in Metz, where Drogo's copy of the *Handbook of 809* was made.

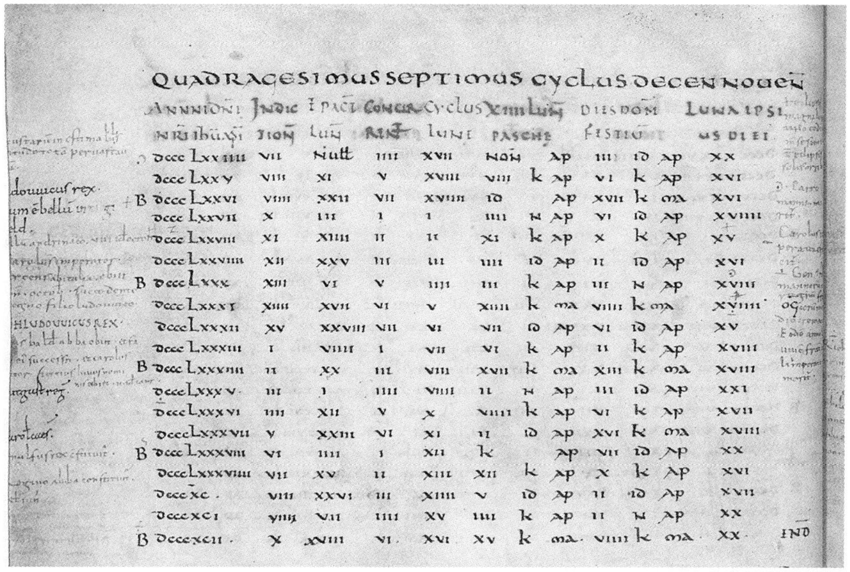

FIGURE 15

Nineteen-year table 47. Madrid, Biblioteca Nacional de España, MS 3307, fol. 18v, top (detail). Photo © Biblioteca Nacional de España.

Wilhelm Koehler has upheld an observation about Hand A advanced in the 1940s by Neuß. Hand A registers the entries for eclipses in the nineteen-year tables of Drogo's copy of the handbook through the lunar eclipse of 820. Since another table of eclipses in the *Handbook of 809* includes a list of eclipses through the year 812 (located on folio 68v in Drogo's book), Drogo's personal copy must have been made after 820.[113] Since by all accounts Hand A entered data at the time of the book's manufacture, this date of 820 supplies a *terminus post quem* for the manuscript's creation in Metz.

Neuß also believed that the entries added by a later hand working in Prüm (Boschen's Hand H, discussed below) included the entry for 828 concerning the succession of Tancradus (d. 829) by Marcuuardus (d. 853) as abbot of Prüm. Neuß thus suggests a *terminus ante quem* of 828 for the creation of Drogo's copy of the *Handbook of 809*. But as Boschen and Koehler have each suggested, Neuß based this hypothesis on the assumption that the list of the abbots of Prüm could not have been added retroactively. In fact, information of such importance to the community in Prüm could have been added at any later time.[114] Koehler guardedly argues instead that Drogo's copy (Madrid 3307) was made some time between 820 and 840 and came to Prüm by the end of the ninth century.[115]

The presence of Hand A in Metz and of Hand B/C in Prüm in fact explains the glaring omission of any entry for Drogo's 855 death in the *Annales Prumienses* in Madrid 3307.[116] Since Hand B/C records an event in the year 882, its presence also precludes

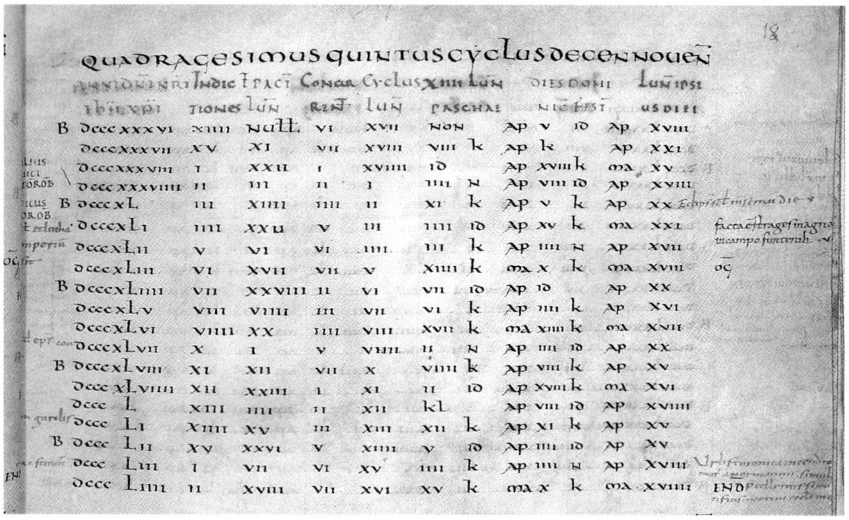

FIGURE 16

Nineteen-year table 45. Madrid, Biblioteca Nacional de España, MS 3307, fol. 18r, top (detail). Photo © Biblioteca Nacional de España.

the possibility, maintained by José Janini and José Serrano, that Drogo's copy remained in Metz until 887.[117] Moreover, if Drogo donated his copy of the *Handbook of 809* to Prüm when his nephew Lothar I sought his final resting place there, and the additional entries in Hand B/C were made around 882, then the lack of an entry noting Drogo's death makes perfect sense. Drogo was no longer of real significance in Prüm, whereas the king, who had died in the abbey's community and was buried in their midst, was worthy of note and recorded in Prüm by Hand B/C in the entry for 855.[118] In fact, the book was probably taken from Metz, where it was in Drogo's possession, and offered to Prüm in gratitude for the medical treatment the dying Lothar I had received there. Perhaps the book was also intended as something of a propitiatory gift to aid the king in his quest to atone for his sins and thereby gain entry into the monastic community at the end of his life.

Scribes in a scriptorium at Metz, where the lavishly illustrated service book for the Mass, the Drogo Sacramentary, had been made, would never have failed to document the death of their beloved Bishop Drogo. The necrology, which is included on folio 128r of the Drogo Sacramentary, lists Drogo's name in honorific gilded uncials. The date for his death is given as "VI ides decembris" (8 December 855). The year 855 provides a significant point of rupture between the campaigns of additions to the annals in the nineteen-year tables. If the *Handbook of 809* was offered to Prüm at that time, as seems highly likely on the paleographic and historic grounds presented here, then the lack of a reference to Drogo in it makes sense. Once the decision had been made to send the handbook to Prüm with Lothar I when he sought admittance to the monastic community there, there was no way to anticipate Drogo's death or retroactively enter its date. It is also relevant that Drogo and Lothar I were not linked by kinship alone: Drogo had served his nephew as archchaplain in Aachen in the years 842–55,[119] and Lothar I had reinstated Drogo in the office of *archicapellanus,* just as his father, Louis the Pious, had done in 834.[120]

From all this, we may reasonably conclude three things: Drogo's donation of his copy of the *Handbook of 809* (Madrid 3307) to Prüm may have been his final act of deference to Lothar I; the manuscript would have pleased the intellectual community of learned men in the monastic school of Prüm; and Drogo's copy probably reached Prüm shortly before Drogo drowned in 855. Drogo had been appointed papal vicar north of the Alps in 844 by Pope Sergius II while on a mission to Rome with Lothar I's eldest son, Louis II (d. 875).[121] For Drogo, the donation of his astronomical-computistical-pedagogical handbook would have been an act of ecclesiastical diplomacy and a generous gift to the Carolingian family into which he had been born as an illegitimate son of Emperor Charlemagne.[122]

The allocation of the hands using Caroline minuscule in the *Annales Prumienses* in the handbook is more complex. Rather than subject the reader to an exhaustive review of Boschen's research on the subject, only a few emendations of importance for the dating of Drogo's copy of the *Handbook of 809* are offered here. Boschen believes that at least four scribes added notes to the annals in the handbook for dates ranging from 841 to 887. The attributions to Hands D, F, and G seem reasonable enough, as illustrated by the following examples: Hand D, "Facta est strages . . ." for the year 841, folio 18r (fig. 16); Hand F, "Obiit Hlotharius iuvenis" for the year 870 (Lothar II, d. 869 actually), folio 18r (fig. 17); and Hand G, "Obiit Hludouvicus rex . . ." for the year 876, folio 18v (fig. 15).[123] Extremely cautious attention to detail has led Boschen to attribute an *emendatio* to the entry for 855 to the work of a novel hand. There is no reason not to group the work of Boschen's Hand E with the work of the prolific and more important figure referred to as Hand H, whose contributions Boschen dates to after 888. The clustering of the letters, especially the way that the letter *e* carries over into adjacent letters of script, suggest that the scribe who wrote "senior" over the entry for 855 (originally written by Hand B/C) on folio 18r, "HLUDHARIUS senior REX OBIIT III. KL. OCT" (LOTHAR the elder DIED 29 SEPTEMBER), is the same person who composed the numerous annal entries, identified by Boschen as Hand H. This would include the information added to the entry for the year 855, "Lotharius filius eius in

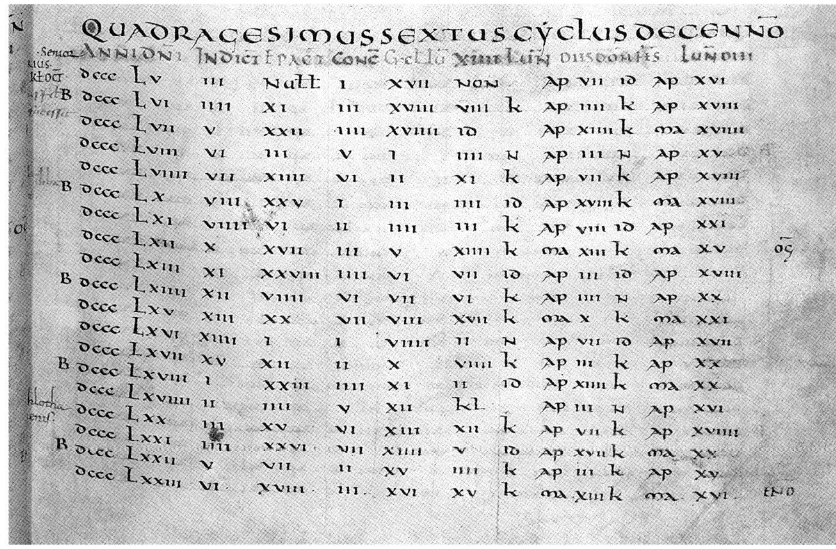

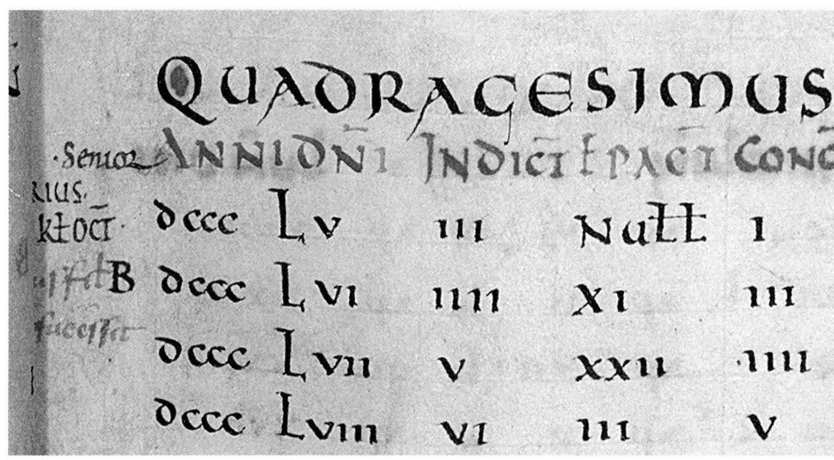

FIGURE 17

Nineteen-year table 46. Madrid, Biblioteca Nacional de España, MS 3307, fol. 18r, bottom (detail). Photo © Biblioteca Nacional de España.

FIGURE 18

Nineteen-year table 46. Madrid, Biblioteca Nacional de España, MS 3307, fol. 18r, bottom (close-up of "senior"). Photo © Biblioteca Nacional de España.

regnum successit" (Lothar his son succeeded as king). The corrections or putative improvements were added later in both cases, as is evident by the very fact that they are clarificatory *changes* to the original annal entry for 855 written by Hand B (figs. 17–18).

Over time, the number of hands adding indices to the annals increased by accretion, demonstrating long-standing engagement with and interest in Drogo's copy of the *Handbook of 809*. The manuscript was a handbook, but it was also something along the lines of a workbook, or a perpetual work-in-progress. It could be changed, tailored, and adapted to the needs and desires of the local library where it was stored. When the book changed hands, new entries could be added to the nineteen-year tables. The commemoration of meaningful events from the political or sacred realms could be added to the tables in perpetuity. That this practice took place in Prüm reveals just how important Drogo's gift was to the monastery. The local abbot would have selected this book of sacred science as the appropriate repository for the history of the Franks and of his local community, thus bringing the confinement of the cloister into alignment with extant Frankish royal authority. In records such as the *Annales Prumienses*, the lines dividing court and cloister blurred and an interpenetration of the sacred and secular aspects of the *Handbook of 809* was achieved.

Since so many entries were added by Hand H, and the entries supplied to the nineteen-year tables by Hands A, B, D, F, and G preceded the work by H, the date for the additions supplied by H must be later than 887, the year of the last of the entries by the other hands.[124] In the case of the "senior" added to the entry for 855, Hand H corrects earlier entries—which must, of course, have been available to correct. Between the entry for 876 on folio 18v, "Obiit Hludouvicus rex . . ." (Louis the German), which is the work of Hand G, and the entry for 882 below it, "OBIIT HLUDOVUICUS REX" (Louis the Younger) by Hand B (in my view, or Boschen's Hand C), the later Hand H squeezes, in a small and speedy script, five additional entries of various lengths on both the left- and right-hand sides of the nineteen-year tables in Drogo's copy of the *Handbook of 809* (fig. 15). These entries made by Hand H were added after Drogo's copy had already left Metz. In fact, since the entries by Hand H extend to 923, they must have been added shortly before or after the book was taken to Liège, as suggested by the entry supplied by Boschen's Hand H for the year 922 on folio 19r: "922. Richarius Rome in sede Tungrensi episcopus ordinatur II. Non. Novb., et Hildradus eodem anno in Prumia abba efficitur III. Non. Marc." (At Rome Richarius was ordained bishop of the see of Tongres-Liège on 4 November 922, and in the same year Hildradus was made abbot in Prüm on 5 March 922).[125] Either Abbot Richarius of Prüm took Drogo's copy of the *Handbook of 809* along with him to Liège in 922 or he requested it from his old monastery, at which time the final updates were inserted into the nineteen-year tables. Whether the important scribe of the actual *Annales Prumienses*, namely, Hand H, added the text in Liège or Prüm is not a vital point. As Boschen prefers, these entries probably were added in Prüm before the book was sent either to or with Richarius upon his elevation to the see of Liège.[126]

Mariana recognizes that the list of abbots of Prüm from 828 through 922 was added in the tenth century in Prüm, before the book was transferred to Liège.[127] Since the records of the abbots of Prüm begin with 828, there is no reason to postulate an early relocation of Drogo's book from Metz (or anywhere else) to Prüm in the first half of the ninth century—unless Hand A had made the entries for the early abbots of Prüm. Boschen attributes the entries for the abbots of Prüm, and this seems right, to Hand H: 828, on folio 17v, for the death of Tancradus and subsequent succession of Marcuuardus; 853, the succession of Eigil, on folio 18r; 860, the succession of Ansbaldus, on folio 18r; 886, the succession of Farabertus after the death of Ansbaldus reported on folio 18v; 892, the succession of Regino, on folio 18v; in 899, on folio 18v, Richarius attains the abbacy; 922, on folio 19r, Hildradus replaces Richarius after Richarius is designated bishop of Liège-Tongres. Presumably, all these historical notes about the abbots of Prüm, rendered in a uniform style of penmanship, were added more or less together before the book came to be with Abbot Richarius in Liège.[128] There also does not appear to be any good reason to delay the transfer of Drogo's copy of the handbook to Prüm, as Mariana would have it, until the "beginning of the tenth century" (aunque nuestra opinión es que no debió realizarse hasta entrado el siglo X).[129]

The *Annales Prumienses* provided a reminder of the community Richarius had served as abbot of Prüm (899–922), following the unusual departure from Prüm to Trier of Regino (d. 915), whose removal from office Richarius had engineered in 899.[130] According to Boschen, additional entries were made until 1044 by a series of hands labeled I and K–R. These entries seem different enough to warrant so many different hands. In addition, there is a long time span covered by these entries, which adds credence to Boschen's desire to posit such widespread individual participation.[131] This suggests that a relatively lively engagement with Drogo's copy of the *Handbook of 809* continued throughout the Ottonian period, until a final addition to the book was made in 1044.

This digression into the conditions of creation for the *Annales Prumienses* supplies more than information about the later history of Drogo's handbook. On the historical and paleographic grounds presented in the annals, there is ample support for the assignment of the book to Metz during the tenure of Bishop Drogo. Nevertheless, it will be necessary to address the question of the book's putative relationship to Murbach Abbey below, after returning to a review of the contents of the *Handbook of 809*.[132]

Books and Topics Covered in Madrid, Biblioteca Nacional, MS 3307

Book II of Drogo's copy of the *Handbook of 809* includes texts that were essential to time reckoning. Practical explanations of the information supplied in the nineteen-year tables and basic computistical techniques enabled a knowledgeable reader to understand the terms and pragmatic application of medieval computistics. For example, there were notes on how to calculate the age of the earth from the creation of the world, *ab initio mundi*, according to an admixture of medieval computistical wizardry and Christian mythologizing. Numerous references to the year of the Lord 809 accompany its corresponding age of 4,761 years since the beginning of the world in book II. These references link Drogo's book directly to the textual transmission of the *Handbook of 809*. The dates can be found in book II.1 and II.7–12.[133]

In book III.1 of the *Handbook of 809*, there is a discussion of hand counting derived from the first chapter of Bede's *De temporum ratione*, and likewise named, traditionally, *Secunda Romana computatio*.[134] According to the authority of the ancient Roman commentator Macrobius, who studied Cicero, the entire universe returns to its original starting position once every 15,000 years.[135] This is interesting because Bede had already calculated in his *De temporum ratione* the need for 532 years to pass before the Easter feast would fall once again on a particular day. Clearly, there was a perceived disconnect between the repetition of the Easter feast day and the recommencement of the cyclical movements of the heavens. In other words, the *Handbook of 809* addressed the discrete projects of determining the feast day of Easter and understanding the paths of the fixed stars. Such notions reinforce the handbook's link to Isidore of Seville's taxonomies in his *Etymologies* and simultaneously offer a glimpse into the non-geometric effort at system-building advocated by Carolingian stargazers. Frankish prelates clearly expected that turning their minds to contemplation of the night skies would lead to the kind of intellectual mastery that a liberal arts education promised young prelates. This is why Carolingian astronomy and its pagan classical roots can sometimes be studied more fruitfully by students of philosophy and culture than by narrow historians of science.[136] Book III.10 of the *Handbook of 809* refers to the year of the Lord 793 ("III. X, Anni ab incarnatione domini anno praesenti sunt 793") and therefore demonstrates a link to the first (Borst's type A) version of the astronomical-computistical-pedagogical handbooks, such as the Verona compilation discussed above.[137]

Book IV provides texts on the planets, including the sun and moon, according to the list of planets commonly adopted in the Middle Ages by proponents of the Ptolemaic, geocentric model of the solar system. As noted above, the orbit of the moon played an important role in the establishment of the feast of Easter, and consequently the earth's satellite—believed to be a planet—is featured in these texts.[138]

The proposed lost archetypal copy of the *Handbook of 809* included an extended cycle of lavish illustrations accompanying book V, and the best version of these images was painted in a workshop affiliated with Drogo for his personal copy.[139] The four diagrams of planetary configurations in book V derive from texts excerpted from Pliny, but the Carolingian encyclopedists had them added to their corresponding texts as pedagogical supplements.[140] Another unfinished diagram can be found on folio 67v in Drogo's copy of the *Handbook of 809*. There is the faint suggestion of a circle on folio 68r as well, although it is difficult to determine whether this circle provided the schematic beginnings of yet another diagram.

Hubert Le Bourdellès has convincingly identified two of the texts from book V as Carolingian contributions to the *Handbook of 809*. They were probably composed in Corbie and transported to the synod of 809: the *Excerptum de astrologia* (hereafter the *Excerptum*) and the *De ordine*. The *Excerptum* describes the constellations in the heavens following the model of Aratus in his *Phaenomena*, which describes the placement of the constellations relative to each other in 276–274 B.C.E.[141] Le Bourdellès argues that an eighth-century translation from Greek into Latin of the Aratus poem, the so-called *Aratus Latinus*, was carried out in Corbie under Abbot Grimo of Corbie (served 723?–48), who was also archbishop of Rouen beginning in 744.[142] On stylistic grounds, Le Bourdellès argues that the *Aratus Latinus* surpasses the clumsy, early poetic Latin literature attributed to Corbie circa 700, yet does not rival the stylistic improvement of compositions written during the early stages of the Carolingian *renovatio* (ca. 750–60).

Le Bourdellès also suspects that it was during Abbot Grimo's tenure that the Greek texts needed for the translation initiative arrived in Frankish Gaul from Rome. This argument is reasonable enough. One need only recall, as Le Bourdellès rightly stresses, the successful pleas voiced by Theodore and Hadrian, who sought the transport of Greek codices from Rome to Canterbury in the seventh century.[143] For that matter, the study of Greek and the preservation of Greek manuscripts at Saint-Gall in the ninth through eleventh centuries also document a legacy of Frankish interest in Greek literature and of limited interaction with the Byzantine Empire.[144]

In the thirteenth and fourteenth centuries, the Carolingian text *Excerptum* became linked to the work of Hyginus and was renamed the *Yginus philosophus, De imaginibus caeli*. The mistake was subsequently treated as fact in an early nineteenth-century edition of Hyginus texts prepared by L. W. Hasper.[145] Neuß apparently relied upon this early blunder as justification for his claim that C. Iulius Hyginus made a second-century Greek (not Latin) version of the Aratus text, which was only translated later by a forgotten intermediary. The Latin text, according to Neuß, was preserved in two different redactions. One was allegedly composed by Bede, while the other was to have been the work containing the two texts under discussion from book V of the *Handbook of 809*: the *Excerptum* and the *De ordine*.[146] Le Bourdellès corrects the lacunae in this account.

The text called the *Excerptum* contains revelatory clues about its history, sources, and subsequent textual transmission. These suggest that there was not a direct relationship between the *Aratus Latinus* and the Carolingian description of the constellations supplied by the *Excerptum*. Interestingly, Le Bourdellès has argued for a direct connection between the

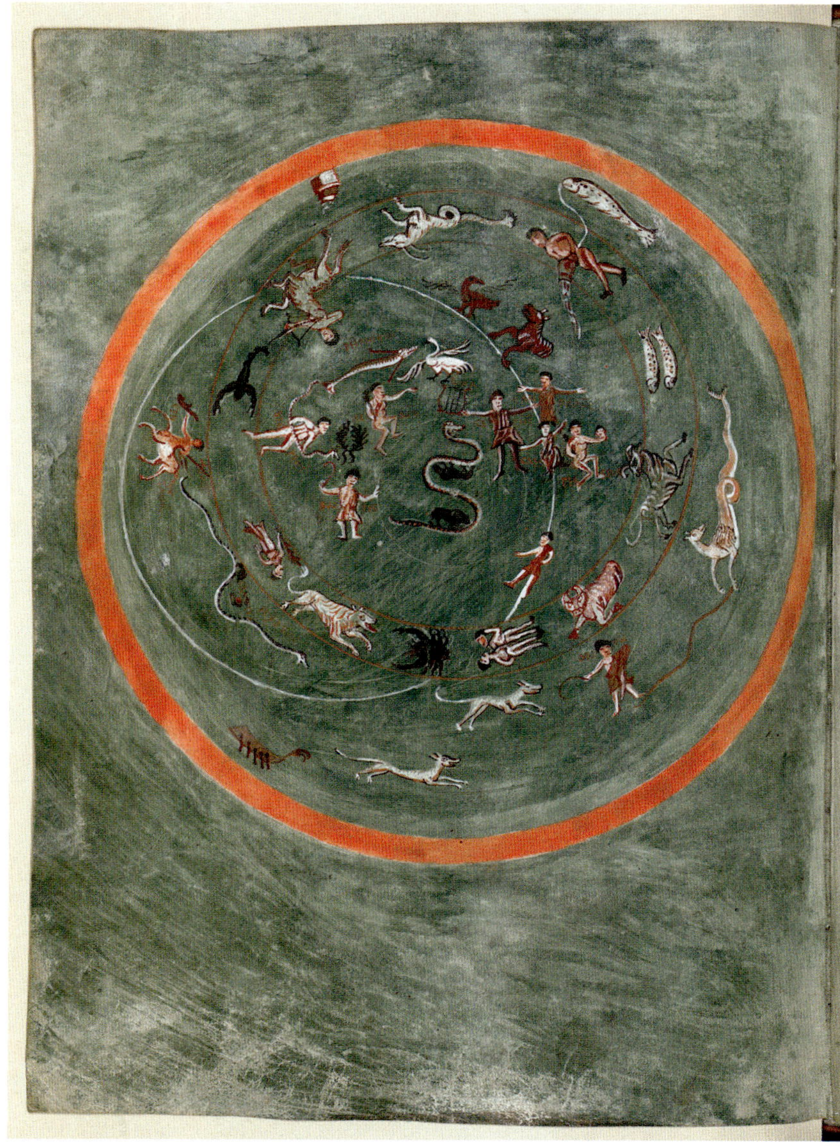

FIGURE 19

Celestial map. Munich, Bayerische Staatsbibliothek, Clm 210, fol. 113v. Photo: Bayerische Staatsbibliothek München.

Excerptum and a celestial map, such as the one found in Munich, Bayerische Staatsbibliothek, Clm 210, folio 113v (fig. 19; hereafter Munich 210), or perhaps a celestial sphere. The description of the Milky Way in the *Excerptum* cannot be found in any other textual redactions of the Aratean traditions. Hyginus described the course of the Milky Way but did not use the Milky Way as a means to situate the constellations relative to one another, as in the *Excerptum*.[147] In particular, the *Excerptum* held that two constellations appear at the intersection of the Milky Way and the band of the zodiac. The two constellations at this juncture are Gemini and Sagittarius, and the usage of *commissura* in this context to designate the locus of intersection harks back to a terminology formulated by Pliny, according to Le Bourdellès.[148]

Pliny's texts feature prominently in the handbook, especially in the texts illustrated with diagrams (book V.3–6; figs. 3–6), so this, too, is significant. Given these discrepancies, Le Bourdellès argues that the *Excerptum* is a Carolingian text derived from a second Latin translation of the *Phaenomena*, likewise made in Corbie. This intermediary source was known as the *Aratus revisé*, or *Revised Aratus*. More precisely, the *Excerptum* derives from the textual variant found in the so-called *Scholia Sangermanensia* (a view already advanced by Ernst Maass, but referred to in his text and in much subsequent literature as the anonymous *Recensio interpolata*).[149]

Although more will be said about the *Aratus revisé* in chapter 2, for now it is vital to understand that the *Revised Aratus* resulted from early medieval efforts to parse the convoluted passages of the *Aratus Latinus*. The most important manuscript of the *Revised Aratus* is a copy from Corbie that was made sometime between 810 and 840. Many manuscripts from the monastery of Corbie were transferred to Saint-Germain-des-Prés in a seventeenth-century effort to preserve them. This resulted in the more common name of the text in question, the *Scholia Sangermanensia* (Paris, Bibliothèque nationale de France, MS lat. 12957). This manuscript contributed a version of the *Catasterisms*—star stories associated with apotheosis, and therefore constellation formation—to the *Revised Aratus*. The *Scholia Sangermanensia* looked to earlier versions of Aratean texts in its revisions of the *Aratus Latinus*'s impenetrable text and scholia, while drawing upon miscellaneous works by other authors as well.[150] The *Revised Aratus*, according to Le Bourdellès, is an early Carolingian effort to revise and distill the astronomical information remaining from antiquity into a format appropriate for early medieval Christian use, including among its revisions selections from Fulgentius, Isidore, and perhaps even Bede. The text of the *Revised Aratus*, which spawned almost as many variants as there are stars in the sky, was probably a product of the second half of the eighth century.[151]

The second text from book V, the *De ordine*, includes forty-two paintings of the constellations, which immediately follow each text passage describing the relative placement of stars within the constellation (see plates 1–17). Unlike many astronomical manuscripts of the ninth century, however, in Drogo's copy of the *Handbook of 809* the miniatures of the constellations do not demarcate the locations of the stars. In the Reg. lat. 309 copy, in contrast, the miniatures of the constellations such as Hercules and Corona do display the positions of the stars (fol. 92r; fig. 20).[152] More famously, depictions of the constellations with stars accompany the images of the constellations from the Leiden *Aratea* (fol. 52v; fig. 21).[153]

This suggests that although the stars were not considered necessary aspects of astronomical depictions in the early ninth century, they were quickly recognized as useful additions to pedagogical handbooks treating the heavens. The text of the *De ordine* derived, curiously, from the tradition of texts stemming from the Germanicus translation of Aratus's *Phaenomena* poem, as did the text included in the Leiden *Aratea*.[154] The so-called Germanicus translation of the Aratus poem from Greek into Latin was originally made around 15 B.C.E.–19 C.E. by Germanicus Julius Caesar, the Roman emperor Tiberius's child through adoption. The Leiden *Aratea* relied directly upon the Germanicus text.[155] The *De ordine* should be considered a novel Carolingian summary of the descriptions of the constellations that formed around the subsequent textual transmission of the Germanicus Latin translation and corresponding scholia.

Le Bourdellès also notes the relationship between the new Carolingian text prepared for the *Handbook of 809* and a collection of supplemental additions, the so-called *Scholia Basileensia* (Basel, Universitätsbibliothek, MS An. IV 18; hereafter Basel *Scholia*).[156] This important connection situates the illustrations in the *Handbook of 809* at the nexus of multiple contributing sources from the classical world and late antiquity. Mechthild Haffner has studied the origins of Aratean star pictures extensively, and she posits a new critical Latin translation of the *Phaenomena*, ca. 300 C.E., that relied closely on the old Germanicus version. She also considers red-framed images of the constellations in blue fields (akin to those in the Leiden *Aratea*; see fig. 21) to reflect, from a remove, the visual tradition associated with copies of the Germanicus translation. As a result, Haffner finds that in this revised Germanicus Latin translation of the early fourth century, philological and art-historical considerations converged, resulting in exquisite individual manuscript illuminations of the constellations

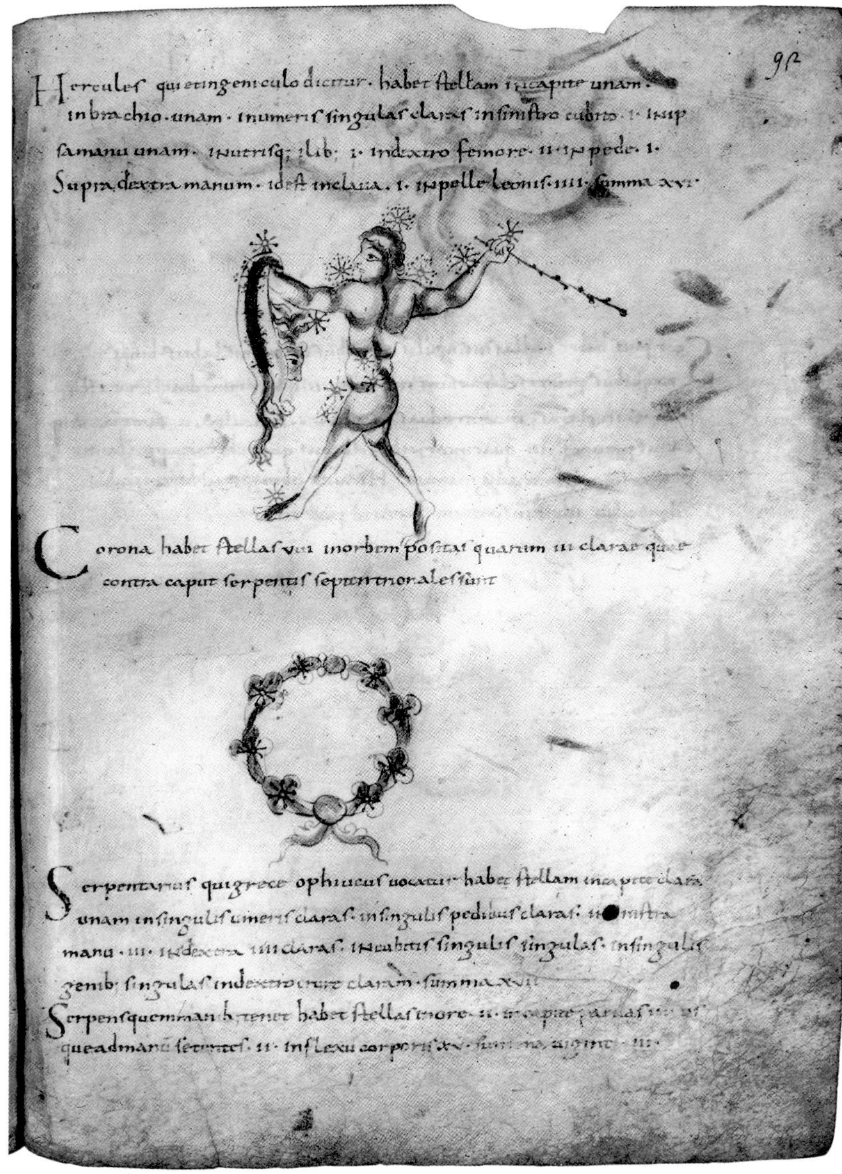

FIGURE 20

Hercules and Corona. Vatican City, Biblioteca Apostolica Vaticana, Reg. lat. 309, fol. 92r. Photo © 2017 Biblioteca Apostolica Vaticana. Reproduced by permission of Biblioteca Apostolica Vaticana, with all rights reserved.

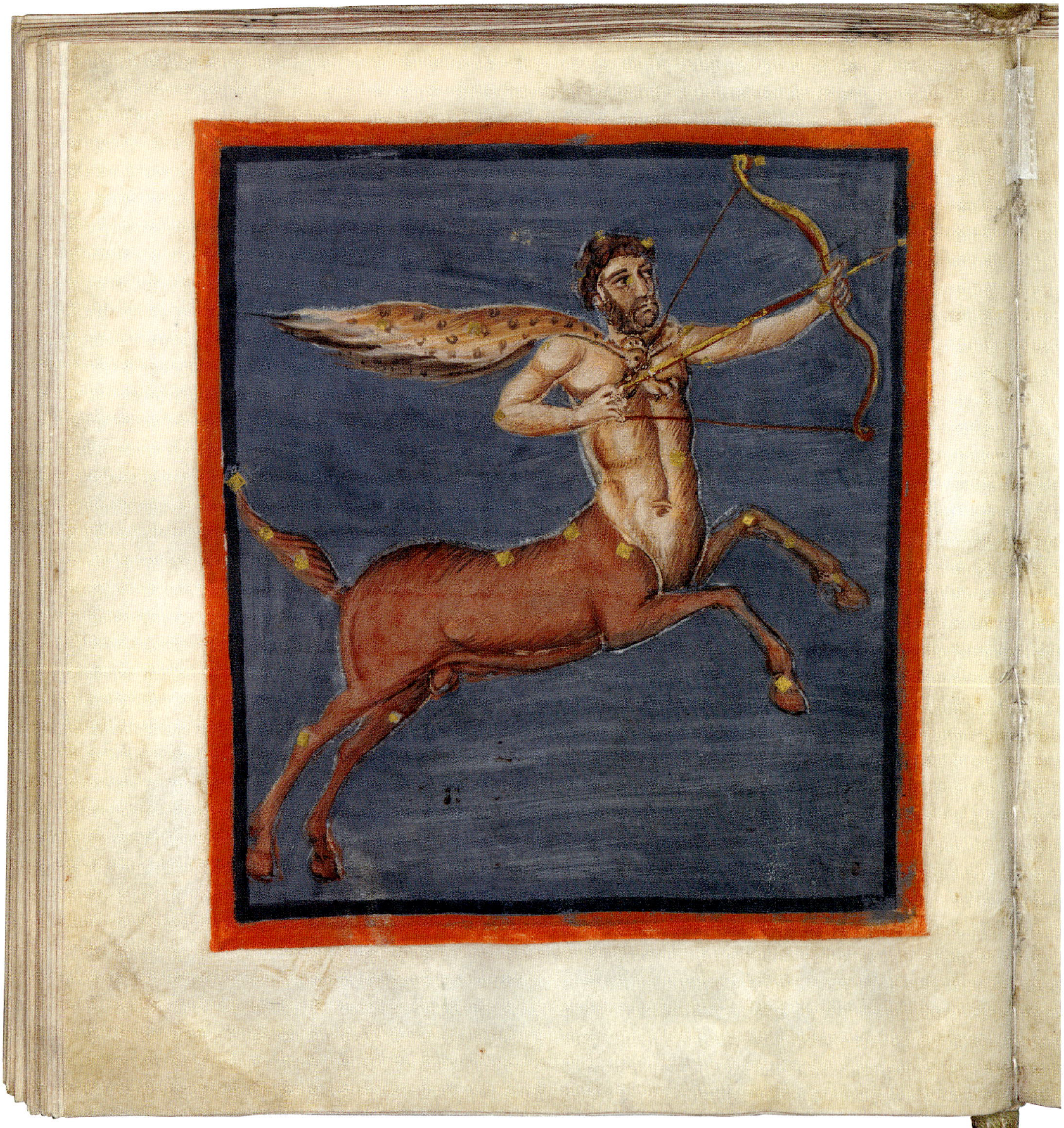

and an authoritative Latin edition of the Greek notes linked to the *Phaenomena*, the Basel *Scholia*.[157]

Drogo's copy of the *Handbook of 809* (Madrid 3307) provides an incomplete record of the two final books from the original *Handbook of 809*. Book VI is missing altogether from Madrid 3307, whereas in the archetypal copy of the *Handbook of 809*, book VI would have contained discussions of cosmic measurements involving the earth, sun, and moon.[158] Book VII is present only in fragmentary form in Madrid 3307, whereas in the archetypal copy book VII offered the entire text of the *De natura rerum* by Bede, as explained in the introduction. In Drogo's copy there is only a fragment of Bede's text, followed by several additional texts on weights and measures derived in part from Isidore of Seville.[159]

It is important to recognize that the third and final (Borst's type C) version of the astronomical-computistical-pedagogical handbooks in three parts attempted to simplify this organizational structure. In 818, Archbishop Arn of Salzburg culled 112 texts from the first two versions of the handbooks and arranged them in three parts. The revised, three-part version of the *Handbook of 809* was intended to be more useful than its predecessors. It has come down to us in two *de luxe* copies, which were both made in or near Salzburg roughly contemporaneously with the original redaction of the compilation: Vienna, Österreichische Nationalbibliothek, Cod. 387, folios 1r–165v (fig. 22), and Munich 210, folios 1r–163v (figs. 23–24). This three-part handbook was truly encyclopedic. It attempted to supplement all that was useful in the two previous handbooks with more educational material on arithmetical calculations that would hone the reader's skills at computistical analysis. Consequently, the third version of the handbook is also known as the *Liber calculationis*, yet among the three versions of the handbooks, only it would deserve to be called an early medieval encyclopedia.[160]

Metz or Murbach?

Given the rich variety of resources that contributed to the creation of Drogo's copy of the *Handbook of 809*, it is not surprising that some confusion remains about the book's site of manufacture and the origins of certain components. Wilhelm Neuß first proposed the link to Metz that has fostered all subsequent debate. On folio 6v in the martyrologium calendar leaf for August, the feast day on 16 August (or XVII KL septembris) celebrates: "Depositio sancti Arnulfi confessoris."[161] This entry apparently refers to the day in 641 on which the relics of Saint Arnulf from Remiremont (in the Vosges region) were transferred to the abbey church of his monastery in Metz.[162] Saint Arnulf (d. 640) was one of the most venerated of the Carolingian saints precisely because he was a forebear of the Frankish kings.[163] Charlemagne used devotion to Arnulf's cult as a political technology demonstrating the spiritual anointing of the Carolingian royal family, who adopted the names and typology of sacred Old Testament kings, such as David or Josiah, in text and art. Charlemagne linked himself to the salvific significance of the spiritual king David, while in the Vivian Bible the image of his grandson, Charles the Bald, is related to that of David, the musical shepherd-king.[164] This theme was reinforced by other Frankish kings who presented themselves as new Frankish avatars of Josiah, resurrecting the holy writ as a sacred salve enabling spiritual renewal of the mind through erudition.[165] Claiming Saint Arnulf only bolstered an exaggerated hagiographic justification of the Carolingian kings' right to rule the Franks.

FIGURE 21

Sagittarius. Leiden *Aratea*, Bibliotheek der Rijksuniversiteit, MS Voss. lat. Q. 79, fol. 52v. Photo courtesy of Universiteitsbibliotheek Leiden.

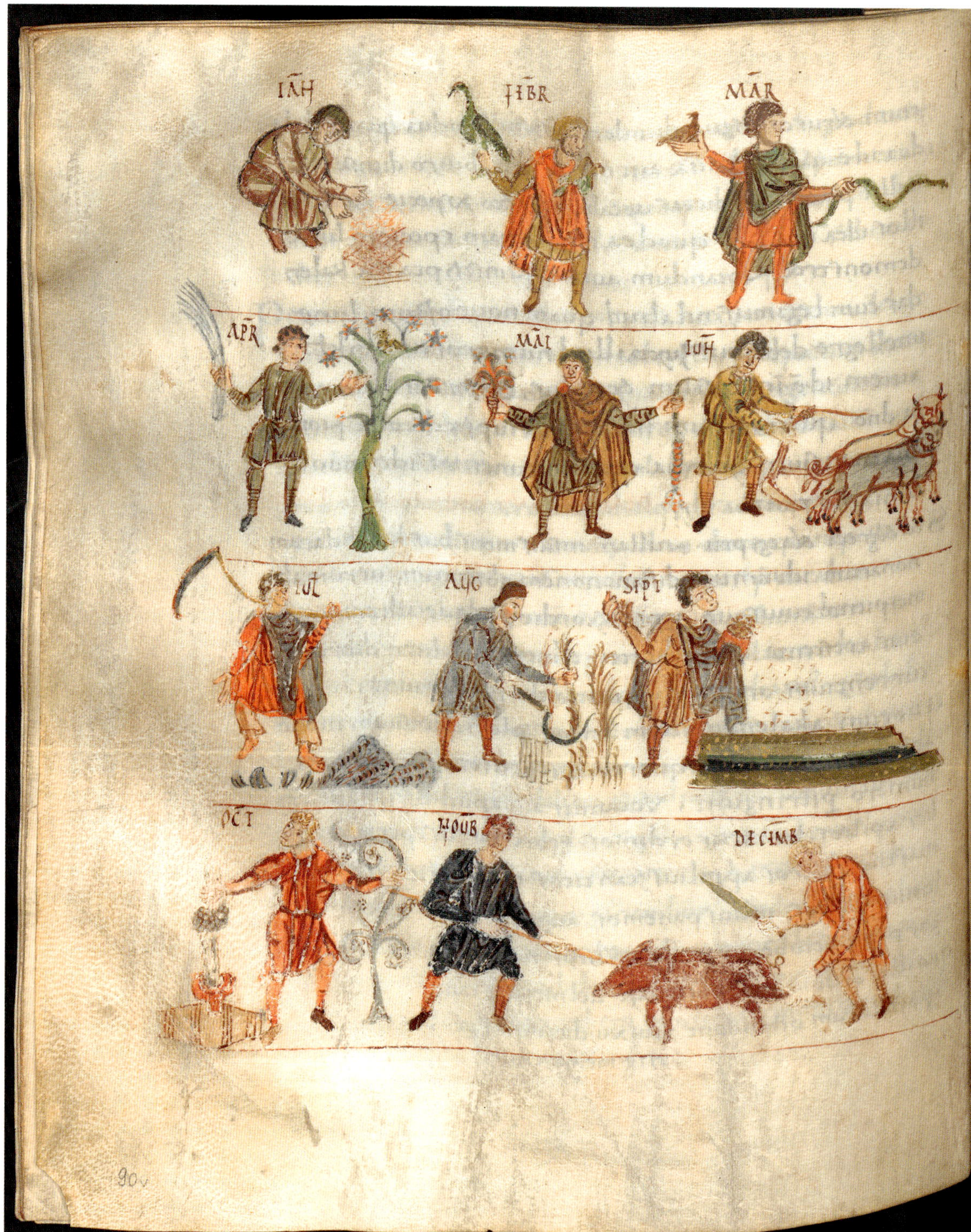

FIGURE 22

The twelve months. Vienna, Österreichische Nationalbibliothek, Cod. 387, fol. 90v. Photo: ÖNB/Wien.

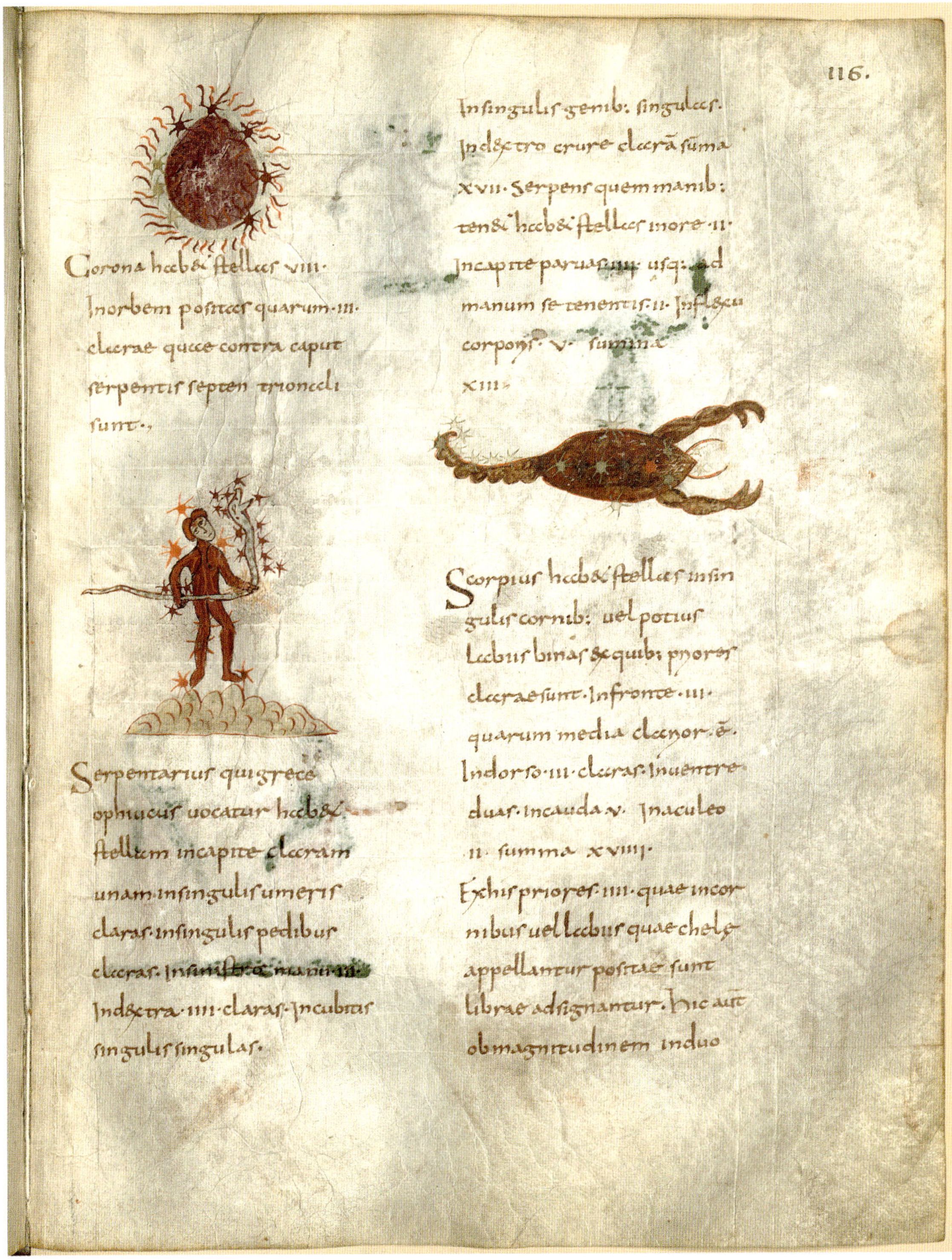

FIGURE 23

Corona, Serpentarius/Ophiuchus, and Scorpio. Munich, Bayerische Staatsbibliothek, Clm 210, fol. 116r. Photo: Bayerische Staatsbibliothek München.

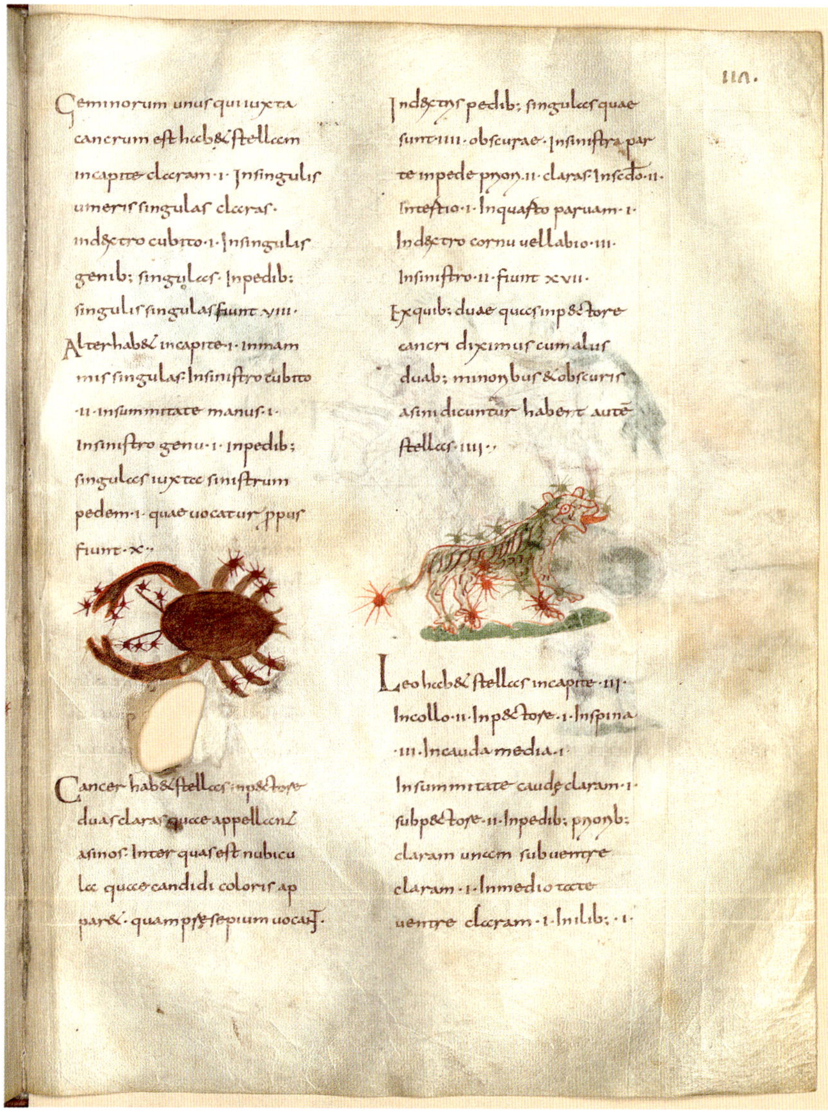

FIGURE 24

Cancer and Leo. Munich, Bayerische Staatsbibliothek, Clm 210, fol. 117r. Photo: Bayerische Staatsbibliothek München.

A brief history of the Carolingian city of Metz, the Austrasian capital city, can clarify certain points about Drogo's copy of the *Handbook of 809*. Michel Parisse has established the contours of such a history in his various publications. The Frankish city of Metz was famously ransacked by Attila and the Huns on the eve of Easter, 7 April 451, according to Gregory of Tours. At that time the cathedral of Saint Stephen survived from the late antique city. After the death of Clovis in 511, the kingdom divided into four smaller ones, and the eastern portion, based in Reims, was allotted to the direct heir, Theodoric (d. 534). This kingdom encompassed sections of Aquitaine and the Rhineland, and the other three capitals would likewise later play an important role in the Carolingian realm: Paris, Orléans, and Soissons. Subsequently, in the sixth century Clotaire I consolidated the territories of his heirless cousins before designating his son, Sigisbert, king of Reims in 561. The kingdom of Reims left its king in a vulnerable position relative to the kingdoms of Orléans, Paris, and Soissons. Sigisbert strategically relocated the royal seat to Metz, whence the city was elevated to an early medieval stronghold and became the courtly heart of the kingdom, renamed Austrasia. Nevertheless, this historical connection between Reims and Metz remained a vital thread in the series of ties binding the two cities together. At Metz, and in Drogo's book in particular, the disparate trends and styles of painting cultivated at these twin cities remain artistically conjoined.[166] (The painterly styles exemplified by Drogo's copy of the *Handbook of 809* are discussed in chapter 3.)

The historical Arnulf joined the team of noble youths who around the year 600 were working and studying with Gondoul, the royal mayor of the palace in Metz.[167] The somewhat apocryphal account of how Arnulf's dynastic lines of consanguinity altered sacred Frankish political history attests to a dramatic ideological shift throughout greater Austrasia.[168] Formerly, the bishops of Metz had come from late antique Roman families of dying significance, but as political power gravitated toward the Frankish kings, a preference for Frankish prelates, too, emerged. Prior to his appointment as bishop of Metz by King Chlotar II (d. 629), Arnulf entered into marriage with Doda, resulting in the births of two important princes, Anschise/Ansegisel (d. after 662) and Chlodulf (d. after 670).[169] Ansegisel continued the familial line after his father became a cleric, marrying Pepin's daughter Begga (d. ca. 693). Arnulf accepted the role of tutor to the young Merovingian king of Austrasia, Dagobert I (d. 639), at the behest of the boy's father, Chlotar II. Before his death in 640, Arnulf became a hermit, finishing his days with his lifelong friend and spiritual companion, Romaricus, atop the Vosges, where they ran a leper colony. Arnulf's bones were translated to a Messine church once dedicated to the Holy Apostles on 16 August 641 (as recorded in the martyrology of the *Handbook of 809*), resulting in the subsequent rededication of the abbey church to the locally venerated ecclesiast Saint Arnulf ca. 700.[170]

As can be seen from these brief historical notes, Arnulf was no mere saint with local significance, but the personal tutor to a king and a powerful player in the development of the seventh-century kingdom of Austrasia. Furthermore, he played a role in the creation of the Carolingian bloodline. For these reasons, Wilhelm Koehler[171] and Lothar Boschen are right to criticize Neuß's original intuition. In his reconstruction of the original calendar accompanying the lost *Handbook of 809*, Boschen argues that the reference to Arnulf was important to Carolingian dynastic pretensions and was therefore included in the archetypal calendar introduced into the handbooks, such as Drogo's personal copy.[172] Consequently, references to Arnulf or any other particular saint in the archetypal calendar of the *Handbook of 809* cannot supply any information about a *particular* manuscript's site(s) of manufacture.

Bernhard Bischoff has assigned Drogo's copy of the *Handbook of 809* to Murbach in two documented contexts. Two asides have resulted in a false dilemma based on benighted methodological beliefs about early medieval workshop practices. As the following review reveals, there is no reason to locate the manufacture of Drogo's copy of the *Handbook of 809* in Murbach rather than Metz. At best, one or more scribes from Murbach may have participated in the production of Drogo's handbook (Madrid 3307) in Metz. In any case, it is important to recall that one interesting result thus far of the investigation into the copies of the *Handbook of 809* is that *de luxe* copies tend to be linked to important Carolingian prelates. No such candidate from Murbach presents himself as a replacement for Drogo ca. 830, although this alone does not discredit any connection to Murbach and its rich tradition during the Carolingian period.[173]

First, at some point prior to 1972, Lothar Boschen and Bernhard Bischoff exchanged opinions about the origin of the copy of the *Handbook of 809* presently in Madrid (MS 3307), and Bischoff reportedly suggested Murbach as the site of its manufacture.[174] Claudia Höhl has pointed out that Boschen's note was one source of this attribution to Murbach, before other references appeared in subsequent secondary literature without grounding or argumentation.[175] Arno Borst and Bianca Kühnel likewise adopted Boschen's position, which for Boschen was derived from Bischoff without justification or an art-historical argument to buttress Bischoff's aside, which eventually became treated as an accepted fact.[176]

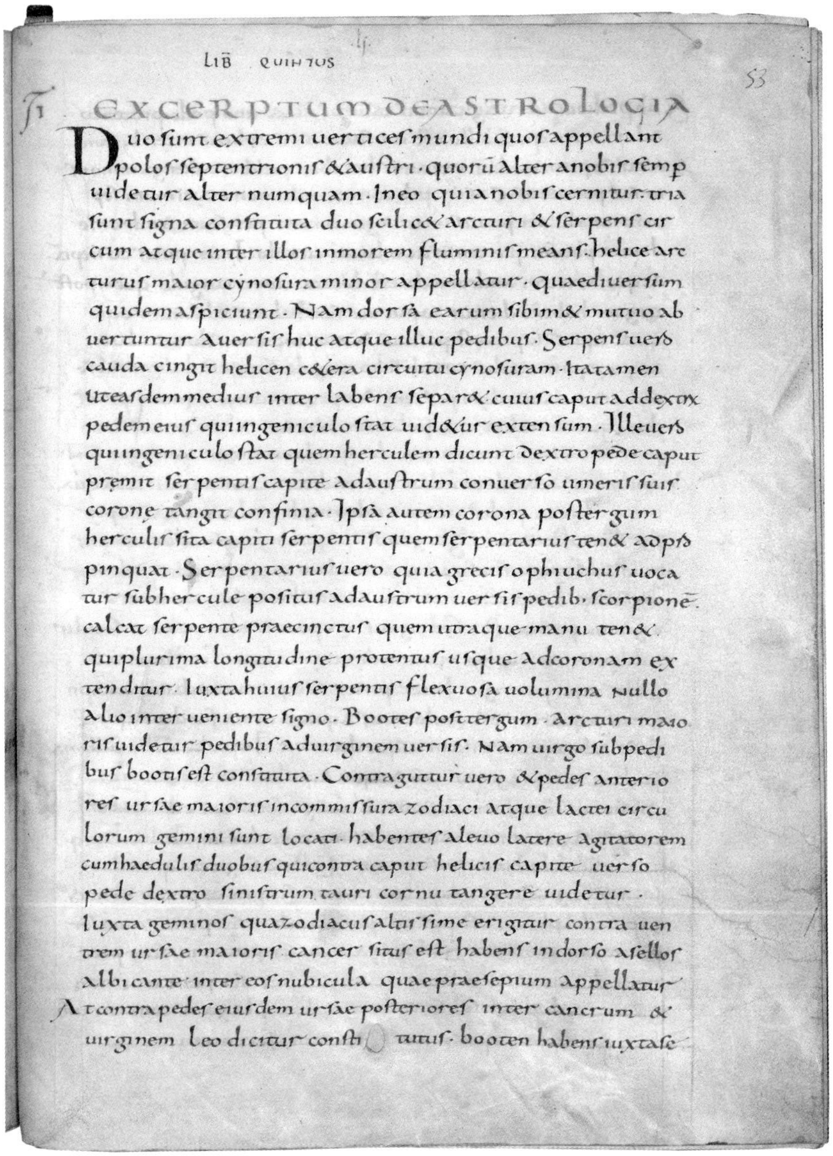
LIB QUINTUS
53
EXCERPTUM DE ASTROLOGIA
Duo sunt extremi uertices mundi quos appellant
polos septentrionis & austri. quorū alter a nobis semp
uidetur alter numquam. In eo quia nobis cernitur. tria
sunt signa constituta duo scilic& arcturi & serpens cir
cum atque inter illos in morem fluminis means. helice arc
turus maior cynosura minor appellatur. quae diuersum
quidem aspiciunt. Nam dorsa earum sibi m& mutuo ab
uertuntur auersis huc atque illuc pedibus. Serpens uero
cauda cingit helicen c&era circuitu cynosuram. Ita tamen
ut eas dem medius inter labens separ& cuius caput ad dextra
pedem eius qui in geniculo stat uid&ur extensum. Ille uero
qui in geniculo stat quem herculem dicunt dextro pede caput
premit serpentis capite ad austrum conuerso umeris suis
coronę tangit confinia. Ipsa autem corona post tergum
herculis sita capiti serpentis quem serpentarius ten& adpro
pinquat. Serpentarius uero quia grecis ophiuchus uoca
tur sub hercule positus ad austrum uersis pedib. scorpionē
calcat serpente praecinctus quem utraque manu ten&
qui plurima longitudine protentus usque ad coronam ex
tenditur. Iuxta huius serpentis flexuosa uolumina nullo
alio interueniente signo. Bootes post tergum arcturi maio
ris uidetur pedibus ad uirginem uersis. Nam uirgo sub pedi
bus bootis est constituta. Contra guttur uero & pedes anterio
res ursae maioris in commissura zodiaci atque lactei circu
lorum gemini sunt locati. habentes a leuo latere agitatorem
cum haedulis duobus qui contra caput helicis capite uerso
pede dextro sinistrum tauri cornu tangere uidetur.
Iuxta geminos qua zodiacus altissime erigitur contra uen
trem ursae maioris cancer situs est habens in dorso asellos
albicante inter eos nubicula quae praesepium appellatur
At contra pedes eiusdem ursae posteriores inter cancrum &
uirginem leo dicitur consti tutus. booten habens iuxta se

FIGURE 25

Excerptum de astrologia. Madrid, Biblioteca Nacional de España, MS 3307, fol. 53r. Photo © Biblioteca Nacional de España.

Bischoff somewhat hesitantly considered Madrid 3307 a possible example of both script and illumination from Murbach, noting, "I do not believe that I deceive myself that the splendorous Madrid Codex of an astronomical textbook was also written and illuminated there" (Ich glaube mich nicht zu täuschen daß auch der prächtige Madrider Codex eines astronomischen Lehrbuchs dort geschrieben und illuminiert worden ist).[177] Before addressing Bischoff's unfounded claim about the style of painting in the miniatures, it is interesting to consider the role that scribes from Murbach might have played in the development or manufacture of Drogo's copy of the *Handbook of 809*.

There is an important record of the holdings at the ninth-century monastic library in Murbach, but it lacks any description of a manuscript that might be a copy of the *Handbook of 809*. In 1464, Sigismund Meisterlin prepared a paper facsimile of this inventory, which is preserved in Colmar (Archives départementales du Haut-Rhin, Cartulaire Abbaye Murbach no. 1, paginated 86–96). One of the catalogue's curiosities is its extensive Carolingian "wish list" of books for acquisition, which also lacks any mention of an astronomical-computistical-pedagogical handbook.[178] Yet if Drogo's copy of the *Handbook of 809* (Madrid 3307) was manufactured in Murbach and relocated to Prüm before midcentury, researchers would have to accept that the monks at Murbach considered their prized *Handbook of 809* worthy of lavish embellishment and careful copying but inessential to their library holdings. Under this interpretation, the best remaining copy of the *Handbook of 809* (i.e., Drogo's personal copy) was thus neither recorded by its makers upon completion nor deemed worthy of a working replacement after it was conferred, by whatever route and whichever intermediaries, upon Prüm. Still, scribes from Murbach were celebrated throughout the network of monastic foundations established by Pirmin, including Hornbach, a monastery from the diocese of Metz.[179]

To assess Bischoff's hunch in favor of a Murbach origin for Madrid 3307, it is useful to look for any paleographic connections between the Germanic style of early or pre-Caroline minuscule scripts cultivated at Murbach and a sample folio from Drogo's personal copy of the *Handbook of 809*.[180] This will help distinguish any traces of the handwriting at Murbach that could have remained as vestigial influences on the more standard and regular Caroline minuscule script penned during the reign of Louis the Pious.[181] For this purpose, the opening text of book V, the *Excerptum*, on folio 53r suffices (fig. 25). Normative features of the early Murbach house style derived from an Alemannic paleographic impulse throughout the scriptoria of the Bodensee (Lake Constance) region, emanating from Reichenau.[182] Key features of the *Excerptum* folio include the general use of a Carolingian *a* with a downward-sloping crest in lieu of the earlier double-*c* ("*cc*") style in line 2; pronounced extended "clubbed ascenders"[183] in the letters *b*, *d*, and *l* in lines 2–5; a variant of the Carolingian *g* with a smoothly turning tail reminiscent of a figure 8 in line 5; the nicely aligned *nt*, without ligature, in line 9; ampersand as "et" on its own (line 3) and in the middle (line 10) or at the end of a word (line 5); and a decorative majuscule *A* formed by the reduplication into a lozenge at the apex of the left-hand shaft in line 1.

Early books that likely hail from Murbach share certain features. A related (albeit not identical) *g* with a nicely curved tail can be found in a fragment from Freiburg-im-Breisgau (Pelagius, *Expositio in Epistulas S. Pauli*, Stadtarchiv, B1/330; fig. 26).[184] A putative model for the Carolingian capital *A* in

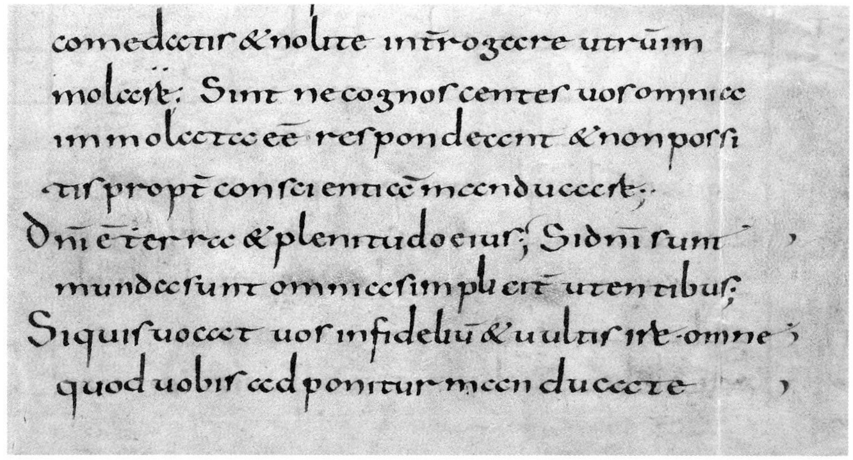

FIGURE 26

Fragment of Pelagius, *Expositio in Epistulas S. Pauli*, early Caroline minuscule (eighth–ninth centuries). Freiburg im Breisgau, Kulturamt, Stadtarchiv, B1/330, fol. 2r. *CLA*, vol. 8, no. 1193. Photo: Stadtarchiv Freiburg.

Madrid 3307 could have derived from Murbach manuscripts such as the Colmar copy of Iohannes Cassianus (*Collationes*, Bibliothèque municipale, MS 82, fol. 44v; fig. 27).[185] The bloated ascending shafts of the letter *l* in a Colmar computistical manuscript, found in the Murbach library inventory under works by Isidore of Seville, recall the script form in Drogo's copy of the handbook as well (Pseudo-Isidorus, *De ortu et obitu Patrum, De Numeris, Computus Paschalis*, Bibliothèque municipale, MS 39, fol. 128v; fig. 28). It is important to remember that this Colmar manuscript, unequivocally linked to the library at Murbach, nevertheless has affinities with a writing style utilized throughout a wider network of monastic centers from modern Alsace to northern Switzerland.[186] Although there is insufficient evidence to reassign the origin of Drogo's book to Murbach, there is ample justification for Bischoff's intuition that

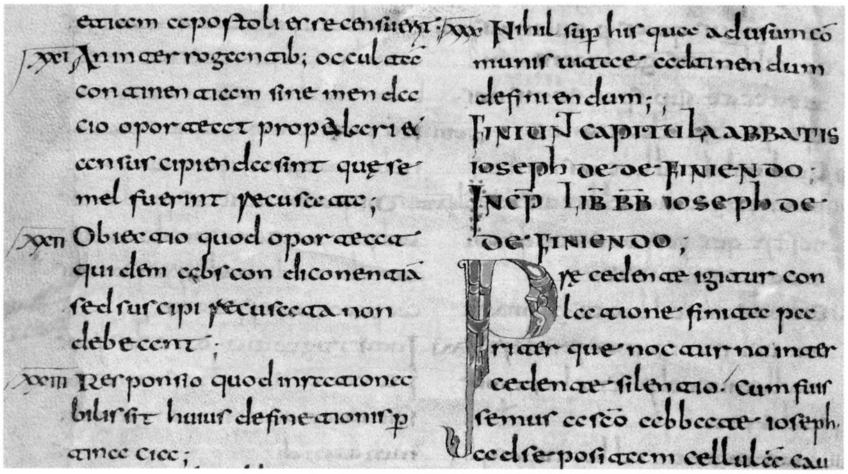

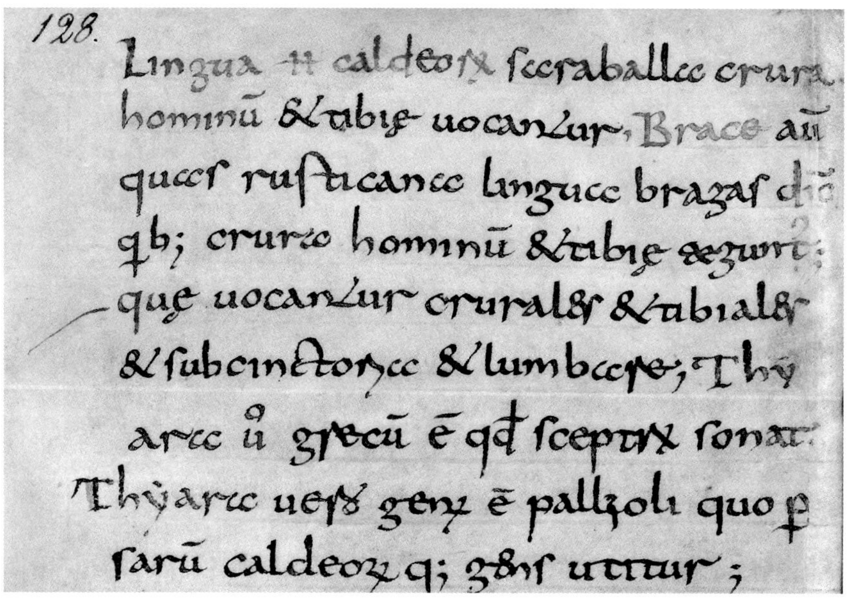

FIGURE 27

Iohannes Cassianus, *Collationes,* German pre-Caroline minuscule (eighth–ninth centuries). Colmar, Bibliothèque municipale, MS 82, fol. 44v. *CLA,* vol. 6, no. 753. Photo courtesy of the Bibliothèque de Colmar.

FIGURE 28

Pseudo-Isidorus, *De ortu et obitu Patrum, De Numeris, Computus Paschalis,* German pre-Caroline minuscule (eighth–ninth centuries). Colmar, Bibliothèque municipale, MS 39, fol. 128v. *CLA,* vol. 6, no. 751 right. Photo courtesy of the Bibliothèque de Colmar.

Drogo's copy has some paleographic connection to Murbach's scriptorium through prevailing patterns of monastic scribal influence.

Those who have mounted art-historical arguments predominantly concur that Neuß's poorly grounded attribution of Drogo's copy of the *Handbook of 809* to Arnulf's Metz is nevertheless correct: these include Koehler (1960), the great Aachen exhibition (1965), Janini and Serrano (1969), Mütherich and Gaehde (1976), McGurk (1981), Mariana (1993), Höhl (1996), the exhibition devoted to the Utrecht Psalter (1996), and the recent exhibition of Carolingian, including Messine, codices at the Bibliothèque nationale de France (2007).[187] Having pinned down the date of transfer of Madrid 3307 to Prüm in the section on the *Annales Prumienses* (i.e., in 855 before the death of Drogo), it is now possible to clarify some of the remaining mysteries about its manufacture. Koehler's paleographical analysis of the Caroline minuscule script in Drogo's copy of the *Handbook of 809* has determined that its script is compatible with a date for Metz manuscripts of ca. 820–40, thus providing evidence for an early variant of Caroline minuscule script that was cultivated at Metz. The manuscript must postdate 820, since the list of eclipses added to the nineteen-year tables by Hand A at the moment of the manuscript's manufacture ends then, as discussed above.[188]

Koehler considers the style of painting displayed by the miniatures in Madrid 3307 to reflect the classicizing tendencies of the problematic Vienna Coronation Gospels (Weltliche Schatzkammer; fig. 29). Since the archetypal *Handbook of 809* arose in the court at Aachen in the years following the synod, 809–12, so the argument goes, the style of the archetypal paintings would probably have resembled this classicizing tendency. Koehler even suggests that the archetypal manuscript may have traveled to Metz, where its classical paintings were copied by the artists working for Drogo. This would situate Drogo's copy of the *Handbook of 809* at the outset of the following sequence of artistic development in Metz:

> Drogo's copy of the *Handbook of 809*, made ca. 830
> Gospel book, Paris, Bibliothèque nationale de France, MS lat. 9383, ca. 835–45 (figs. 30–31)
> Gospel book, Paris, Bibliothèque nationale de France, MS lat. 9388, ca. 845–55 (fig. 32)
> Drogo Sacramentary, Paris, Bibliothèque nationale de France, MS lat. 9428, ca. 845–55 (fig. 33)[189]

In fact, Marie-Pierre Laffitte's recent reconsideration of the MS lat. 9383 gospel book also argues that its use of purple folios should bring it into alignment with the conservative tendencies of the Coronation Gospel group emanating from Aachen.[190]

The artistic style of the miniatures and diagrams in Drogo's book is discussed in chapter 3. For now, however, it is vital to recognize that a working date of ca. 830 respects Koehler's original intuition about the script without pushing the date of the paintings too late, and it builds upon the detailed analysis of the *Annales Prumienses* above. However, Bischoff's contribution to this debate suggests that the scribes who penned the text of Drogo's copy of the *Handbook of 809* included Metz within a regional network of scribal influences that benefited from the older political connections fostered by Merovingian or Austrasian kings and palatial mayors. In addition, the inclusion of Hornbach within the diocese of Metz underscores the ecclesiastical importance of artistic exchange—both scribal and painterly—between monastic foundations. It is best to consider the paleographic style cultivated at Metz in books such as

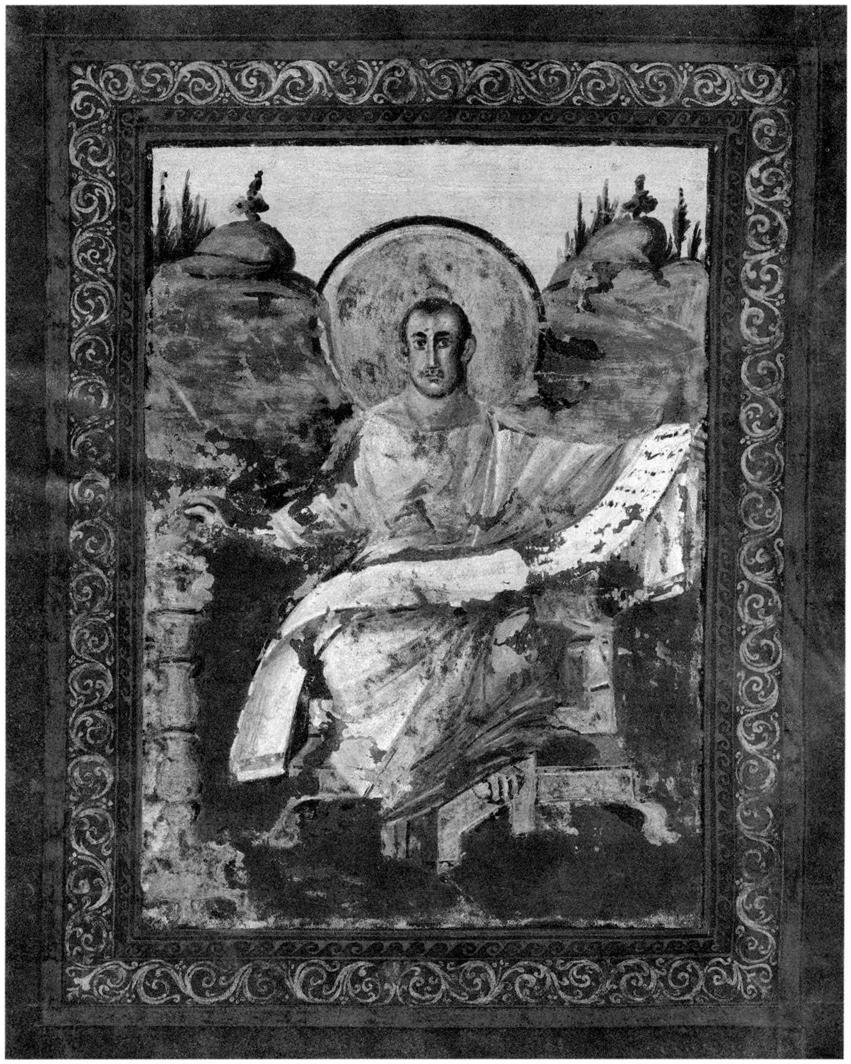

FIGURE 29

Gospel-author portrait of Mark. Coronation Gospels (SK XIII 18), Vienna, Weltliche Schatzkammer, fol. 76v. Photo: KHM-Museumsverband.

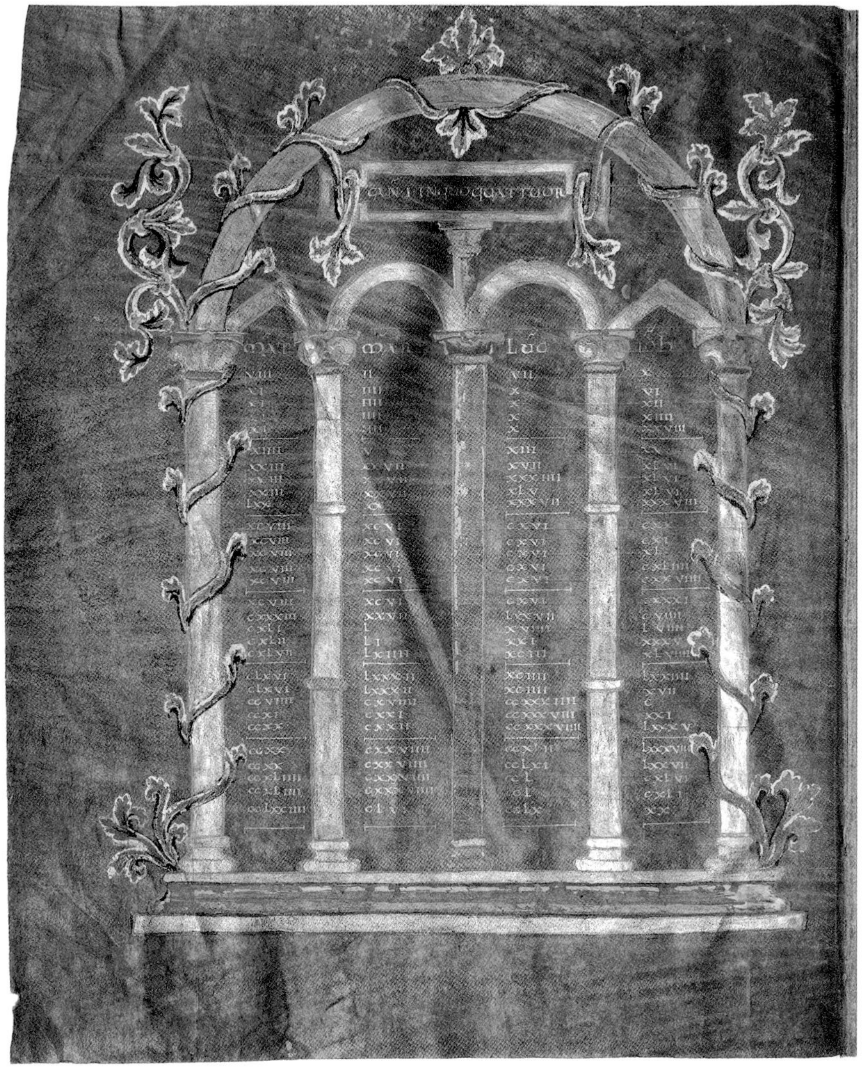

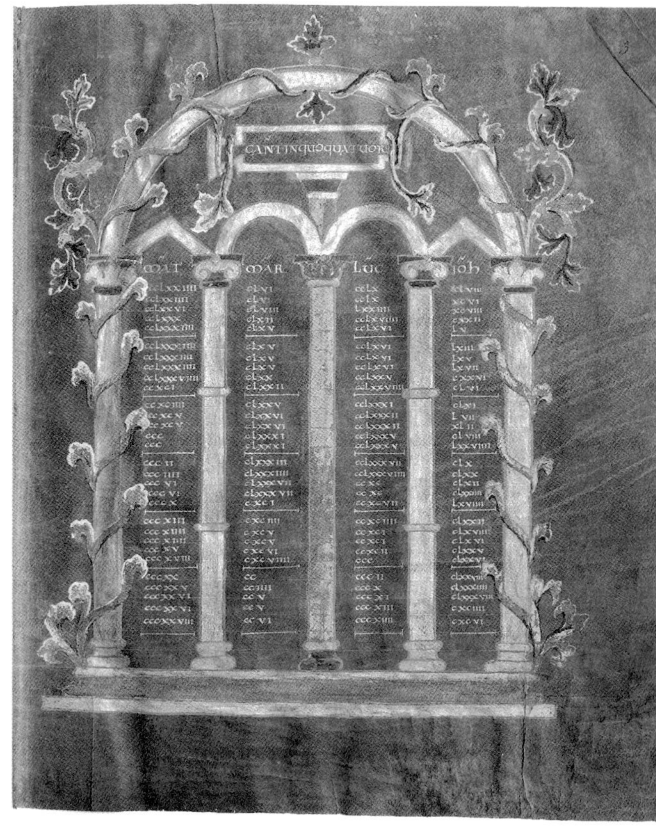

FIGURE 30

Canon table. Gospel book, Paris, Bibliothèque nationale de France, MS lat. 9383, fol. 2v. Photo: BnF.

FIGURE 31

Canon table. Gospel book, Paris, Bibliothèque nationale de France, MS lat. 9383, fol. 3r. Photo: BnF.

the Drogo Sacramentary as the fulfillment of a house style already nascent in Drogo's copy of the *Handbook of 809* and maturing over a generation, under the influence of the Alemannian scribal traditions associated with Murbach and the Bodensee.

Detailed observation of the paintings produced in the gospel books strongly supports this attribution. In the end, these *comparanda* provide the best argument for the link between Madrid 3307 and Metz. Koehler adduces a list of salient comparisons between paintings in Drogo's copy of the *Handbook of 809* and other books from Metz. One of the most compelling examples is the common style of modeling the serpentine form in the handbook painting of Serpens on folio 54v (see

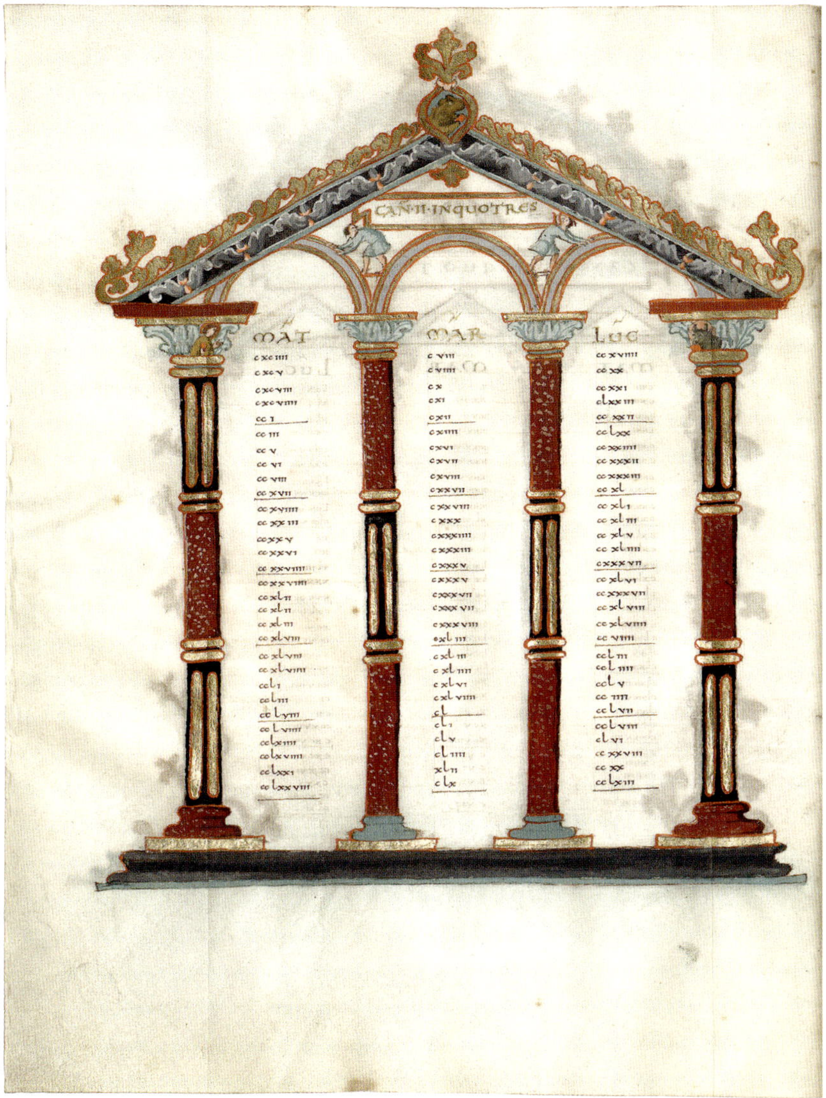

FIGURE 32

Canon table. Gospel book, Paris, Bibliothèque nationale de France, MS lat. 9388, fol. 8v. Photo: BnF.

FIGURE 33

The three women at the empty tomb of Christ. Drogo Sacramentary, Paris, Bibliothèque nationale de France, MS lat. 9428, fol. 58r. Photo: BnF.

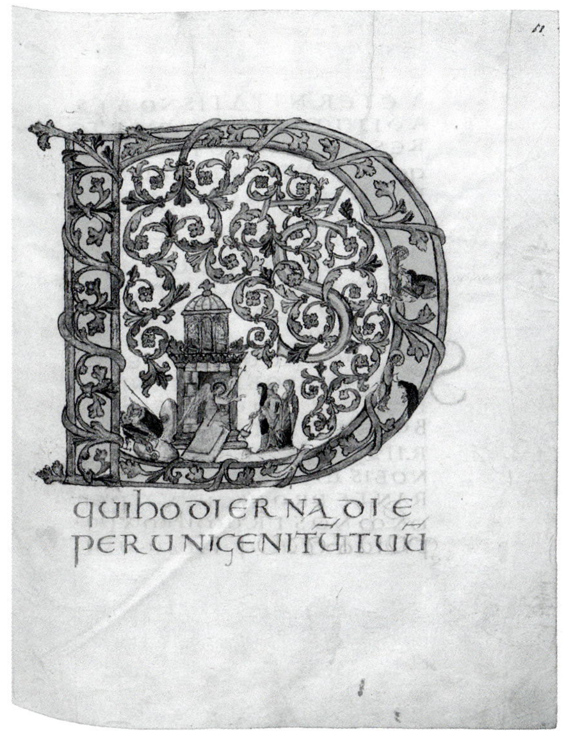

plate 1) and the snake clutched in the talons of the eagle accompanying the gospel initial author portrait for John on folio 150v of Paris, MS lat. 9388 (fig. 34). Both works show a three-toned modeling of the reptile's flesh and a stippling effect evoking a scaly texture.[191]

It is not clear what sort of artists were at work in Metz. Certainly they need not have constituted a stationary, isolated workshop in the permanent employ of Bishop Drogo. As discussed in the introduction, Lawrence Nees has argued that at least some Carolingian miniatures were the work of itinerant specialists.[192] Whoever the painters were who began their work at Metz, honing and plying their trade on Drogo's copy of the *Handbook of 809*, they began

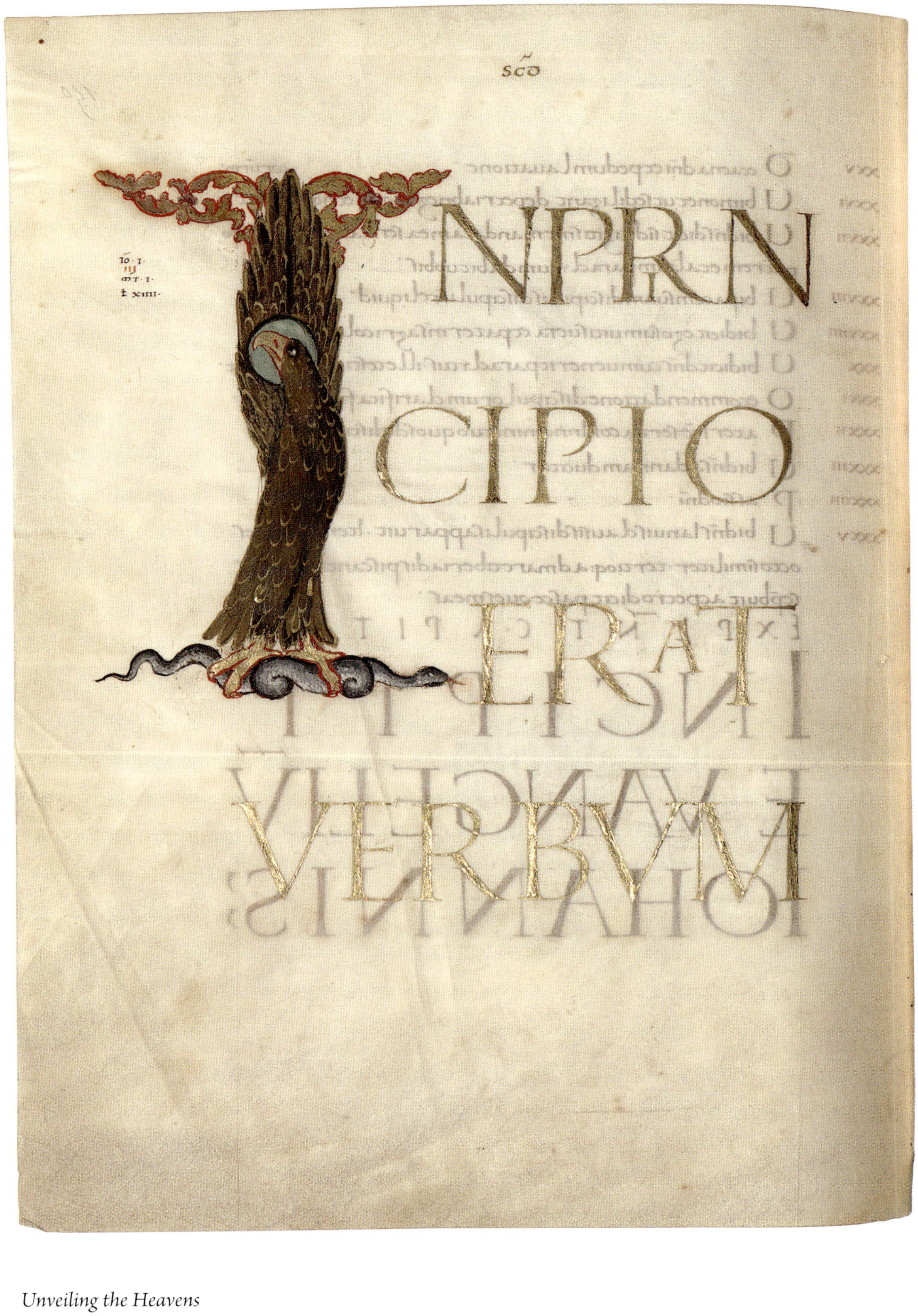

FIGURE 34
Evangelist symbol for the Gospel of John. Paris, Bibliothèque nationale de France, MS lat. 9388, fol. 150v. Photo: BnF.

much more than the creative work on the book. They stoked Drogo's desire for more luxury codices in early medieval Metz.

Later History of Drogo's Copy of the *Handbook of 809*

The later history of Drogo's copy of the *Handbook of 809* can be summarized succinctly, because the scant evidence in the manuscript itself attests above all to its antiquarian value as a symbol of imperial appropriation. As already noted, Richarius received the manuscript, complete with updates, in 922, when he assumed his new see of Tongres-Liège. The scribe whom Boschen identifies as Hand H added an entry in the annals of the nineteen-year tables describing the event on folio 19r.[193] A calendar entry on folio 80r for 17 September (XV kalendae octobris) confirms the time Drogo's book spent in Liège: "Lamberti epi. et ma." (Feast of Saint Lambert, Bishop and Martyr). The historic Saint Lambert (d. 698–701) had served as bishop in Maastricht and remains a patron saint of Liège.[194] Janini and Serrano have perspicaciously identified four little lines of script at the top right corner of folio 77v as poetry linked to another local bishop, Notker of Liège (d. 1008), who hailed originally from Saint-Gall: "Hoc Teotoche tuo Nodkerus vovit honori Lambertoque suo" (German-speaking Notker dedicates this text to Your honor and to his Saint Lambert).[195]

Subsequently, as Neuß had noted in the 1940s, an inscription on folio 17r informs the reader that the book had come to Sicily by the sixteenth century: "Anno 1543 ego Fr. Franciscus Monachi minorita hunc librum dono accepi" (In the year 1543 I, Brother Francis of the Monachi, a minorite, received this book as a present).[196] Franciscus Monachi likely traced his ancestry back to the Sicilian Monachis of the northwest Trapani region.[197] This may be the same Franciscus who composed a new text for Archbishop Octavian Praeconius of Palermo in or shortly after 1562, the *Epistola de situ orbis ac descriptione eiusdem*. Within three years' time, Withagius of Antwerp had already decided to disseminate the *Epistola* more widely in an accessible quarto edition dated 1565.[198]

Drogo's copy of the *Handbook of 809*, then, offers a witness to the fraught relationship between Spain and Sicily during the seventeenth century, moving about like a vulnerable piece of illuminated flotsam on the troubled Mediterranean Sea. Toward the end of its stint in Sicily, Drogo's copy of the *Handbook of 809*, like many other codices formerly housed within the royal libraries at Palermo, fell prey to the acquisitive bibliophilic tendencies of the duke of Uceda and viceroy (r. 1687–97) Don Juan Francisco Pacheco Téllez Girón, as argued by Mariana, whose thorough treatment of the subject is closely followed here.[199] A 1692 inventory of the viceroy's collection, compiled by its librarian, Johannes Silvester, identifies the shelf location of Drogo's book as III B, as Mariana reports: "De cyclis decemnovenalibus et alii tractatus de astrologia et de rebus naturalibus; aiunctus est in principio duernio impressus et intitulatus *Thesaurus eruditionis pro sole*; in folio."[200] The treatise mentioned in the entry, the *Thesaurus eruditionis pro sole per duodecim Zodiaci signa discurrente*, is a short printed text of ca. 1628 drafted by the Jesuit astronomer Francisco de Macedo. The ca. 1628 printing resulted in a fascicule from Madrid being inserted in the *Handbook of 809* at the time of an approximately dated seventeenth-century Sicilian rebinding.[201] These four leaves were removed when Drogo's copy of the *Handbook of 809* was restored in 1995, but the commonly used foliation (likewise adopted here) continues to number the manuscript leaves as though the short printed material still occupied the first four folios of the manuscript.

When Téllez Girón returned to Spain in 1697, he brought his newly expanded library with him. Naturally, his collection included the manuscripts he and his predecessor had confiscated in Sicily. A similar fate awaited his own library during the War of the Spanish Succession (1701–14). Téllez Girón eventually backed Archduke Charles of Austria (1685–1740, later Charles VI of Austria, r. 1711–40), and as part of the ongoing struggle, Philip V's foot soldiers confiscated his library in 1711. At that time, Drogo's illustrious book of star pictures passed into the nascent collection of the Real Biblioteca (olim L. 95); the manuscript was later designated olim Vitr. 13-3 and MS 3307 in turn, once it was accessioned into the Biblioteca Nacional of fin-de-siècle Madrid.[202] There, to this day, Drogo's copy of the *Handbook of 809* provides one of the best records of Carolingian confrontation with the pagan roots of classical culture—and of its study by ninth-century Christian Franks. That syncretistic engagement with the past resulted in the images of the constellations. An iconographic study of those images is the subject of chapter 2.

CHAPTER TWO

DROGO'S COPY OF THE *Handbook of 809*

A Simulacrum of Celestial Order

The celestial host surrounding medieval Metz maintained an otherworldly permanency. On Bede's own authority, the realm of the fixed stars and the planetary wanderers would survive the Apocalypse, bathed in the radiant splendor of the Almighty, who purged the unclean with fire. In chapter 70 of *De temporum ratione*, Bede had informed Drogo and all learned Franks that prior to the donning of "everlasting bodies," "the airy heaven will shrivel up in fire, [but the heaven] of the stars will remain undamaged. In fact, the heavenly bodies will be darkened, not by being drained of their light, but by the force of a greater light at the coming of the Supreme Judge."[1] The texts and images of the fixed stars in a book such as Drogo's copy of the *Handbook of 809* were signs of stability in a world of transience. Even the hope and promise of the Apocalypse, which played a fundamental role in the evangelistic mission and doctrine of renewal within the early medieval church, could not undermine these integral components of God's observable creation. Time was consistently restored through the annual and daily cycles of the heavenly bodies, regulated by the sun and moon.[2]

For the Frankish princes who owned copies, the combined moments of meditative reflection upon the texts and star pictures of the *Handbook of 809* were part of the unfolding salvation history of the Franks. Individual episcopal study by a prelate such as Drogo indirectly helped fulfill the soteriological mandates of the early medieval Christian church that called for sacred study of the liberal art of astronomy.[3] Drogo's personal copy of the *Handbook of 809* was thus an ephemeral portal—bound for destruction within the Apocalyptic flames of penitence—that nevertheless opened his mind and spirit to the eternal mysteries of the heavens.

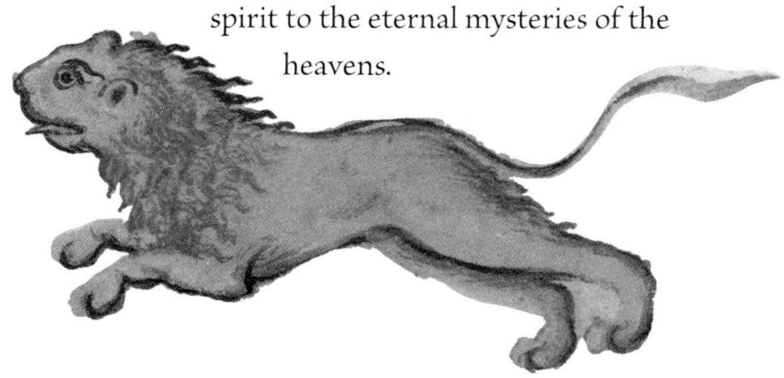

Paul Edward Dutton has argued in this context that courtly encomia likening worthy bishops to stellar vicars of Christ had more than a literary significance, operating as a topos for a poet such as Sedulius Scottus (fl. ca. 840–60) within the Frankish realm. Frankish kings such as Charlemagne or Louis the Pious achieved a veritable "Christianized Carolingian apotheosis" that could be extended by royal blood to their episcopal sons, nephews, and cousins, legitimate or otherwise, in service to the church.[4] As Dutton notes, Sedulius, apparently an itinerant Irish monk, celebrated Bishop Hartgar (served 840–55), under whom he had served in Liège, perhaps achieving the rank of *scholasticus* (roughly equivalent to a headmaster) within Hartgar's cathedral school dedicated to Saint Lambert.[5] It is highly likely that Sedulius also composed a number of lyrical panegyrics in Metz for Bishop Adventius (served 858–75).[6] Sedulius specifically makes Hartgar "earth's most radiant star," handpicked by Aurora in one poem,[7] and in another compares the physical ascent of a tower to the expansion of the prelate's mind, making the fulfillment of his episcopal and homiletic duties to the people of Liège possible:

The daughter of Zion rejoices in such a shepherd,
and the rich and poor exult with joy.
He builds a lofty tower, a hundred cubits high,
so that he may ascend above the stars.
Climbing the stairway rising towards heaven,
he instructs his flock with sage words and examples.[8]

Drogo, similarly preparing himself through study of his personal copy of the *Handbook of 809*, thereby "wards off wolves and rescues his lambs."[9] It is within this framework, joining personal spiritual study to the establishment of a celestial Carolingian hagiography, that the text-image relationship in the *Handbook of 809* must be addressed, and Drogo's personal book is the best example of an official spiritual-yet-scientific manual disseminated throughout the Frankish realms for such use. Stephen McCluskey has said of such "astronomical and computistical anthologies," otherwise known as "star catalogues," that "their goal was not astronomical observation but artistic and mythological edification."[10] More important, study of the stars brought Carolingian bishops into the heavenly realms, where the planets and fixed stars served as stepping stones to intellectual freedom and spiritual enlightenment, purifying the hearts of these prelates while simultaneously preparing them to accomplish their spiritual mission on earth by bringing their fellow Franks along with them.[11]

In this chapter, the central theme of part 1 of the present volume continues to unfold: the art-historical significance of Drogo's personal copy of the *Handbook of 809* requires an investigation into the iconographic history of classically inspired but notably reinterpreted images of the heavens. This chapter focuses primarily on a detailed review of the pictorial program of constellations in Drogo's handbook. After an initial explanation of Aratean traditions, I offer a complete iconographic analysis of the Greco-Roman traditions that informed the symbolic properties of the miniatures in Drogo's handbook. In addition, specific attention is given to images such as Chiron the Centaur in order to supply a cogent example of Carolingian exegetical emendation and strongly advocate for Carolingian painterly creativity, as argued in the introduction. As seen in chapter 1, a thorough examination of Drogo's copy of the *Handbook of 809* permits a celebration of the ways in which students of astronomy and computus created a dynamic record of their spiritual science.

Here in chapter 2, the profound appreciation for classical iconographic traditions manifested by the Carolingian prelates and painters who created Drogo's copy of the *Handbook of 809* is contextualized, in terms of both the mythological narratives informing image types and the scientific soteriological benefits of studying the star pictures along with their corresponding texts. In every viable instance, Frankish painters demonstrated an appreciation of the traditions they transformed, creating a novel Christian presentation of the heavens one constellation at a time. At the end of this review, some closing remarks concerning Panofskian iconographic traditions build upon the comments in the introduction and conclude part 1.

History of the Aratean Pictorial Cycles and Their Texts

This novel spiritual justification for the Carolingian star catalogues supplied the impetus for the original description of the constellations offered to readers of the *De ordine*, which was systematically derived from an admixture of mythological lore and the presentation of the constellations modeled after star globes. The textual and visual considerations of vital significance for the *De ordine* hark back to Aratus of Soli, who came to Pella in 276 B.C.E. and composed the *Phaenomena* at the Macedonian court of Antigonus Gonatas, ca. 276–274 B.C.E.[12] According to a somewhat apocryphal legend, Aratus did not derive his poetic description of the constellations, their stellar utility for discerning temporal changes, or their meteorological significance directly from the heavens.[13] According to Cicero, Aratus was a well-trained literary dilettante from Athens without professional astronomical skills: "Eudoxus of Cnidos, who was a pupil of Plato's . . . marked on the globe the stars that are fixed in the sky. Many years after Eudoxus, Aratus adopted from him the entire detailed arrangement of the globe and described it in verse, not displaying any knowledge of astronomy but showing considerable poetical skill."[14]

Douglas Kidd has valiantly defended Aratus against such Ciceronian naysayers, arguing persuasively that Aratus neither imported wholesale nor uncritically endorsed Eudoxus's interpretation of Greek astronomy. On the contrary, Aratus fundamentally altered the perspective from which he drafted his 1,154 lines of polished verse, adopting the standpoint of the stargazer and earthbound student of the heavenly host.[15] This radical move reflects more than authorial repositioning. The novel voice of Aratus harnessed the heavens as a cosmological realm populated by mythopoetic heroes and celestial wonders replete with signposts for knowledgeable stargazers to discern. The abundance promised to farmers who augured well the agrarian signs, and the safe journeys vouchsafed to sailors who navigated auspicious times for travel, had been established by none other than attentive Zeus. The stars and their constellations were a road map to Greek or Frankish survival, and arguably even to individual success, according to Aratus's proem:

> Filled with Zeus are all highways and all meeting-places of people, filled are the sea and harbours; in all circumstances we are all dependent on Zeus. For we are also his children, and he benignly gives helpful signs to men, and rouses people to work, reminding

> them of their livelihood, tells when the soil is best for oxen and mattocks, and tells when the seasons are right both for planting trees and for sowing every kind of seed. For it was Zeus himself who fixed the signs in the sky, making them into distinct constellations, and organized stars for the year to give the most clearly defined signs of the seasonal round to men, so that everything may grow without fail.[16]

It is interesting that the providential role of divine oversight in the opening verses of Aratus's Greco-Macedonian *Phaenomena* (or proem doubling as a panegyric to Zeus) also provides a pagan foil to a kindred idea of Christ the Good Shepherd that was appropriated by the early Christian church. The later Latin derivatives of the Aratean textual traditions, culminating with the *De ordine* in the *Handbook of 809*, attest to this development, which further enhanced the individualistic turn in Aratus's transformation of the earlier information about the stars reported by Eudoxus, his fourth-century predecessor (fl. 368–365 B.C.E.). Other notable deviations from the lost text of Eudoxus include Aratus's effort to align the advent of the constellations appropriate to seasonal zodiac signs with their respective positions on the solstitial or equinoctial colures.[17]

In part 2 below, chapter 3 examines the stylistic concerns of importance for a revisionist history of Carolingian painting in Metz, while chapter 4 details the significant transformations of long-standing Aratean traditions in the service of a Carolingian scientific soteriology. (Students of philology rather than art history may prefer to read chapter 4 before returning here, recognizing that an alternate trajectory through the text will affect the presentation of ideas and the overarching argument of this book: namely, that Drogo's copy of the *Handbook of 809* documents in its texts and imagery a comprehensive Carolingian effort to preserve a simulacrum of celestial order.) In the present chapter, a survey of the star pictures in Drogo's copy of the *Handbook of 809* permits an exhaustive iconographic review of the origins and nature of the forms of the zodiac and constellations that were available to ninth-century painters.

Zeus's array of constellations, which resulted in part from the legendary catasterisms of Greek mythology, were stolen in Promethean fashion from the pagan deity and set within the minds and hearts of Frankish students of the liberal art of astronomy. The *De ordine* recounts in general terms the mythic heritage of the constellations and thereby announces from the outset the twin themes of the text:

> Here is the order and the placement of the constellations, which were fixed in the heavens by the grouping of several stars into signs. The kinds of forms are believed to have been admitted into the heavens from some models or fables. The arrangement of the natural forms did not yield their names, but human conviction gave the numbers and names to the stars. But, in accordance with Aratus the number of stars belonging to each sign has been appointed; the description should follow according to the order, which is his own.[18]

Not Zeus but the pagan ancestors of eighth- and ninth-century Christians thus clustered the stars and then gave these arbitrary groupings the names by which they remain known to this day. Advancing a personal outlook, the Frankish author of the *De ordine* relied upon the same powers of imagination and intellect that earlier people had used to designate the constellations. In so doing, the author of the *De ordine* followed Aratus's lead, but in keeping with standard interpretations of the Frankish renewal: the individual early medieval Christian's position within the cosmos (and the concomitant placement of his microcosmic soul) required a novel celebration of Aratean tradition. The Franks reoriented their understanding of the heavens and pointed Carolingian astronomical study in a joint scholarly and spiritual direction, one that permitted revision of classical narratives that had come down via lines of textual transmission discussed here and in chapter 4. The emphasis on the individual person's observation of the heavens, inherited from Aratus and coupled with a sincere desire to reinterpret tradition to fit the needs of a new ruler and polity, were therefore parlayed into an unexpected Christocentric, soteriological focus as a driving motivation for the Carolingian study of astronomy and the stars. As explained in the introduction and argued in this chapter, the images of the constellations themselves attest to the scientific sacralization of antiquity and make possible the Frankish scientific soteriology witnessed by novel presentations of the constellations.

Given these preliminary considerations, it is not surprising that a learned translator of astronomical texts from Corbie adopted and adapted the order of the star pictures belonging to Aratean textual traditions for his Christian presentation of the constellations. This supplied a meaningful structure for the *De ordine*, transported by Adalhard of Corbie to the Aachen synod of 809. In all Aratean texts, the standard way to adduce the constellations, including the zodiac, was to begin with the celestial hemisphere north of the ecliptic and continue with those constellations found south of the ecliptic.[19] Hence the depictions of the constellations in Drogo's copy of the *Handbook of 809* will be described in the order in which they appear, adhering to and revealing the standard text of the *De ordine*. This ordered presentation of the constellations recurs in other texts derived from the Aratean *Phaenomena*.[20] Since there is such a high degree of standardization in the sequence of the cycle, the idiosyncrasies in discrete pictorial programs and alterations associated with various textual recensions become significant. Before returning to a discussion of Adalhard, it is useful to review precisely how he and others envisioned the structure of the celestial sphere.

The *ecliptic* is the oblique that circles the celestial sphere on which are permanently situated the fixed stars, composing the constellations. The entire celestial sphere rotates daily in a clockwise direction about a stationary earth. This line receives its name because it defines the path of the sun throughout the course of the year. Since all planets—including the moon and sun, according to the Ptolemaic model of the solar system—revolve about the earth in a counterclockwise direction, the only place the sun, moon, and earth can overlap is on the ecliptic, at the center of the geometrically distributed twelve-part zodiac belt, causing eclipses.[21]

According to a Carolingian understanding of Plinian and classical astronomy, the zodiac belt extends 6 degrees north and south of the ecliptic, defining a 12-degree diagonal swath of space through which the planets move latitudinally, according to their inclination, relative to the ecliptic. The equator defines the central circumference of the earth and is complemented by the so-called celestial equator exactly parallel to it.[22] The parallel tropics of Cancer to the north and Capricorn to the south of the equator identify zones of roughly 23.5 degrees of geographic latitudinal space, respectively. The oblique ecliptic intersects the tropic of Capricorn at its southernmost point in the region once occupied by the constellation Capricorn, whereas it intersects the tropic of Cancer in the former region of the zodiac constellation Cancer at its northernmost point. This also explains the origin of the tropics' names.[23] Because of the phenomenon of precession, or the perpetual shift of all constellations on the celestial sphere from west to east (i.e., counterclockwise) over time, the geometric 30-degree sections of the ecliptic defined as the signs of the zodiac do not align perfectly with the constellations that bear their names and through which the sun passes annually. In other words, stars from the constellation Pisces are actually located within the spatial zone identified by the geometric zodiac sign of Aries; Aries, the constellation, has entered Taurus's territory; and so forth.[24]

Again, according to the geocentric model of the universe, the fixed and wandering stars (planets) rotate around the earth. There are two conflicting motions, which typically cause great confusion for those trying to understand a medieval view of the heavens. On the one hand, the sun and all planets make their way through the zodiac signs along a counterclockwise trajectory (ignoring for the moment the phenomenon of apparent planetary retrograde motion discussed in chapter 3) that follows the order on the diagram "The Apsides of the Planetary Orbits Within the Zodiac" in Drogo's copy of the *Handbook of 809* (fol. 65v; see fig. 5). Thus we see Aries at 9:00, designating the beginning of solar springtime renewal, followed by Taurus, Gemini, Cancer, Leo, Virgo, Libra, Scorpio, Sagittarius, Capricorn, Aquarius, Pisces. Simultaneously, every heavenly body within the geocentric cosmological model, including *a fortiori* the celestial sphere of the fixed stars, rotates about the stable earth once daily in an apparently alternate clockwise motion.[25] These two aspects of the medieval celestial model derived from Ptolemy permit the daily passage of the sun and moon as well as the seasonal shifts linked to the liturgical calendars, regulating lives on earth through their rapport with the celestial bodies surrounding them on all sides.[26] In the words of Aratus, "The numerous stars, scattered in different directions, sweep all alike across the sky every day continuously for ever. The axis, however, does not move even slightly from its place, but just stays for ever fixed, holds the earth in the centre evenly balanced, and rotates the sky itself."[27]

Hubert Le Bourdellès has emphasized the significance of Adalhard of Corbie's involvement in the synod of 809. Adalhard was, as noted in the introduction, named in the report that followed the convocation of the computistical subgroup that assessed the state of the knowledge of computus during the synod of 809.[28] It is highly likely that Adalhard brought two vital records of Frankish interest in astronomy with him to the Aachen synod. Those two new Carolingian creations from Corbie became the key illustrated components of book V in the *Handbook of 809*. Le Bourdellès has fully identified the textual sources for these two records, and this information is key to the story of the creation of Drogo's copy of the *Handbook of 809* undertaken here. The *Excerptum* (text V.1) supplied an abstract, as it were, of the version of the *Aratus Latinus* known as the *Revised Aratus*. The *Aratus Latinus* was a translation into Latin from the ancient Greek poem described here, the *Phaenomena* of Aratus. The earlier translation probably took place at Corbie around 735 and then provided the foundation for the *Revised Aratus*, which is also likely to have been made in Corbie, around 790.[29] The most important example of the *Revised Aratus* is the third-generation copy from Corbie presently located in Paris (Paris, Bibliothèque nationale de France, MS lat. 12957; hereafter the Paris *Aratus*). Created in the ninth century (probably ca. 810), this copy of the *Revised Aratus* includes a set of miniatures (fig. 35). The line drawings suggest to Le Bourdellès that both the original copy of the *Aratus Latinus* translation from Greek into Latin and the lost original copy of the *Revised Aratus* itself were illustrated with Aratean star pictures.[30]

These are contentious points, although Le Bourdellès is probably accurate in his assessment. An explanation of the reasons he is apt to be correct is essential for an understanding of the origin of the star pictures found in the *Handbook of 809*, and *a fortiori* for those in Drogo's copy. In addition, for this art history of Carolingian astronomical manuscripts, it is equally important to assess whether lost, hypothetical cycles of star pictures included drawings alone or miniatures as well. It is highly likely that there was an originally illustrated copy of the *Revised Aratus* made in Corbie. The evidence for an originally illustrated *Aratus Latinus* is sketchier. There were no illustrations in the early copy of the type A version of the astronomical encyclopedia (according to Borst's nomenclature) discussed in chapter 1, the Verona compilation.[31] In the late eighth century, medieval scribes therefore did not consider astronomical compilations in need of artistic illustration. In fact, some later copies of the *Aratus Latinus* lack illustrations (e.g., Paris, Bibliothèque nationale de France, MS lat. 7886, ca. 850–75; Brussels, MS 10698, twelfth century). But other copies of the *Aratus Latinus* were probably illustrated (see fig. 7) with detailed line drawings depicting the stars, like those accompanying the

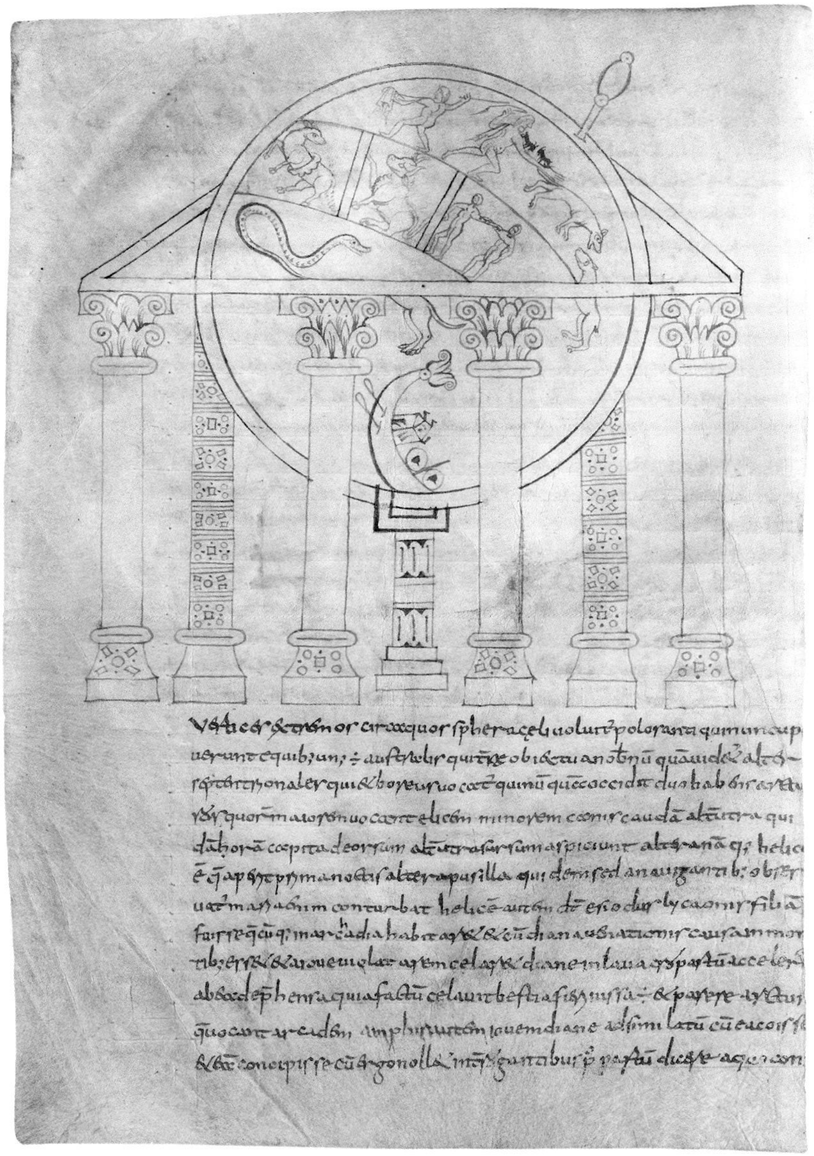

FIGURE 35

Celestial globe. Paris, Bibliothèque nationale de France, MS lat. 12957, fol. 63v. Photo: BnF.

portion of the Basel *Scholia* devoted to the Germanicus translation with scholia.[32]

In any case, Aratean manuscripts were not always illustrated. In the case of the *Handbook of 809*, however, miniatures constituted an integral component of the early luxury copies of the handbook. The ninth-century pictorial cycle in Paris could have been a later addition to the lost original copy of the text, although this is unlikely. The introduction of pictures into the *Revised Aratus* could have taken place during the manufacture of the hypothetical intermediary copy, postulated by Le Bourdellès, that probably supplied the immediate precursor for the copy from Paris and another presently in Cologne (Cologne, Dombibliothek, MS 83-II, hereafter the Cologne *Aratus*; fig. 8).[33]

The Cologne *Aratus* includes a date from the incarnation of the Lord of 798 on folio 14v and was probably completed on art-historical grounds in 805. Hildebald, who was archbishop of Cologne (before 787–818) and archchaplain to Charlemagne, compiled the manuscript.[34] Charlemagne decreed that clerics should undertake the mastery of the computus with the Capitulary of Diedenhofen (Thionville) in 805. The coincidence between the completion date of work on the Cologne manuscript (805) and the capitulary has led Anton von Euw to hypothesize a direct connection between Charlemagne's desire for an astronomical-computistical-pedagogical handbook and Cologne.[35] In other words, the Cologne copy of the *Revised Aratus* was an early attempt to create an astronomical and computistical teaching manual before the *Handbook of 809*.

Two observations about the Cologne *Aratus* are noteworthy for the history of star pictures during the Carolingian era. First, the completed miniatures in Cologne, finished by 805, are truly painterly, such as the image of Hercules and the serpent on folio 156v (fig. 8). By comparison, the drawings in the Paris copy of the *Revised Aratus* supply linear contours (fig. 35). A parallel disparity exists between the paintings in Drogo's copy of the *Handbook of 809* (plate 2) and the linear forms belonging to a set of suitably related copies of the *Handbook of 809* that justify their assignment to a specific recension: Monza f–9/176 (figs. 36–37) and Vat. lat. 645 (fig. 38). Painted miniatures were actually the exception rather than the rule among the Carolingian manuscripts with star pictures. The inclusion of painted miniatures, as opposed to linear contour drawings, was clearly intended to convey the superiority and excellence of a *de luxe* manuscript such as Drogo's princely copy. Astronomical-computistical compilations of great refinement, replete with painterly miniatures, fulfilled more than a pedagogical purpose. They were also expressions of a patron's power and influence. Given this conclusion, there is little doubt that the original copy of the handbook compiled in Aachen under the direction of Adalhard of Corbie also included a lavishly painted cycle of miniatures. It was made for Charlemagne.

Second, the Cologne and Madrid manuscripts were both made for patrons who were at one time archchaplains of the Carolingian kingdom, as discussed briefly in chapter 1. Hildebald of Cologne ascended to the archchaplaincy after the death of Angilram of Metz in 791.[36] Bishop Drogo served both Louis the Pious (after 834) and Lothar I (after 842) as archchaplain, too.[37] Since Drogo's copy of the *Handbook of 809* was made around 830, it is reasonable to see in the painterly production of the miniatures an effort to surpass the quality of painting in Hildebald's proposal for a handbook.

In any case, the pictorial cycles of the Paris *Aratus*[38] (fig. 35) and Cologne *Aratus* (fig. 8) are similar

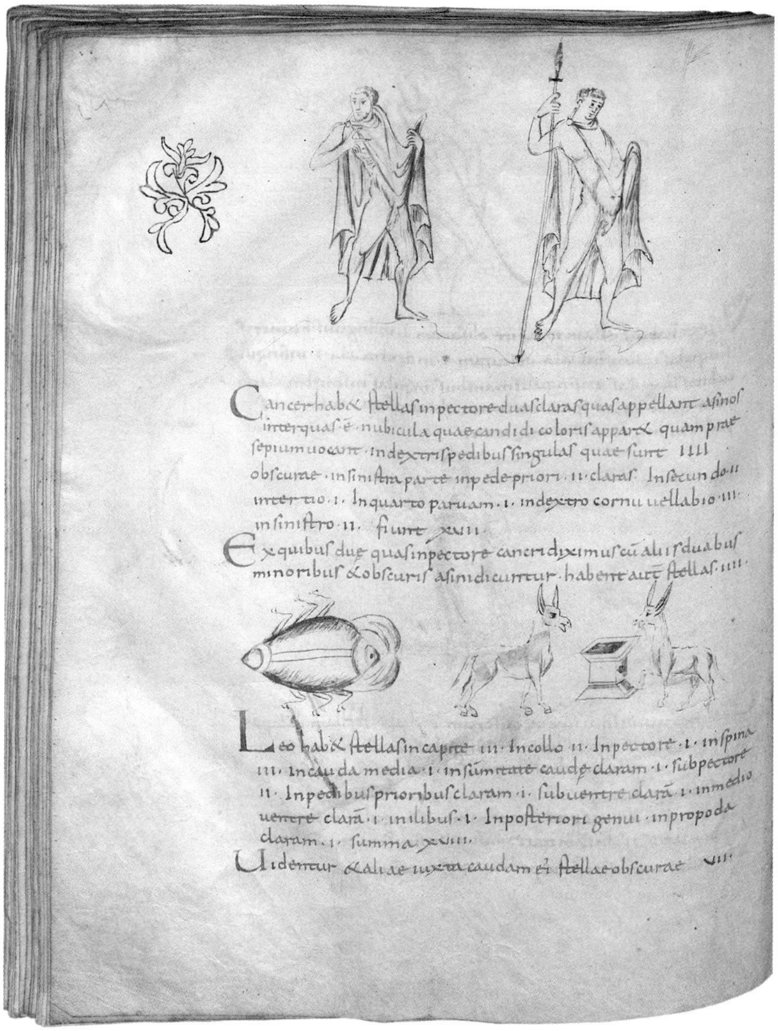

FIGURE 36

Gemini and Cancer. Monza, Biblioteca Capitolare, f-9/176, fol. 63v. Photo © Biblioteca Capitolare del Duomo di Monza.

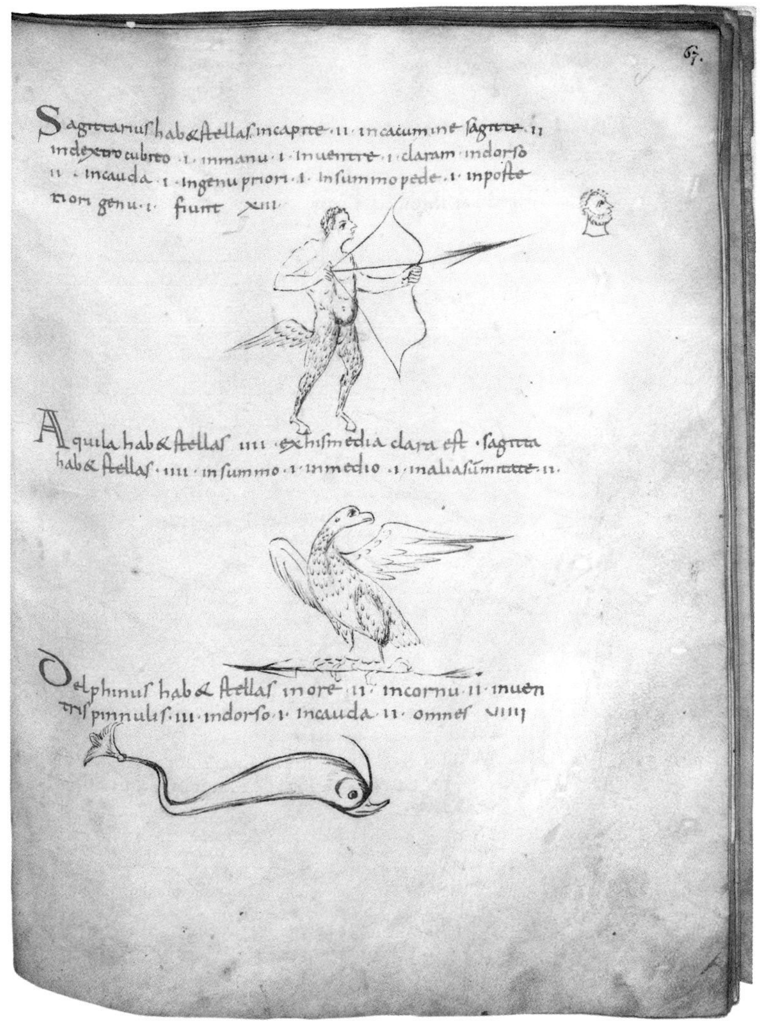

FIGURE 37

Sagittarius, Aquila, and Delphinus. Monza, Biblioteca Capitolare, f-9/176, fol. 67r. Photo © Biblioteca Capitolare del Duomo di Monza.

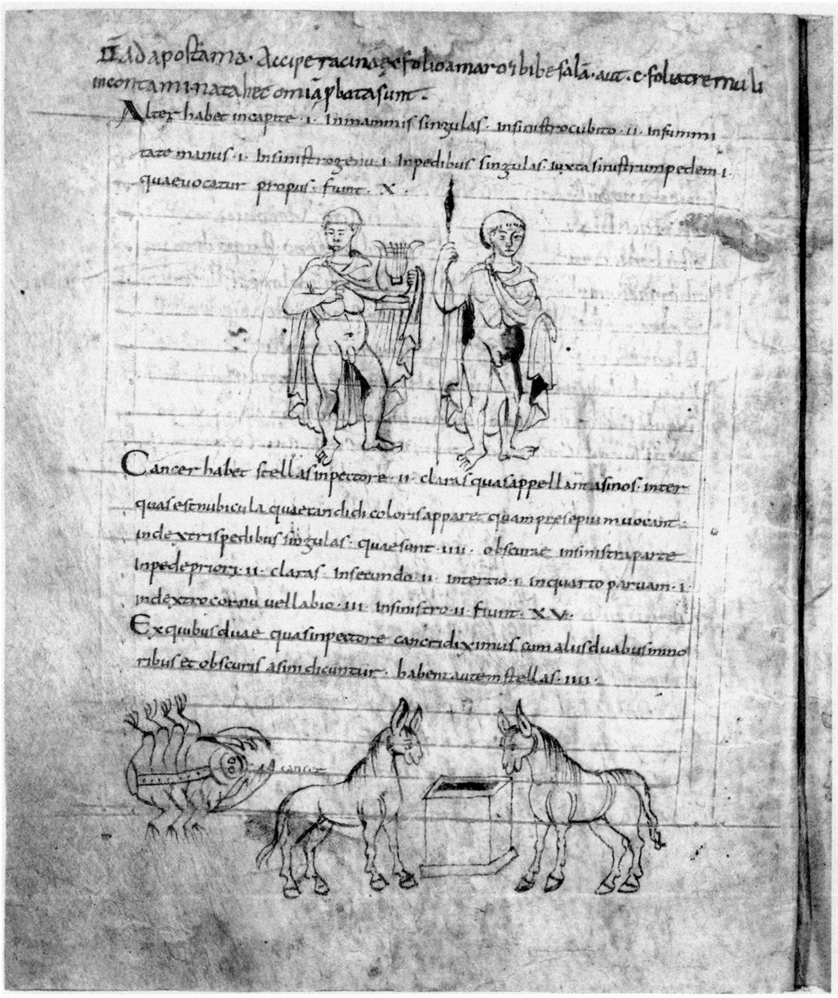

FIGURE 38

Gemini and Cancer. Vatican City, Biblioteca Apostolica Vaticana, Vat. lat. 645, fol. 58v. Photo © 2017 Biblioteca Apostolica Vaticana. Reproduced by permission of Biblioteca Apostolica Vaticana, with all rights reserved.

enough to warrant the suggestion that their predecessor influenced the order and the arrangement of the pictorial cycle in these manuscripts. The pictorial program in these two manuscripts is also relatively similar to the various copies of the *Revised Aratus* belonging to the second branch of the bifurcated stemma provided by Le Bourdellès. This further supports his contention that the original copy of the *Revised Aratus* was indeed illustrated. For example, one of the copies from the other branch of *Revised Aratus* manuscripts is Saint-Gall, Codex 902 of the mid-ninth century (Stiftsbibliothek Sankt-Gallen, Cod. 902, hereafter Saint-Gall 902; figs. 39–40). Saint-Gall 902 includes the illustrated portion of the text referred to as the *Recensio interpolata* in earlier literature.[39] The image of Aries on page 90 (fig. 39) resembles the basic linear form of the painted image in Drogo's copy of the *Handbook of 809* (plate 9), but in reverse.

This iconographic similarity with manifest variety is a standard phenomenon found throughout the individual Carolingian manuscripts containing star pictures. The similarities derive from the long-standing Aratean precursors under discussion here, but the individual differences in specific codices permit additional meanings beyond straightforward classical references. These novel insights are the true legacy of Carolingian book painters to pictorial art. That does not compromise the significance of the Aratean texts transmitted for posterity. On the contrary, along two intrinsically intertwined trajectories, the value of Carolingian scribes to the accurate preservation of texts can be celebrated equally with the interpretative efforts of Frankish painters who built visually on the work of their predecessors and turned tradition into new ninth-century erudition, replete with the awe and wonder the heavens can induce.

The uniformities displayed by depictions of the constellations in the *Revised Aratus* manuscripts supply the best reason to argue backward for an originally and similarly illustrated *Aratus Latinus*. For example, the image of Arcturus Maior (The Great Bear) on page 82 of Saint-Gall 902 resembles the basic linear form of the related images in the Paris *Aratus* (fol. 64r) and Cologne *Aratus* (fol. 155r). Moreover, in all three cases, the quadruped faces to the left. Naturally, the general forms of the zodiac and constellations can be traced back to Hellenistic precursors. According to Mechthild Haffner, the finalization of the relatively standard pictorial program of the illustrated copies of the Germanicus translation of the *Phaenomena*, which influenced the order and form of the pictures in the *Aratus Latinus* and the *Revised Aratus*, took place in the late third century. These images were attached originally to a hypothetical Greek copy of the *Phaenomena* with scholia, referred to in the scholarly literature as Φ.

Haffner has noted that certain miniatures in Aratean manuscripts can only be explained by the scholia introduced into the textual redactions. This suggests that the iconography of the star pictures changed over time or that the pictures were added to the text after its composition (at a potentially much later date). For example, a kneeling Hercules only makes sense in the context of the conflict between the hero and the serpent, described in scholia, which later influenced the creation of the Basel *Scholia* supplementing the Germanicus translation into Latin of the *Aratus* poem (fig. 7). In other words, Haffner believed that the definitive set of illustrations that influenced medieval astronomical and astrological cycles of pictures was added to the Aratus tradition roughly five centuries after the composition of the original poem by Aratus, and the textual analyses of Le Bourdellès have corroborated this result.

Under this interpretation, the advent of the codex roughly coincided with the creation of a series of illustrations for the Aratean manuscripts—a series that would later become comprehensive. In the third century, the images adhered to a pictorial logic appropriate for a rotulus in Alexandria. During the transition period from the rotulus to the codex, Haffner has argued, there were linear star pictures without frames accompanying a Greek edition of the *Phaenomena* with scholia. Interestingly, many of the putative earlier examples of the constellations tended to display the dorsal sides of the constellations, as though they were viewed outside the ring of fixed stars on an ideal celestial globe. Haffner's ingenious solution to the problem of the transmission of the Aratean text and star pictures was to postulate a new Latin translation of the *Phaenomena* in the late third or early fourth century. This edition relied upon the Germanicus translation for a model, while the translators also addressed the Greek scholia, thereby creating the accompanying Latin scholia, namely, the Basel *Scholia*. The miniatures accompanying the copies of the Germanicus translation from around that period onward supposedly displayed true codex illustrations, with blue fields and red frames (fig. 21), recalling the pictures in the Carolingian ninth-century Leiden *Aratea*.[40]

Colin Roberts and T. C. Skeat have identified the late third and early fourth century as the point at which bound books generally replaced rotuli as the vehicle of choice for the transfer of written information.[41] This finding buttresses Haffner's hypothesis. Here, however, in assessing the origins of the astronomical miniatures in Drogo's copy of the *Handbook of 809*, we must recognize the plausibility of an

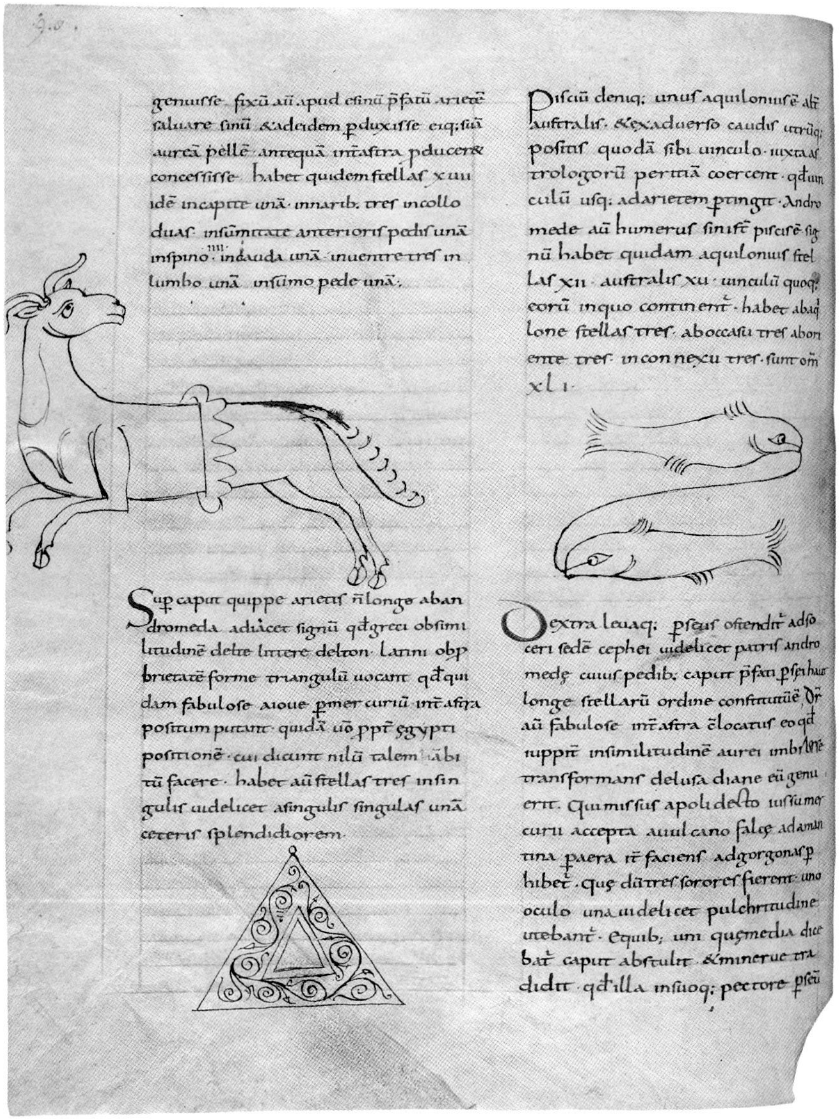

FIGURE 39

Aries, Deltoton, and Pisces (Aquilonalis and Australis). Stiftsbibliothek St. Gallen, Cod. Sang. 902, page 90. Photo: Stiftsbibliothek St. Gallen.

originally illustrated *Revised Aratus*. That pictorial cycle probably harked back to an original program created in the third century in Alexandria for the Greek edition of the *Aratea* with scholia. That edition, as Le Bourdellès proposed, probably provided the model for the pictures introduced into the *Aratus Latinus* and then into the *Revised Aratus*, recognizing that with every copy there were opportunities for creative aesthetic decisions.[42]

Like Drogo's copy of the *Handbook of 809*, the Cologne *Aratus* included miniatures in open fields between single-column text blocks. These *de luxe* painted astronomical manuscripts preferred to return to the more ancient graphic style of unframed images, which was actually associated with a rotulus. On the one hand, these aspects of the pictorial cycles in early medieval illustrated astronomical manuscripts are emblematic of beliefs about early manuscript illustration espoused by Kurt Weitzmann.[43] On the other hand, the pictures in these two codices have received uncommon emphasis. In the Paris *Aratus* (fig. 35) and Saint-Gall 902 (fig. 40), the linear renderings of the constellations have been squeezed into the available allotted spaces, pushing up against the margins. The linear forms seem to be inconsequential diversions from the dominant text blocks in those manuscripts. Yet not only did the *de luxe* copies of ninth-century astronomical treatises receive painted miniatures, but those miniatures were also placed in free-standing, unframed open spaces, highlighting the independence and artistic refinement of the forms. It is important to note here that the highly developed tinted drawings of the sister manuscript to the Saint-Gall 902 (fig. 40), the Saint-Gall Cod. 250 (Stiftsbibliothek Sankt-Gallen), were also set apart, according to the design strategy for an exceptionally refined cycle of miniatures. The monastic draughtsmen inserted their highly finished drawings into Saint-Gall 250 in open spaces between single text columns in the last quarter of the ninth century, following the procedure for a more luxurious copy of an illustrated astronomical text.[44] The one-column format was in no way reserved for the painted manuscripts alone, although the presence of paintings and a single-column format in the Madrid and Cologne manuscripts clearly conveys their exceptional refinement and splendor. This actually opposes one tenet of Weitzmann's history of early manuscript illustration. Weitzmann has argued for a steady progression from early cycles of illustration in scrolls to the emancipation of gilded images, which resulted in framed manuscript illuminations.[45] Notably, early medieval artists could accentuate the value of their miniatures without the use of frames.

In summary, it is highly likely that a pictorial program enlivened the original edition of the *Revised Aratus,* and that this pictorial cycle had a significant impact on the development of future star pictures. The *Excerptum* from the *Handbook of 809* (V.1) was not illustrated, but it was the prelude to the text that included the illustrations, namely, the *De ordine* (V.2). For this reason, it is highly significant that the *Excerptum* was derived from the *Revised Aratus,* as argued by Le Bourdellès, and that the *Revised Aratus* originally included a set of star pictures like those in the Cologne *Aratus.*

Le Bourdellès also identified the second text from book V of the *Handbook of 809* as an original contribution, which Adalhard brought from Corbie to the synod meeting in 809. The text provided a summary of the Basel *Scholia.*[46] The latter name derived from the scholia included within a manuscript likewise containing the fragmentary copy of the *Aratus Latinus* (and not *Revised Aratus*) attached to an edition of

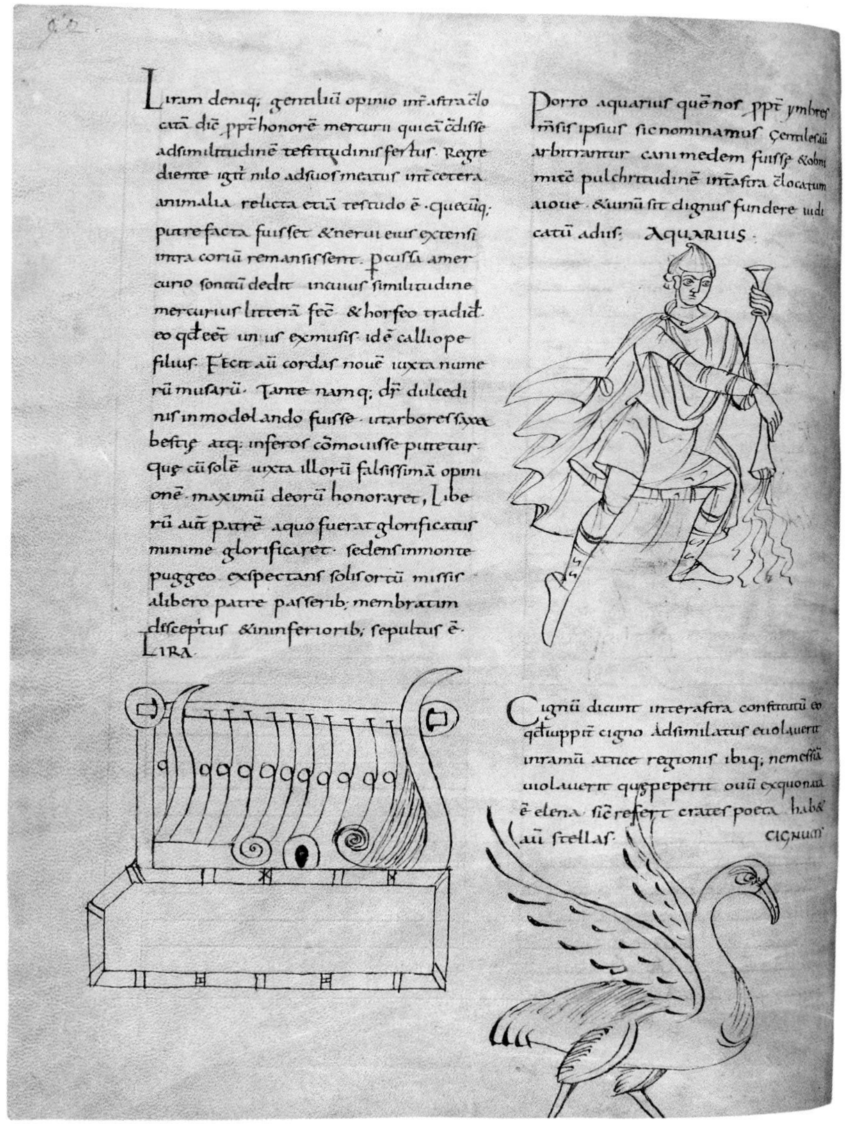

FIGURE 40

Lyra, Aquarius, and Cygnus. Stiftsbibliothek St. Gallen, Cod. Sang. 902, page 92. Photo: Stiftsbibliothek St. Gallen.

the Germanicus translation in Basel, as mentioned above. It is known that the scholia in Latin could not have been added to any putative version of the Aratus poem that arose before the imperial period of classical Rome, since the scholia included references to the writings of Publius Nigidius Figulus (active ca. 98–45 B.C.E.).[47] The Latin scholia were translated ca. 300 C.E. from the commentaries to the lost Alexandrian redaction of the *Phaenomena* by Aratus, as discussed already. Le Bourdellès followed J. Martin when he argued that the hypothetical Alexandrian text combined the Aratus poem with an introductory text summarizing the ideas of Eratosthenes (ca. 276–194 B.C.E.) on terrestrial or celestial globes. In any case, the commentaries had been added to the Germanicus translation by the fourth century, as Le Bourdellès noted, because the Christian exegete from Trier, Lactantius (240–ca. 320 C.E.), alluded to the commentary in his *Divinae Institutiones* (I.11).[48] All of Le Bourdellès' findings, therefore, corroborate the positions advanced by Haffner.

In the third-century Alexandrian corpus, the Aratean descriptions of the constellations also included mythological tales about the constellations and star pictures. These supplemental mythological texts provided an account of the fictional reasons for which each of the constellations appeared in the heavens as the handiwork of the gods. Whenever a person, such as Hercules/Engonasin, or a creature, such as Cancer the Crab, was magically preserved for posterity by the introduction of a constellation in its form and honor, a *catasterism* had taken place. These catasterisms, described in a text derived from Eratosthenes, also catalogued the stars visible within a depiction of a constellation. The Latin translation of the *Catasterisms* from the Alexandrian compilation provided a supplemental list of the stars in the heavens added to the Germanicus translation of the Aratean *Phaenomena*. These insertions into the Germanicus text adduced mythological references and a list of stars, one constellation at a time. This complement constituted most precisely the so-called *Scholia Basileensia*.[49] The descriptions formed the basis of the new Carolingian catalogue of the stars within the constellations, the *De ordine*, albeit without the mythological references.

In 1898, Georg Thiele had already recognized the link between the group of manuscripts containing the *De ordine* and the *Catasterisms* in the most important early contribution to the study of medieval astronomical illustration, the *Antike Himmelsbilder*. Thiele labeled this set of manuscripts the "Phillippicus Class," because there is another Carolingian copy of the *De ordine* closely related to Drogo's copy of the *Handbook of 809* in Berlin (Preußischer Kulturbesitz, Staatsbibliothek zu Berlin, MS Phill. 1832; hereafter the Berlin *Ordine*). Thiele linked, for example, some of the pictures with curious iconographies, such as the satyr form of Sagittarius (fig. 41), to an earlier book of *Catasterisms* on these grounds.[50] Whenever possible, the text of the *De ordine* and the imagery in the Carolingian *Handbook of 809* emphasized meteorological details or made references to antiquarian forms of the constellations associated with zodiac cycles and celestial globes. These decisions were motivated by a zealous desire for scientific accuracy and an unshakable, fundamental belief in the intellectual sophistication of classical learning throughout the Frankish lands. The Franks could only achieve the scientific sacralization of the zodiac and the Greco-Roman history of the constellations by first fully immersing themselves in the historiographic tradition.

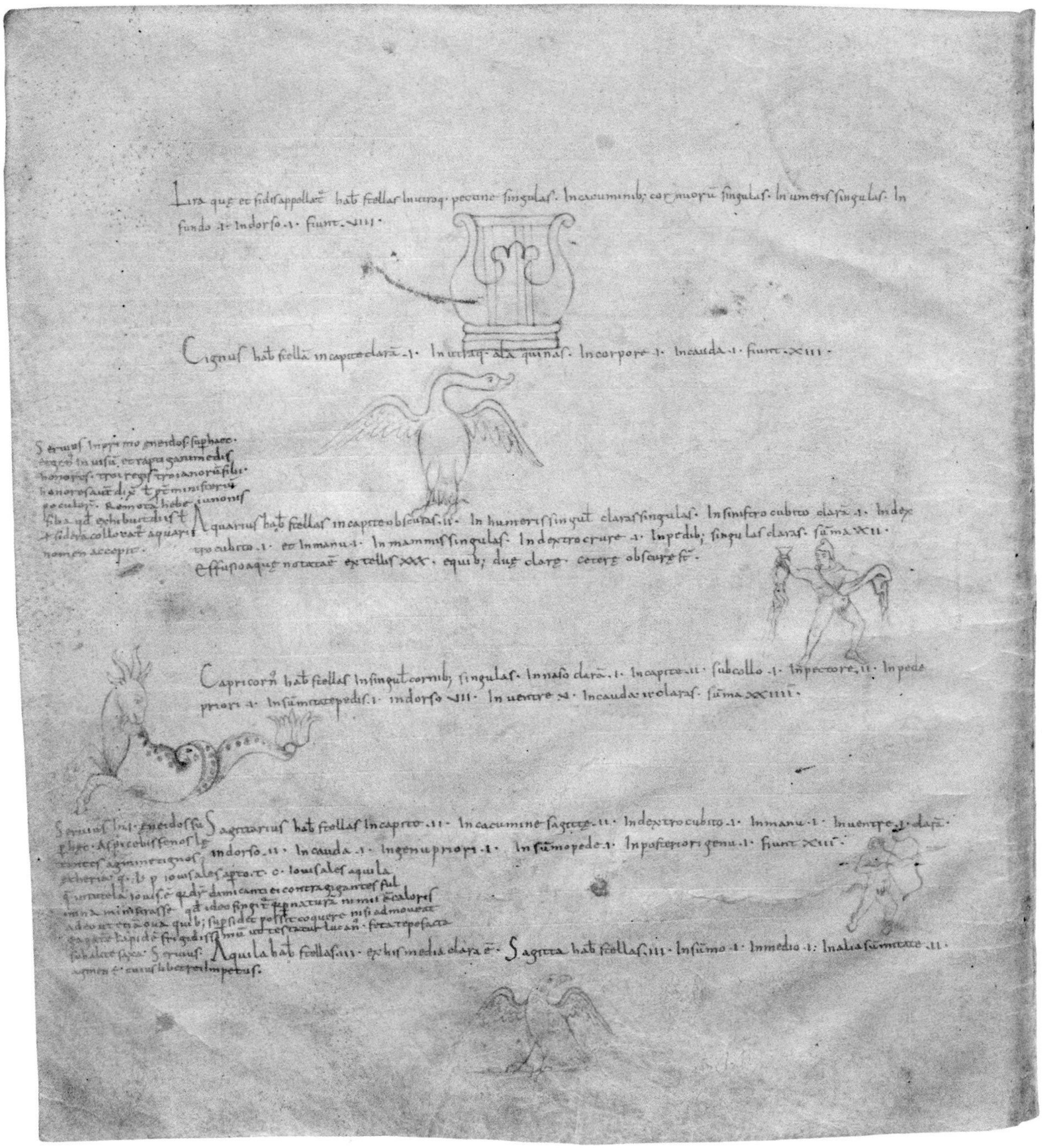

FIGURE 41

Lyra, Cygnus, Aquarius, Capricorn, Sagittarius, and Aquila. Staatsbibliothek zu Berlin, Stiftung Preußischer Kulturbesitz, Hs. Phillipps 1832, fol. 84v. Photo: bpk, Berlin / Staatsbibliothek zu Berlin / Art Resource, New York.

Christian Pictures of Pagan Constellations North of the Ecliptic

In Drogo's copy of the *Handbook of 809*, forty-two descriptions[51] of the constellations in single-column format are each followed by a free-standing Carolingian miniature. Drogo's personal reading reenacted the passage of the celestial bodies, meditating upon the information about the stars presented in the trimmed-down, palatable Christian text, the *De ordine*. The paintings document the Carolingian reaction to painterly styles of modeled forms, which Frankish viewers could access in late antique mosaics, extant frescoes, and manuscripts such as the *Calendar of 354*, known to have existed in a Carolingian copy.[52] The Carolingian star pictures, along with their linear style of execution, perhaps derive in part from such models, or from a lost late antique art of celestial globe production. Carolingian artists are known to have employed a polycyclic strategy of image selection when producing new pictorial cycles for codices that were intended to relay conceits and concepts linked to science and the liberal arts. The roots of the star pictures are nevertheless unequivocally linked to the history and textual transmission of the illustrated Aratean texts just discussed.[53] It is important to underline here exactly what is meant by an innovative or original pictorial cycle during the Carolingian period, especially vis-à-vis star pictures.

To make a general point, it is impossible to label anything even vaguely reminiscent of a crab a thoroughly original depiction of the zodiac sign Cancer. That is precisely because the form belongs to a standard type. The standardization of the form of the sign had been defined by human observation, in agreement with how people had long since clustered the stars in the sky into recognizable patterns. By the ninth century, the form of the constellation Cancer had been fixed for all of human time. The selection of the crab as an appropriate form for Cancer was no longer an interpretative artistic decision in the ninth century; its use provided an accurate scientific report of the facts of the matter. It would be just as meaningless to evaluate a chemistry student's sculptural creativity with respect to her model of a water molecule, rendered accurately with three colored wooden balls and floppy springs. *Mutatis mutandis*, the same applies to contemporary digital imaging software, rendering discrete components of coded molecular structures more readily apparent and discernible than the blades of grass on the lawn beyond one's window. In all these cases, the presentation conforms to an accurate model grounded in scientific observation and theory. Various star pictures become significant for the history of art when they are rendered with powerful individuality. The way in which the Carolingian artists in Metz represented the ninth-century images of the constellations in Drogo's personal copy of the *Handbook of 809* (Madrid 3307) turned the manuscript into a superior creative achievement. It is possible to evaluate Carolingian star pictures for their relative adherence to the prototypes derived from Aratean traditions and ancient star globes. It is equally possible to celebrate the Carolingian star pictures in Madrid 3307 for their innovative painterly forms of the constellations.

Beginning with the northern hemisphere, at the north pole, the serpent, Serpens, intertwines with the forms of the two bears: Ursa Major, otherwise known as Haelice or Arcturus Maior, and Ursa Minor, also known as Cinosura or Arcturus Minor. The forms of the miniatures in Drogo's copy of the handbook derive in part from those found in the Aratean manuscripts, which were consulted in the composition of the *De ordine* in Corbie. The similarities and dissimilarities among the Carolingian star pictures become readily apparent when we compare the individual depictions of the bears and serpent (plate 1) with the single image of the trio found on folio 3v in the celebrated Leiden *Aratea,* which belongs to a set of manuscripts associated with the court of Louis the Pious (fig. 42).[54]

The text begins with a description of the stars in the Great Bear: "Haelice, the Great Bear, has seven stars in its head. There is one star in each visible shoulder, one in the arm, one in the chest, two bright stars altogether in the front paw, one bright one in the tail, one bright one in the belly, two in the back leg, two in the rear paw, and three more in the tail. This makes a total of twenty-two stars."[55] Immediately thereafter, an artist painted the Great Bear (plate 1). The descriptions of the constellations in Drogo's book, derived in part from the descriptions in the *Catasterisms* attributed to Eratosthenes, had been stripped of their original mythological content.[56] In the text of the *De ordine* prepared for the *Handbook of 809,* the paramount concern was the composition of a text that would fulfill Charlemagne's mandate for a pedagogical tool. Adalhard of Corbie was probably personally responsible for the removal of this mythological content from the Basel *Scholia* to rid the astronomical information of pagan associations.[57] This was perhaps the most important innovation made by the Carolingian author(s) of the *De ordine,* and it continued a practice that had already played an important role in the development of the *Revised Aratus.*

As Le Bourdellès has argued, the new Latin edition of the Aratean text from the late eighth century provided an ideal opportunity to render it up-to-date with the latest in Christian education—a goal that was in complete accord with widespread

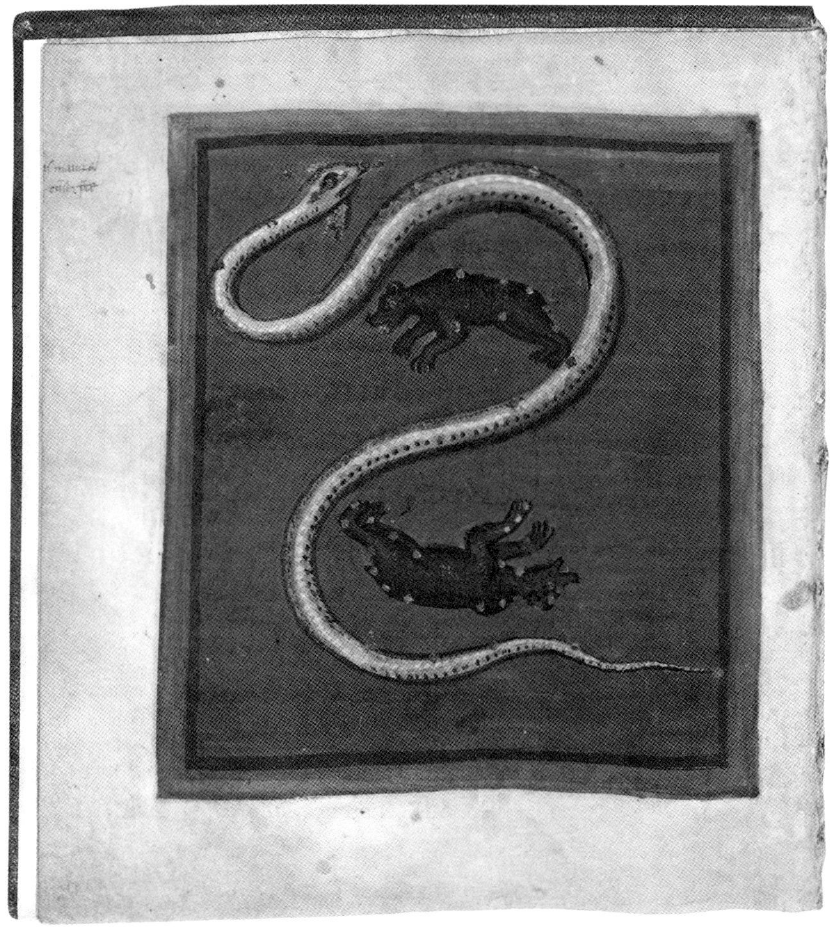

FIGURE 42

Ursa Major, Ursa Minor, and Draco from the Leiden *Aratea,* Bibliotheek der Rijksuniversiteit, MS Voss. lat. Q. 79, fol. 3v. Photo courtesy of Universiteitsbibliotheek Leiden.

Carolingian strategies for scientific sacralization. Classical ideas in the *Revised Aratus* were intentionally juxtaposed with information culled from Isidore, and inspiration for at least one passage even harks back to the recently deceased Venerable Bede.[58] The suppression of the mythological material emphasized the condemnation of paganism by the prelates who

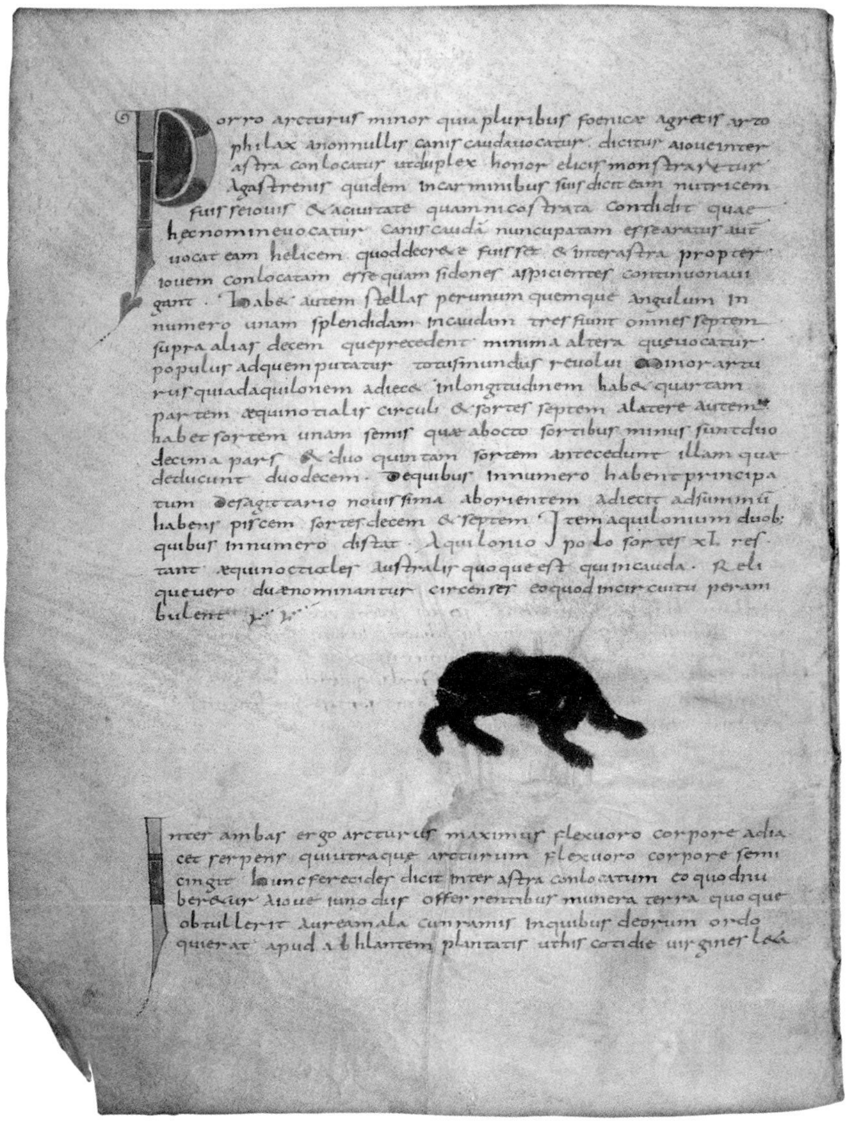

FIGURE 43

Arcturus Minor / Ursa Minor. Cologne, Erzbisch. Diözesan- und Dombibliothek, Cod. 83-II, fol. 155v. Photo courtesy of Erzbisch. Diözesan- und Dombibliothek.

had attended the synod of 809. The refusal to include or to examine the secular content traditionally associated with the constellations also underscored their rejection of late classical Roman paganism and the indigenous Germanic religious practices with which such beliefs intertwined in the ninth century.[59]

Nevertheless, it is equally important to underscore that this Christian revisionist interpretative bent likewise permeated the artistic portrayals of the constellations in Drogo's book.[60] For this reason, the requisite classical references are included in the discussion that follows, primarily to elucidate the areas in which Christian artists from Metz foisted an eighth- and ninth-century Carolingian outlook upon their star pictures and advanced a creative pictorial vision of the heavens in their novel reinterpretations of long-standing Aratean imagery.

Intriguingly, the miniatures in Drogo's copy of the *Handbook of 809* include neither representations of the adumbrated lists of stars nor a monumental format. The scale of the handbook's miniatures is intimate, and the artistic refinement expressed by the painterly compositions was probably intended for the purposeful and pleasurable observation of its owner. Alternatively, Drogo's copy may have been intended to be enjoyed in small group settings by a select number of elite students, either in the bishop's company or in that of his academic masters at the cathedral school. It is easy to see how full-page versions of the same cycle of illustrations would have better fulfilled this purpose; however, it is equally meaningful to envision the use of Drogo's manuscript in tandem with other astronomical observational tools appropriate to the ninth century, such as a star globe or a star clock (*horologium nocturnum*). Indeed, without such instruments, the lack of accurate stellar displacement in the images of the constellations calls into serious question their pedagogical efficacy.[61] This is, however, to confuse the *Handbook of 809* with a typical field guide for stargazers. The images are not arranged or presented in the same manner. The intellectual work for students of the liberal art of astronomy was different. Careful observation and analysis of the images of the constellations in Drogo's copy of the *Handbook of 809* on site in Madrid could not confirm Dieter Blume and his colleagues' claim that the stars were once included in the constellation pictures of Drogo's handbook.[62]

The handbook fulfilled Charlemagne's pedagogical desire for a treatise on astronomy and computistics in the service of reforms within the Frankish lands. The miniatures provided basic forms that could serve as mnemotechnic teaching aids for younger stargazers.[63] The lavish pictorial program presented in Drogo's manuscript, however, surpasses in quality the linear drawings found in other copies of the *Handbook of 809* (figs. 36–38). Even the simpler forms are adequately suited to fulfill their purpose as a teaching tool. The presence of such lavish miniatures in Drogo's copy truly sets it apart as a royal commission worthy of an illegitimate son of Charlemagne. This heightened aesthetic quality gives Drogo's book its apparent distinction as a definitive expression of the *Handbook of 809*. That does not mean that Drogo's book is a more accurate reflection of any putative model; on the contrary, the profound difference in quality between Drogo's book and the others underscores the unique nature of this creative undertaking by superior Frankish illuminators.

This lack of stars in Drogo's handbook differs from the image cycle in the Leiden *Aratea*, which includes an original Germanicus Caesar translation. An additional Latin translation of the *Aratea* by Rufus Festus Avienus (fourth century) supplemented the text of the Germanicus translation in the Leiden manuscript.[64] In the image for the Great Bear, Little Bear, and Serpent on folio 3v of the Leiden *Aratea*, the three constellations are clustered into one picture without true concern for the text beyond the framed border (fig. 42).[65] This can be considered a general design strategy, which arranged visual information analogously to the semiotic construction of hypotactic subordinated sentences in Latin literature. This can be considered a form of visual hypotaxis, emphasizing integrated relational placement and expressing pictorial information with pristine clarity.[66] In a text displaying parataxis, such as a work by Sallust, ideas are linked together in clusters without the subordination of clauses.[67] In the descriptions and illustrations of the constellations from Drogo's copy of the *Handbook of 809*, the *De ordine* presents a progression of text-image pairs. This sequence of syntactic text-image pairs remains subordinated to the overall structure of the *De ordine*, but follows the order of the constellations established by Aratus in a concatenated series of semiotic units, adduced in sequence like a paratactic history from Sallust. The imagery in Drogo's book therefore separates the Bears from the Serpent in three discrete pictures (plate 1). The arrangement of the Bears and Serpent into three sequential text-image syntactic units emphasizes the form of each constellation and asserts the individuality of each one within the whole list of constellations in the *De ordine*.

This paratactic strategy was also adopted for the representations of the bears and serpent on folios 155r–156r of the Cologne *Aratus* (fig. 43). Comparisons of the types of images to be found in both manuscripts can tell us much about the kinds of lost late antique precursors that were harbingers of celestial understanding, relaying the classical traditions to

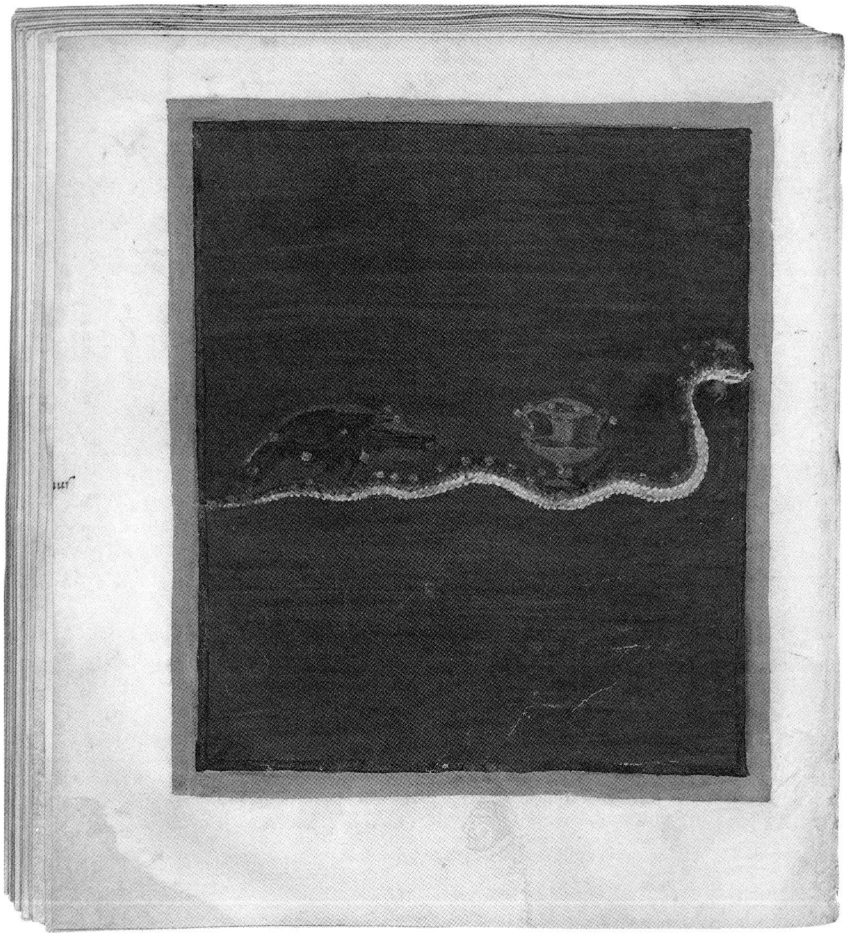

FIGURE 44

Hydra, Crater, and Corvus from the Leiden *Aratea*, Bibliotheek der Rijksuniversiteit, MS Voss. lat. Q. 79, fol. 76v. Photo courtesy of Universiteitsbibliotheek Leiden.

Carolingian artists who then interpreted the ancient forms for themselves.

For example, three discrete images depict the Hydra, the Raven, and the Crater on folios 62r–v in Drogo's copy of the *Handbook of 809* (plates 16–17).[68] The Leiden *Aratea* represents the three clustered together in the framed miniature on folio 76v (fig. 44), while in the Cologne *Aratus*, the hydra, bird, and vessel are all included within one scene at the top of folio 167r (fig. 45). The artists working in Metz preferred to emphasize the individuality of the constellations and their significance for Carolingian readers by painting three miniatures. In Drogo's book on folio 62r, one image clusters the trio together (plate 16), according to the more standard hypotactic form included in the Leiden manuscript, followed by individual depictions of the Raven and the Crater on folio 62v (plate 17). This conforms to the semiotic strategy of compilation through juxtaposition noted in the introduction. The painters of Drogo's copy of the *Handbook of 809* arguably sought to emphasize the coeval informational status of both text and image in the *Handbook of 809*. The painters in Metz pushed the design strategy of visual parataxis—experimented with in the Cologne copy of the *Revised Aratus*—even further. The upshot was that the pictures in Bishop Drogo's manuscript were arguably more useful references of the forms of the individual constellations for teaching purposes than the miniatures available in the picture book, made for the court of Louis the Pious, contained in the Leiden *Aratea*.

The representations of the constellations of Serpens and the two bears in Drogo's copy of the *Handbook of 809* nevertheless recall the essential features of the corresponding animal forms in Leiden, taken one at a time. This was natural; the two cycles both derived from common pictorial prototypes, and of course depict the same constellations. The two bears look away from each other in opposite directions in both books. In the Leiden manuscript, they stand paw to paw across the divide delimited by the tortuous form of the serpent. In Drogo's handbook (Madrid 3307), the bears both stand on all fours, heads forward in opposite directions. Whereas the bears in Leiden appear to leap through space in an endless, rhythmic circular chase, the two bears in Drogo's book hover within the illusionistic space of the parchment folio. The bears in Madrid 3307 conform to the illusionistic pictorial logic appropriate to landscape painting, for they seem to stand on a missing horizon line. Medieval painters generally adopted the pictorial logic of the groundline but did not include one in most cases; in Madrid 3307, Heridanus on folio 61v may be an exception to this rule (plate 15).

More important, in Drogo's handbook, as in the Cologne *Aratus*, the text block announces a spatial-semantic frontier that runs parallel to the implied horizon line on which the otherwise free-floating bears seem to stand in profile. In this way, the text not only provides the descriptions for the star catalogue but establishes a visual perimeter by blocking off the open space of the pictorial zone within which the ninth-century painters carried out their work. The fact that the paws of the Great Bear in Drogo's book are set slightly above a line ruled in drypoint on the recto side of folio 54 provides additional visual proof that this layout was a design decision rather than an accident.

Unlike the Leiden *Aratea* images the miniatures in the *Handbook of 809* do not have frames. The Carolingian artists who painted the original lost archetypal *Handbook of 809* and the copy made for Drogo (Madrid 3307) dismantled the late antique framed images copied into the Leiden manuscript into the smaller discrete images preferred by the makers of the handbook. At the very least, the painters of Drogo's book saw no reason to add any frames. Frankly, the preference for visual parataxis shattered the frames of the miniatures. This reduction in Madrid 3307 of the pictorial sequence into isolated units subordinated to the rhetorical structure of the *De ordine* was

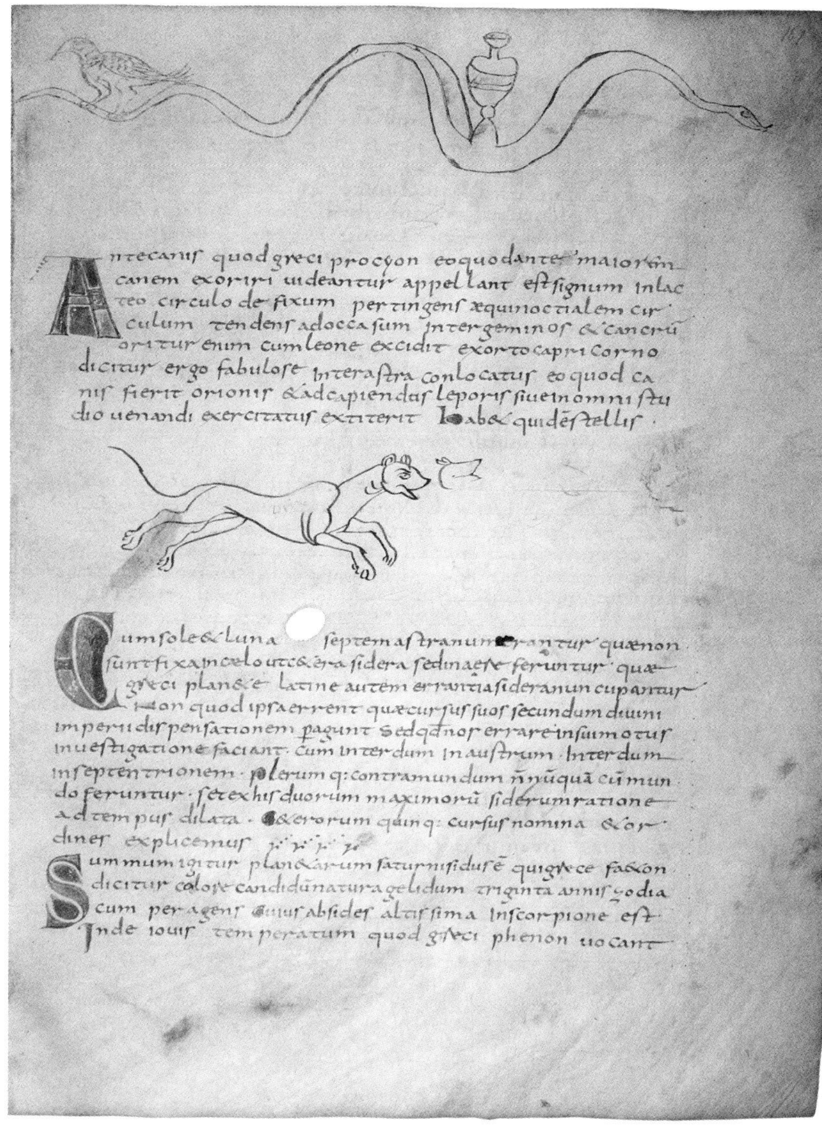

FIGURE 45

Serpens/Hydra and Anticanis/Canis Minor. Cologne, Erzbisch. Diözesan- und Dombibliothek, Cod. 83-II, fol. 167r. Photo courtesy of Erzbisch. Diözesan- und Dombibliothek.

a preferred Carolingian pictorial book-making device, likewise confirmed by the Cologne copy of the *Revised Aratus*. The pictures in the *Handbook of 809* deviate from the pictorial tradition copied by the artists who worked on the Leiden *Aratea*. As a result, the pictorial logic of the *De ordine* adheres to the twin Carolingian desires for (a) an original but simplified textual and pictorial presentation of the astronomical information appropriate to meditative study, and (b) perfect mastery in a pedagogical handbook.

The lack of stars in the paintings of the constellations in Madrid 3307 is an odd solution, however, and calls into question the independent educational value of the miniatures. The simple text of the *De ordine* arguably permitted a general mental reconstruction of the placement of the stars within the painted image of the Great Bear. But such verbal descriptions could have benefited from the sort of visual representations of the stars within the constellations that are included in the Leiden *Aratea*. Alfred Stückelberger is the most optimistic voice about the reliability of the images in the Leiden *Aratea*, noting a strong correlation between the star catalogue of 1,022 fixed stars utilized by Ptolemy in the second century and the illustration of a constellation such as Sagittarius on folio 52v (fig. 21).[69] This probably overstated view has been significantly revised by Elly Dekker, who instead finds the general lack of accuracy in the Leiden *Aratea* star pictures to derive from the joint influence of both the original Ptolemaic star catalogue and diverse literary "descriptive" traditions, including the Basel *Scholia* under discussion here.[70]

The upshot is, of course, that in both Drogo's copy of the *Handbook of 809* and the Leiden *Aratea*, there is only one potentially helpful source for the localization of the stars. There is an emphasis on the semantic potential of the text in Drogo's book, which deviates from the preference for pictorially placed stars in the Leiden *Aratea*. There is no reason to believe, however, that the lack of stars in the depictions of the constellations originated with Drogo's book. Presumably, the pictorial cycle designed for the original archetypal handbook created after the synod of 809 was made the same way. Interestingly, there are also no depictions of stars in the free-standing, single-column miniatures included within the early Cologne *Aratus* (fig. 8). Nor are there any frames around the illustrations on the two-column text pages in the copy of the *Recensio interpolata* (that is, the *Revised Aratus* explained above) manuscript from Saint-Gall, such as the ninth-century Saint-Gall 902 (fig. 40).[71] As a result, the pictures in the original, archetypal *Handbook of 809* probably resembled the cycles of pictures found in manuscript copies of the *Revised Aratus*, not the tradition of pictures associated with the Germanicus translation in Leiden. This only serves to demonstrate that the *Handbook of 809* stood at the nexus of several programmatic efforts throughout the Frankish lands: the book simplified and comprehensively conveyed astronomical and computistical acumen to the schools; the pictorial program was a coeval presentation of period science and not merely a dilettantish fancy for prelates; and the *Revised Aratus* and the *Handbook of 809* were twin efforts at developing a serious Christian presentation of classically inspired astronomical traditions and *not* merely the result of sanguine reclamation projects.

The representation of Ursa Major in Drogo's copy of the *Handbook of 809* follows its textual description on folio 54v (plate 1). The bear has a stern grimace and an upturned snout, as if it is on the defensive. Its paws terminate in black lines, which reduce the form to a stylized rendition of a quadruped, seemingly with hooves. This might simply be the result of the remaining underdrawing, however. A harsh black contour supplies the basic form of the animal. Certainly, some attention was paid to ursine anatomy, such as the convincing and elevated dorsal hump. Arguably, there is a sense of overlapping perspective in the representation of the limbs of the Great Bear. In fact, the underdrawing has bled through the pigment in the right foreleg and hind leg, suggesting that the artist employed underdrawing to sketch in the animal's general form. The painter nevertheless manipulated the basic form of the animal at a second stage, while applying the color to the parchment. This is highly informative about the practices of Carolingian painters working in Metz. Each artist first created an outline drawing before completing a miniature. Since the styles of painting and the styles of the underdrawing always display a high level of coordination, it is highly likely that each painter also provided his own preliminary contour drawing. The sketch expressed the basic form, while the finished figure communicated the mature aesthetic vision of the artist. Such a two-step process argues in favor of a sophisticated Carolingian painterly practice that had already developed by the 830s, when Drogo's copy of the *Handbook of 809* was created.

The almost mirror image of Ursa Major, Ursa Minor (plate 1)—also known as Cinosura, according to the *Epitome Catasterismorum* that is derived from the now-lost star catalogue with myths contained within the original *Catasterisms* of Eratosthenes,[72] and also as Arcturus Minor—reproduces the overall form of its bigger relative in reverse: "Cinosura, Ursa Minor, has four bright stars on one side arranged in a square shape; there are three bright ones in the tail. They total seven. Under these the star appears that is called Polaris, around which it is believed that the whole sphere is turning."[73] In keeping with the Carolingian reformers' aim to cleanse the descriptions of the constellations of pagan mythological associations, the only reference to the role the bears played in saving Zeus from Cronus lingers in their alternate names. According to the legend, Haelice and Cinosura attended to the infant god for a year in a cave near Mount Ida at Lyctus. In addition, the Carolingian author omitted any reference to the shipping practices of the Greeks, who navigated with the assistance of Ursa Major, and removed the reference in the *Phaenomena* to the superior navigational skills of the Phoenicians, who relied instead on Ursa Minor. The little bear reportedly revolves in a tighter orbit around Polaris, yielding a lower relative error for Sidonian sailors.[74]

The loss of the navigational anecdote from the *Phaenomena* is a significant one. Navigation was one of the most celebrated uses of the constellations from antiquity into the medieval period, and there was no obvious reason for the clerics to rid the star catalogues of these pragmatic applications. This removal was a deliberate hermeneutic maneuver. The *Handbook of 809* demanded a list of star pictures. Telling time was a practical concern of Frankish ecclesiasts such as Drogo, whereas the benefits of the constellations to Phoenician seafaring were not a priority within the cloister walk or cathedral precinct.

Like Ursa Major, Ursa Minor is brown with black highlights. Especially noteworthy are the efforts at rendering the paws, and the eye delineated by a heavy lid under a harsh eyebrow and a beady pupil. Although there was an effort made to capture the ursine form in both images of the bears, neither displays an overall attention to the texture of their coats.

The next constellation in the series is Serpens (plate 1), which weaves between the bears near the north celestial pole. This image of the constellation

Draco resembles the style of other serpents painted in Metz during the ninth century, as noted in chapter 1. Since the archetypal *Handbook of 809* arose in the court at Aachen in the years following the synod, 809–12, the style of the archetypal paintings would undoubtedly have reflected the stylistic trends associated with a later phase of the development of Charlemagne's court school. The juxtaposition of three blue tonal values suggests the rounded form of the serpentine flesh. First, a light blue wash laid in the base color of the image, and then the Carolingian painter adapted a technique common to late antique fresco painting to model the forms.[75] The progressively darker shades of blue indicate shadows on the serpent, but also approximate the scaly texture of the reptile: "The serpent which lies in the space between the bears has five bright stars in its head and ten stars throughout the whole body. They total fifteen."[76]

It is important to underscore that although both the text of the *De ordine* and its imagery in Drogo's copy of the *Handbook of 809* hark back to earlier Aratean traditions, the text itself could not provide a painter with the pictorial cycle. For the pictures in the *Handbook of 809*, the artists had to look to illustrated Aratean manuscripts or other illuminated codices available in the ninth century. Since star pictures regularly supplied visual summaries of the information included in their texts, odd iconographic details could be carried over between manuscripts when old pictures were consulted in illustrating new texts. Sometimes artists employed stylistic devices (such as the stippled modeling of the serpent's flesh) that could not have been derived from the text. Texts exist in visual cultures, too, and artists executing their commissions naturally draw on a multiplicity of sources. For this study, I emphasize Carolingian astronomical manuscript illumination for the sake of economy and to clarify meaningful aspects of the Frankish contribution to early medieval astronomical illustration.

The painters in Metz used the scoring on the opposite side of the handbook's folio 54v as a guide for the undulating twists of the serpent's body. This was a common practice in the atelier making these star pictures, and a similar layout recurs throughout the pictorial program. The form of the serpent adheres to a standard twisting shape also seen in the Leiden *Aratea*, despite the tendency of painters in Metz to employ their style of visual parataxis, eliminating the bears. The space remaining at the bottom of folio 54v, however, permitted the introduction of an additional bend in the snake's body. This reveals the extent to which the pictures in Drogo's handbook were made by creative painters who made *ad hoc*, innovative decisions rather than merely introducing all of their images copied in rote fashion from diverse sources.

The next image, on folio 55r, depicts Hercules (plate 2). The painter modeled the figure of Hercules with three ochre flesh tones. The use of three shades resembles the technique for modeling the scaly flesh of the serpent on folio 54v. The hero's protruding, heavy brow and pupils resemble those of Ursa Minor. These miniatures can be attributed, then, to a first hand. The text explains that "Hercules, of whom it is said that he is kneeling, has one star in his head, one in his arm, one bright star in each shoulder, one star in the left elbow, one in the same hand, one in each hip, two in the right leg, one in the foot, above there is one in the club in the right hand, there are four stars in the lion pelt. They total sixteen."[77] The image of Hercules includes a visual reference to the Aratean character known as the kneeling one, or "Engonasin," here dubbed "Ingeniculo." Aratus explained that the figure tilled the earth, like a farmer.[78]

As Florentine Mütherich has noted in her discussion of the miniatures in the Leiden *Aratea*, the connection between Hercules and the kneeling figure was made at an early time and pervades the imagery of the Aratean manuscripts. A *Revised Aratus* manuscript, such as the Cologne *Aratus*, instead represents Hercules attacking the Serpent or Hydra (Draco), who is wrapped about the Tree of the Hesperides (fig. 8). This scene better illustrates certain scholia attached to the Germanicus translation of the *Phaenomena*.[79] The artists who created the pictorial cycle for the *Handbook of 809* removed the mythological battle, while keeping the older visual reference to the "one who kneels." In Drogo's copy of the *Handbook of 809*, the painter decided to portray Hercules from the front rather than in the quasi-dorsal view displayed by the miniature in Cologne.

The visual references to the defeated Nemean lion and the club of Hercules gave the painters in Metz captivating narrative details to portray. In this respect, the image seems at odds with the desire to create a sterilized, Christian view of Hercules. Be that as it may, there remained a need, in compliance with the pedagogical goals of the original handbook, for the artists to render a form recognizable as the standard type of Hercules/Engonasin. The educational aims of the handbook vied with the clerics' desires to sanctify the pictures, precisely because the classical content of a figure such as Hercules could not be separated from its form. The description of Corona precedes an image (plate 2) of a laurel with a jeweled clasp: "The crown has eight stars placed in a circle of which three are bright ones that abut the head of the northern serpent."[80] Curiously, the artists painted the laurel in the same ochre hues as their depiction of Hercules. Green was used later—for the image of Piscis Magnus on folio 61v (plate 15), for example—so this hue was available to the painters in Metz. This suggests that these miniatures with the brownish hues at the beginning of the handbook are all the work of the first painter. The economical reuse of the pigments for the laurel crown and for Hercules (who is stylistically related to the image of Ursa Minor) reinforces the idea that all the paintings thus far are by a single hand. Although stylistic analyses of scientific manuscripts are not a common concern for historians of either art or science, this has led to the inaccurate dismissal of the qualitative richness of certain scientific or philosophical manuscripts, and to ignorance about the sorts of teams that were manufacturing medieval books that were not overtly biblical for meditation and study.

The next image also conforms to the design strategy of visual parataxis (plate 3).[81] It is true that the Serpentarius, "the man wrapped about with a serpent," also known as Ophiuchus, would be meaningless without his reptilian adversary in the picture:

> Serpentarius, who is called Ophiuchus by the Greeks, has one bright star in the head, one bright star on each shoulder, one bright star on each foot, three stars in the left hand, four bright stars in the right hand, a single star in each elbow, a single star in each knee, a bright star in the right leg. They total seventeen.
>
> The serpent, which he holds in his hands, has two stars in its mouth and four dim ones in the head, although it has two stars in the area up to the hand, and it has fifteen stars in the bend of the body. They total twenty-three.[82]

But here, too, the artists deviate from a type of Aratean image that portrays Serpentarius trampling Scorpio, as on Leiden folio 10v, for example.

Serpentarius and Scorpius (plate 3) each have their own miniatures in Drogo's copy of the *Handbook of 809*, according to the pervasive paratactic design preference for a single picture per constellation. The artists who had made the cycle of images for the computistical manuscript in Cologne also separated out the pictures, accordingly, into two scenes on folio 157r–v of the Cologne *Aratus*:

> Scorpius has two stars on each claw, or preferably both lips, before which there are three bright ones, of which the one in the middle is brighter; there are three bright stars on the back, two in the stomach, five in the tail, and two stars in the spine. They total nineteen. Before these stars four more were placed, whether they are considered to be in the claws or the lips, which are called the "chele of Scorpio," and that were assigned under the circumstances to Libra. On account of its size it [Scorpio] was in two houses of the zodiac. So, it [Scorpio] was divided into the space appropriate for two zodiac signs.[83]

The history of the zodiac is divided into two discrete phases. In the early phase, around the late sixth century B.C.E., there were eleven accepted signs of the zodiac. Scorpio and Libra were conflated into a large-clawed creature reminiscent of a lobster. This tradition harks back to such ancient astronomers as Anaximander (d. 546/545 B.C.E.), Cleostratus of Tenedos (active ca. 500 B.C.E.), and Oenopides of Chios (active in the second half of the fifth century B.C.E.). After Alexander the Great (d. 323 B.C.E.), the twelve-part zodiac wheel gradually became normative throughout the Mediterranean.[84]

The introduction of a pithy tidbit from astrological history into the otherwise terse descriptions of the constellations is intriguing. The scientific writers from Corbie apparently found the fact noteworthy. Two points should be noted here. First, the Carolingian author was aware of not just the standard twelve-part zodiac but also of the alternate ancient eleven-part zodiac system. This suggests that the ninth-century writer of the *De ordine* had a sophisticated understanding of the history of astrology. Second, the *Handbook of 809* as a whole was first and foremost an astronomical-computistical treatise. The author of the *De ordine* emphasizes the relationship between the twelve months of the year and the corresponding twelve-part series of zodiac signs associated with the calendar, and is also aware to some degree of the problematic impact of equinoctial precession on such alignments.[85] Depictions of the months and their related signs in Western medieval manuscripts date back at least as far as the *Calendar of 354*, and calendrical illustrations, such as a cycle of the months (consider fig. 22), embellish manuscripts to the end of the late Gothic period.[86]

The image of Scorpius/Scorpio[87] in the handbook includes the long pincers mentioned in the text, and it therefore alludes to both the standard form of Scorpio and arguably to its more ancient prototype (belonging to the eleven-part zodiac cycle). This was also the form of Scorpio adopted for inclusion in the *Revised Aratus* from Cologne on folio 157v. In Drogo's copy of the *Handbook of 809*, the dark brown and black underdrawing yields a sophisticated reticulation of the arachnid's form.

The next painting in Drogo's handbook, on folio 56r, is of Boötes, otherwise known as Arctophylax (plate 4). According to Aratus, Boötes endlessly prods the circumpolar constellations of the Great and Little/Lesser Bears as they circle the north pole, which is why the Carolingian painter presents him with his shepherd's crook: "Boötes, who is called Arctophylax by the Greeks, has four stars in his right hand that do not set, and in his head he has one bright star. He has one star in each shoulder, one star per pectoral muscle with the brighter one on the right, beneath one pectoral muscle there is a light star. In the right elbow there is one bright star. Between both knees there is a bright star, and there is one star per foot. They total fourteen."[88] There are losses to the head of the figure in Drogo's book that probably took place at the time of the seventeenth-century rebinding (discussed at the end of chapter 1), when the early modern bookbinders cropped the folios. Enough of the brownish flesh-tones and blue mantle remain to warrant an attribution to the first painter. The characteristic large eye resembles other eyes found in the miniatures of Hercules and Ursa Minor.

This first artist also painted the angelic vision of Virgo (plate 4): "Virgo has one light star in her head, one star per shoulder, on the left upper arm/wing she has one light star, on the right upper arm/wing she has one star, she has one star per elbow, one star per hand, and on the left there is a brighter star called 'the Spica, the ear of grain,' in the tunic there are six dim stars, there is also one star per foot. They total eighteen."[89] The maiden Dike is referred to by Aratus in the *Phaenomena* and is alluded to in the Carolingian description of Virgo.[90] The winged goddess of Justice, who retreated from earthly life according to the mythological tale, was arguably reinterpreted in an angelic form by the Christian painter. As the winged harbinger of justice, the mythological presentation of Virgo held her ear of grain or sheaf in her left hand, according to the text. This is highly significant, because Carolingian representations of the constellation Virgo differed greatly. She holds three sheaves of grain in her right hand in the Basel *Scholia* on folio 18v, whereas she holds scales in her role as a personification of Justice in the manuscripts of the *Recensio interpolata* such as Saint-Gall 902, page 86.[91]

The conscious removal of the scales, which identified Virgo in her mythological role as winged Justice, permitted the painter of the maiden in Drogo's handbook to sanctify the representation of the zodiac sign. This innovative form removes the pagan mythological content from the image of Virgo, just as the author of the *De ordine* in Corbie made every effort to omit mythological references from the text. This unified editorial vision provides the best evidence for the hypothesis that Adalhard of Corbie was instrumental in the development not only of the text but also of the pictorial program for the original *Handbook of 809*.[92] The lack of scales is also an intelligent scientific reaction to the content of the previous passage treating Scorpio: since the scales *qua* Libra had been removed from Scorpio, it would have been odd to pass them along to Virgo and include them in either her description or her depiction.

The salmon hue used for the overtunic worn by Virgo is also used for the mantles worn by the curiously sexless (censored) Gemini twins (plates 4–5). Granted, gender-neutral figures are common in early medieval art. The relative uniqueness of this decision in Drogo's copy of the *Handbook of 809* becomes apparent, however, when that depiction of the nude Gemini is compared with other copies of the *Handbook of 809* displaying sexual organs (figs. 36, 38). The manner of painting the eye sockets makes it clear that this image was also painted by the first hand. The faces of the Gemini twins, especially the one on the left, recall the linear construction of the face also found in the image of Cepheus on folio 57v (plate 7). This early medieval technique for rendering the face of the figure requires a *z*-shaped line that begins with a heavy brow, descends as it defines the contour of the nose, and then ends with a line that defines the bottom of the nose. Two tick marks frame the philtrum of the upper lip. One or possibly two roughly parallel lines define the lower lip and seemingly dimpled chin of figures like these in early medieval manuscripts. When a figure stands in a three-quarter pose facing left, like the left-hand twin of the Gemini, the *z*-shaped line begins on the proper right side of the figure's face, corresponding to the viewer's left (plate 5). Examples of figures looking toward the right with such conventionally rendered faces include Virgo on folio 56r (plate 4) and the right-hand Gemini twin on folio 56v (plate 5):

> One of the Gemini twins, who is next to Cancer the Crab, has one bright star in his head, and one bright star per shoulder. In the right elbow he has one star. In each of the knees there is one star, and in each of the feet there is one star. They total eight.
>
> The other one has one star in his head, one star per pectoral muscle, and two stars in the left elbow. He has one star at the tip of his hand, one star in the left knee, one star per foot, and one beside the left foot that is called Propus. They total ten.[93]

The image of a standing soldier and musician was atypical for Carolingian depictions of the Gemini, although not a unicum, and the symbolic reference to the depicted set of twins can be explained by appeal to the Leiden *Aratea*. The related image in the Leiden *Aratea* on folio 16v (fig. 46) nevertheless switches the left-right orientation of the twins in the miniature from Madrid 3307 (plate 5). The Gemini twins also look away from each other in Drogo's copy of the *Handbook of 809*, whereas they engage each other within the visual confines of the square frame in the Leiden *Aratea*. As for the image of Hercules, the classical content of the legend provides the iconographic source for the depiction of the twin children of Zeus who erected a strategic wall for Thebes.[94]

Whereas in the Leiden image one of the twins, Zethus, holds a lance resembling a spear and rests upon his club, he only holds the lance in the miniature from Drogo's copy of the *Handbook of 809*.[95] The image in Drogo's copy draws in a conservative way upon the mythological information that provided the distinguishing emblems of the Gemini. In this case, the imagery of the lyre denotes Amphion as the musician, and armed Zethus is the warrior.[96] The text of the Cologne *Aratus* refers back to an alternate and common set of names for the mythical brothers, Castor and Pollux, on folio 158r.[97] This text makes the more common identification of the Gemini twins as the "Dioskouroi." The image in the Leiden *Aratea* also includes an allusion to Castor and Pollux. The brothers' white domical caps are emblems of the Dioskouroi that the painter sacralized in the Leiden *Aratea* picture, setting gold Christian crosses on top of their headgear (fig. 46).[98] This editorial act is symptomatic of the general approach to reform with respect to the images and ideas linked to the liberal art of astronomy. Occasionally, a Christian gloss was added to a classical concept, but equally importantly, the classical content in the original needed to be elevated to a careful, reflective level of significance before it could be revised. Exegetical emendation already embeds a tactical respect for, or fear of, the original into the very idea of revision.

Each time a new manuscript was created, these images, and their lost precursors, were assessed for their potential inclusion in the books associated with the various Aratean recensions. The situation became even more complicated when monks combined or altered certain cycles of star pictures, seemingly *ad hoc*, recombining favorite exempla, or pulling together scant resources, in order to create an astronomical or computistical handbook or encyclopedia anew. It is vital to distinguish between moments of creative interpretation and the occasional effort at straightforward classical emulation.[99]

The first painter in Drogo's copy of the *Handbook of 809* also rendered Cancer the Crab with a convincing anatomical form (plate 5). Tonal washes complement a confident contour line and model its features:

> Cancer the Crab has two bright stars in its chest that are called the Donkeys between which there is a little haze, which seems to be of a bright color and that they call the manger. Cancer has one star per foot on the right side, all four are dim. On the left side, Cancer also has two bright stars in the first foot, two in the second, one in the third, and a little one in the fourth foot. In the right claw or lip Cancer has three stars, but it has two stars on the left. They total seventeen.
>
> From among this group, the two stars, which we set in the chest of Cancer with two small and dim stars, are called the Donkeys. So, they total four stars.[100]

The references to the two donkeys in the text and the miniature seem odd at first. In fact, by including the donkeys in the image of Cancer, the author of the *De ordine* further demonstrated his integral involvement with the creation of all of the texts compiled into book V. One northern and one southern star designated as the donkeys gather at the manger, which lies before Cancer, according to Aratus in the *Phaenomena*. Both stars were considered useful for predicting inclement

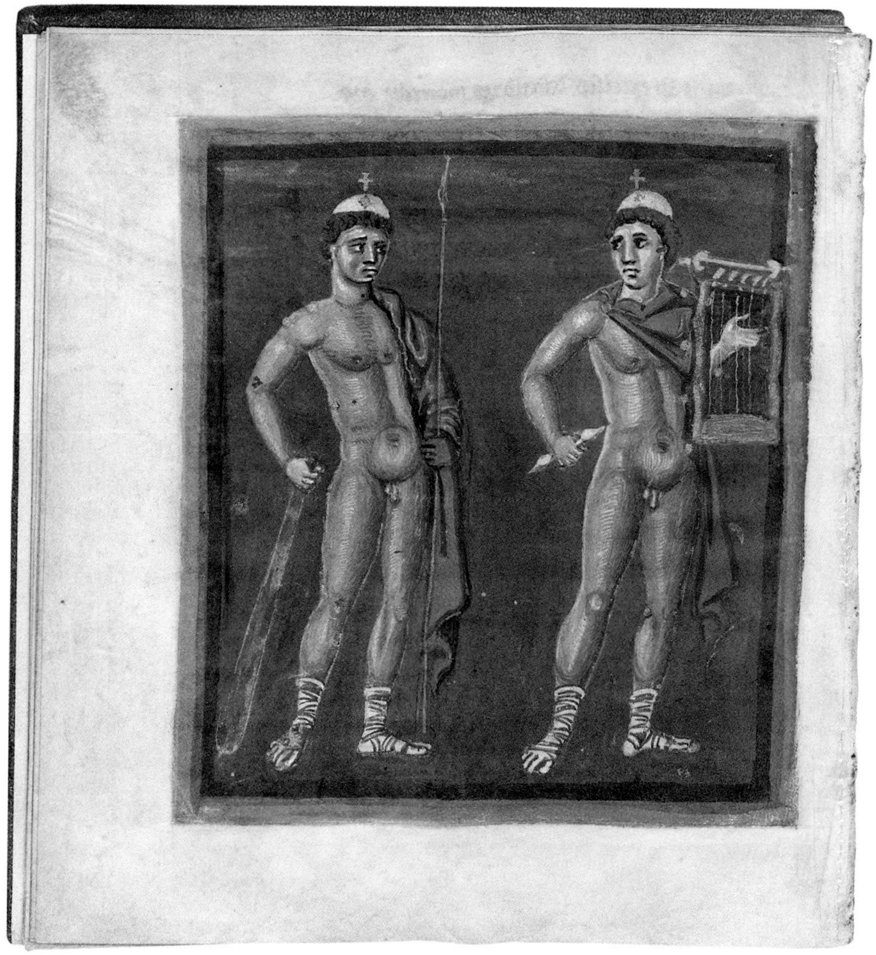

FIGURE 46

Gemini. Leiden *Aratea*, Bibliotheek der Rijksuniversiteit, MS Voss. lat. Q. 79, fol. 16v. Photo courtesy of Universiteitsbibliotheek Leiden.

weather. For example, a fictitious rhyme can easily pass for an old adage: "if the manger grows dark while the stars twinkle, it surely soon will sprinkle." In fact, according to Aratus, the donkeys were useful for making precisely such unimpressive meteorological predictions, which nevertheless would arguably benefit travelers and agriculturalists: "Observe also the Manger . . . if suddenly, when the sky is clear all over, it disappears completely, and the stars that go on either side appear close to each other, then the fields are drenched with no mean storm. If the Manger should darken and the two stars be at the same time recognizable, they will be giving a sign of rain."[101]

Alternatively, when the northern star darkens but the southerly one remains fully visible, the winds rise in the south, and vice versa for the southern star.[102] The information to be gleaned from this is that when winds are blowing clouds and fog in a given direction, the observer notices things getting blurry. This is not particularly helpful, but it does offer an empirically oriented natural explanation for precipitation and changing visibility that does not require a pantheistic view of the environment. To my knowledge, none of the earlier Aratean manuscript traditions that include pictorial cycles show the donkeys feeding at their manger before Cancer the Crab.[103] The image of Cancer in the *Handbook of 809* relies on the ancient Aratean text in the *Phaenomena*, and on Pliny's recapitulation of the same information in book V.12 of the *Handbook of 809*, as discussed further in chapter 4. The creator of this innovative image thereby also demonstrated his astute awareness of late antique and early medieval science. This reinforces the suspicion that Adalhard was involved in the selection of images for inclusion in the original handbook. In fact, Le Bourdellès follows Bernhard Bischoff and has determined that a dismembered copy of Pliny was an early ninth-century copy from Corbie. As Le Bourdellès argues, Adalhard would have been familiar with this material.[104] This further buttresses the likelihood that Adalhard played an important role in the iconographical innovation for the images in the handbook.

Early Christian astronomy had considered certain zodiac signs of meteorological importance for the agrarian and seafaring cultures of the Mediterranean. The significant texts about meteorology (taken in the broadest sense to mean any scientific assessment of the effects of weather) through 400 C.E. included Pliny's *Naturalis historia*, already noted; Columella's *De re rustica* (first century);[105] and the translated Latin versions of the Aratean texts, the *Phaenomena* and the *Prognostica*.[106] Interest in the influences that celestial bodies could exert on meteorological matters waned during the sixth and seventh centuries. Stargazing continued to be useful, however, for computistical calculations, which were necessary for the creation of the church calendar, as addressed in chapter 1. Two texts in particular were instrumental in perpetuating this revised vision of the utility of astronomy for Christian communities: the *Institutiones divinarum et saecularium litterarum* (Institutions of divine and secular learning; hereafter *Institutiones*), by Cassiodorus (d. 585) in the second half of the sixth century, and the *De natura rerum* of Isidore of Seville in the early seventh century.[107] (These texts are also discussed in chapter 4.) In this light, the allusions to meteorological concerns in the Carolingian *De ordine* were actually a creative revival of early Christian interests.

The first painter likewise rendered the image of Leo using his preferred ochre palette (plate 6): "Leo has three stars in its head, two in its neck, one in its chest, three along the back, one in the middle of the tail, one bright star at the tip of the tail, two under the chest, one bright star in the front paws, one bright star under the stomach, one bright star in the middle of the belly, one in the groin, one in the back knee, one bright star in the first paw. They total eighteen. Seven other stars next to its tail seem relatively dim."[108] Nothing mattered to the Carolingian painter here besides rendering a convincing portrayal of the constellation associated with the zodiac sign Leo. All references to the Nemean lion slain by Hercules and to the animal's role as the legendary king of the beasts have been suppressed.[109]

In the next miniature on folio 57r, however, the Carolingian painter included depictions of the Hedi, or goats. They are two stars said to be near the left hand of Auriga, the charioteer (plate 6): "Auriga, otherwise known as the charioteer Ericthonius, has one star in his head and one star per shoulder; the brighter named, the she-goat, is on the left. Each knee has a star. At the tip of the [right] hand are two stars and in the left hand two more that are called the 'Hedi,' the Goats. They total nine."[110] The image from the Leiden *Aratea* also displays the goats in the left hand, but without the team of two horses, on folio 22v. The blue and brown steeds in Drogo's copy of the *Handbook of 809* are fanciful, but they display a generally convincing equine form. The drawing of the goat—which is partly in the margin, rather than over Auriga's left shoulder—actually resembles the donkeys accompanying the image of Cancer. The donkeys of Cancer, the goats of Auriga, and the horses pulling Auriga's chariot all share a robust abdomen and attenuated limbs, which terminate in similarly stylized hooves. These forms were all made by the handbook's first painter. The tunic and mantle of Auriga come from his limited palette: blue, salmon pink, ochre, and brown.

Two interesting details emphasize the desire of the Carolingian author and painter to accentuate the scientific content of the description of Auriga. First, according to Germanicus, the character Ericthonius, referred to in the description, allegedly came up with the idea for the four-horse chariot. Germanicus alternatively suggests that the constellation could refer to the mythical assassin of King Oenomaus, Myrtilus.[111] By focusing on Erichthonius, the Carolingian author and artist of the *De ordine* drew attention to technological advancements in equine transportation rather than infamous examples of regicide. Secondly, the stars known as the goats, like the donkeys accompanying Cancer, were included in the description because they were considered useful for predicting inclement weather. In the *Phaenomena,* Aratus explained that the Olenian goat, which had nursed Zeus, rested on the shoulder of Auriga, while her kids were the two goats on the charioteer's left hand. These two stars appeared dim whenever the sea raged.[112] The wet nurse of the king of the Olympian gods, the Olenian she-goat, is relegated to the margin of Drogo's ninth-century manuscript, while her kids are represented in the center of the leaf without reservation. The reason was twofold: symbolically, the she-goat included a suspect mythological reference, whereas the kids were useful on pedagogical grounds. Practically speaking, however, the composition called for the inclusion of the kids in the center of the miniature. The Olenian she-goat was only useful for the composition insofar as she drew attention to her kids.

The same painter supplied the bucolic and rather lifelike representation of Taurus on folio 57v (plate 7). The eye of the bull resembles the eyes of the other animals in Drogo's copy of the *Handbook of 809*. An interesting departure in this image, however, is the use of the salmon hue as a fleshy highlight. In the

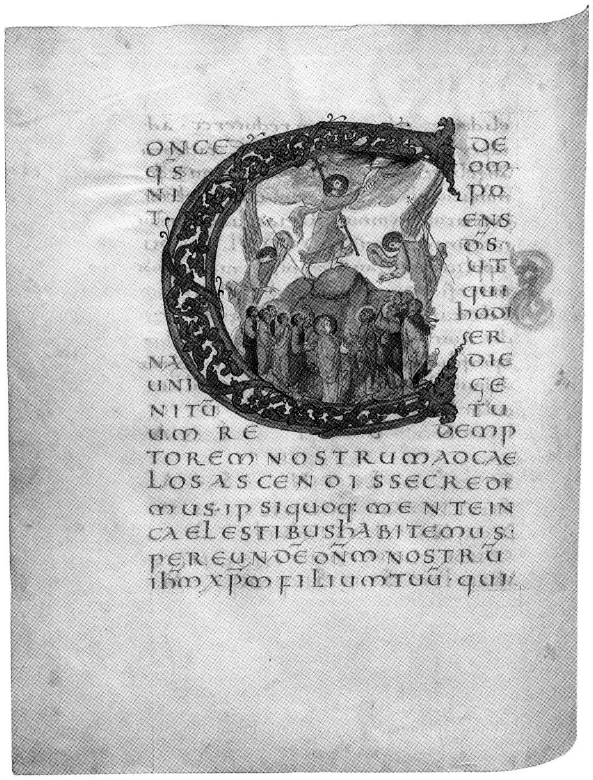

FIGURE 47
Ascension. Drogo Sacramentary, Paris, Bibliothèque nationale de France, MS lat. 9428, fol. 71v. Photo: BnF.

original cycle of the handbook, the full-length reclining image of Taurus deviates from the standard half-figure found in the Leiden *Aratea* on folio 24v and in the *Recensio interpolata* manuscript, Saint-Gall 902, on page 87[113]: "Taurus has one star in each horn, one star in each eye, one star in the nose; these five are called the Hyades. In the hoof Taurus has three stars, two in the neck, two in the back, one final brighter star below the stomach, and a bright star in the chest. They total fifteen. There are seven of the stars, which are called Atlantides or Pleiades; of these six can be seen, but the seventh is faint. It is said that they were placed in the tail of the bull, Taurus."[114] The images painted in Drogo's copy of the *Handbook of 809* and perhaps its archetype were more likely than not either an original appropriation from zodiac cycles of the complete bovine form or a truly innovative attempt at rendering a bull. The reference to the Hyades is another example of a reference to meteorological lore. In the classical world, the ascension of the Hyades in conjunction with the sun was considered a sign of rain.[115] The inclusion of the Pleiades under the description of Taurus demonstrates another authorial effort to omit unnecessary mythological references. While the Leiden *Aratea* includes an image of the seven daughters of Atlas on folio 42v, the originator of the pictorial cycle in the *Handbook of 809* excludes any portrait of the mythological siblings. In the Christian Carolingian *Handbook of 809* meteorological prognostication with the stars was useful, but mythical stories about the daughters of Atlas, the Pleiades, were not. Consequently, the author of the *De ordine* found a way to include the reference to the seven stars that usher in fall or spring while downplaying their mythological content.[116]

That these explanations are not themselves included in Drogo's book is also significant. The labels, names, and pictures of constellations or star clusters in the *Handbook of 809* were mnemotechnic, insofar as they enabled a Carolingian student to create a taxonomic, hierarchical inventory of astronomical information and folk tradition. The cursory descriptive quality of the text of the *De ordine* parallels the uninformative nature of many of the star pictures when they are considered in isolation. The *Handbook of 809* offers introductions for pupils and discussion points for teachers, yielding an elementary synthesis with royal imprimatur.

The next three images in Drogo's copy of the *Handbook of 809* were also painted by the first artist working on that manuscript (Madrid 3307). They depict the characters immortalized by the mythological tale of Andromeda's rescue from the sea monster by Perseus (plates 7–8). None of the descriptions makes this clear:

> Cepheus has two bright stars in his head, one bright star in his right hand, one dim star in his elbow, one bright star in the left hand, one in the left shoulder, three stars arrayed obliquely in the baldric, two stars on the right hip, two stars on the left knee, one star at the tip of the foot, three stars above the foot. They total seventeen.
>
> Cassiopeia has one bright star in the head, one bright star per shoulder, one bright star on her right breast, one bright star on her right thigh, two bright stars on her left thigh, one bright star in the knee, and there are two bright stars in the pedestal of the chair where she sits. They total ten.[117]
>
> Andromeda has one brilliant star in her head, one star per shoulder, one star in the right elbow, one bright star in the hand, one bright star in the left elbow, one star in the forearm, three stars in the girdle, four more above that area, one bright star per knee, two stars in the right foot, one star in the left foot. They total nineteen.[118]

Each figure displays the same pronounced eye sockets and limited range of colors, making the most of blue, salmon, and yellow-brown. These three miniatures

FIGURE 48

Crucifixion. Drogo Sacramentary, Paris, Bibliothèque nationale de France, MS lat. 9428, fol. 43v. Photo: BnF.

betray, however, a quality of color that is thin and diffuse, like washes. The quality of the painting almost tapers off at this point, as though the pace of production had accelerated. Yet this use of painterly washes has often been associated with the style of painting in Metz during the ninth century (figs. 33, 47–48).

This first painter of Drogo's copy of the *Handbook of 809* was adept at rendering animal forms such as that of the fabulous winged horse Pegasus on folio 58r (plate 8). Its eyes resemble those of Taurus (plate 7). The hooves and equine grimace also resemble the second, brown horse in the team pulling the chariot of Auriga (plate 6): "The horse called Pegasus has two bright stars in its face, one star in its head, one in the haunch, a bright star in each ear, four stars along the neck, the next bright star is the [aforementioned] one in the head, has one star in the shoulder, one in the chest, one bright star in the navel, one star in each of two knees, and stars in each of those hooves. They total seventeen."[119] The truncated representation of Pegasus resembles the image found in other manuscripts derived from Aratean precursors.[120] It is interesting that there is a tight relationship between the description and depiction of Pegasus. The text clearly relates to the frontal presentation of a bifurcated animal, since a complete equine form would have called for at least four more stars, for a total of at least twenty-one. In contrast, the Pleiades set within the tail of Taurus necessitated the inclusion of an entire recumbent bovine form. Arguably, the choice of animal picture here derived at least in part from an assessment of the total number of stars and ideas expressed by the star catalogue, the *De ordine*, in Drogo's copy of the *Handbook of 809*. The fact that Pegasus is the only half-length figure in Drogo's *Handbook of 809* confirms this and underscores the book's educational importance for the Carolingian study of astronomy.

The first painter employed the same painterly technique for rendering the hooves of the ram Aries (plate 9) that he used for the hooves of the Olenian she-goat in the Auriga miniature (plate 6). The blue wash coloring the ram's fleece is girded by a green laurel. This symbolizes the ram's role as the inaugurator of a new zodiac cycle and as the zodiac sign initiating spring.[121] In other words, the imagery once again takes full advantage of an extant pictorial tradition to emphasize those aspects of the *Handbook of 809* that clarify the relationship between the constellations and other celestial phenomena, such as seasonal shift, permitting the straightforward Christian association of springtime renewal with Easter. At this point in the text, many of the descriptions become terser: "Aries has one star in the head, three stars in the nostrils, two in the neck, one bright star at the tip of the right foot, four in the back, one in the tail, and one star at the tip of the left foot. They total thirteen stars."[122] The salmon-colored triangle Deltoton follows in sequence (plate 9): "The triangle, which is called Deltoton by the Greeks, has three stars altogether with one in each vertice."[123] The image of Pisces displays a celestial thread of stars linking the mouths of the pair of fish known as Borius/Aquilonalis and Australis (plate 9): "There are two Pisces fish, of which the northern one (Borius) is called Aquilonalis and has twelve stars; the southern one (Notius/Notus), however, is called Australis and has fifteen stars. They are joined together with a line, which has three stars toward the northern part, three to the south, and three stars toward the west. The total number of stars is thirty-six."[124] Pisces is the final zodiac sign through which the sun passes on its annual journey and designates the conclusion of the cyclical progression of the seasons. In some

Aratean images linked to the Germanicus translation, the fish are joined tail-to-tail with a knotted set of chains that restrict their mobility and reportedly cause them to fight for dominance as they swim in opposite directions, such as in the image from the Leiden *Aratea* on folio 38v.[125] The image in Saint-Gall 902 belonging to the *Recensio interpolata* displays instead the fish joined by a cord at the lips on page 90, as they are likewise depicted in Drogo's copy of the *Handbook of 809*. In what remains of the damaged miniature in Drogo's copy, the three-tone technique has been used by the first painter to model the form of the fish. The feathery tail fin lacks a certain anatomical verisimilitude, but salmon-colored highlights added pectoral fins and relatively convincing fleshy sets of gills. Apparently, there was no normative guideline that regulated the placement of the cord linking the fish.[126]

Perhaps the first painter relied on an image of Pisces from a zodiac wheel when painting the representation in Drogo's copy of the *Handbook of 809*. For example, the fish are joined at the mouth in the zodiac from the Utrecht Psalter, folio 36r (Utrecht, University Library, MS 32, fol. 36r; fig. 49), although in that depiction Aries was drawn without its wreath. The belt, wreath, or hoop around the ram's waist (plate 9) indicates the position of the constellation at the equinoctial colure, or the astronomical dividing line, which crosses the ecliptic at the equinoxes and separates the fall/winter from the spring/summer seasons. Aries is the first sign of spring at the vernal equinox, and by jumping through the circular celestial ring, or "hoop," uniting the equinoxes, the ram ushered in the spring and summer signs of the zodiac.[127] This underlines the extent to which the basic forms of the zodiac were established, but iconographic details occasionally supplied additional

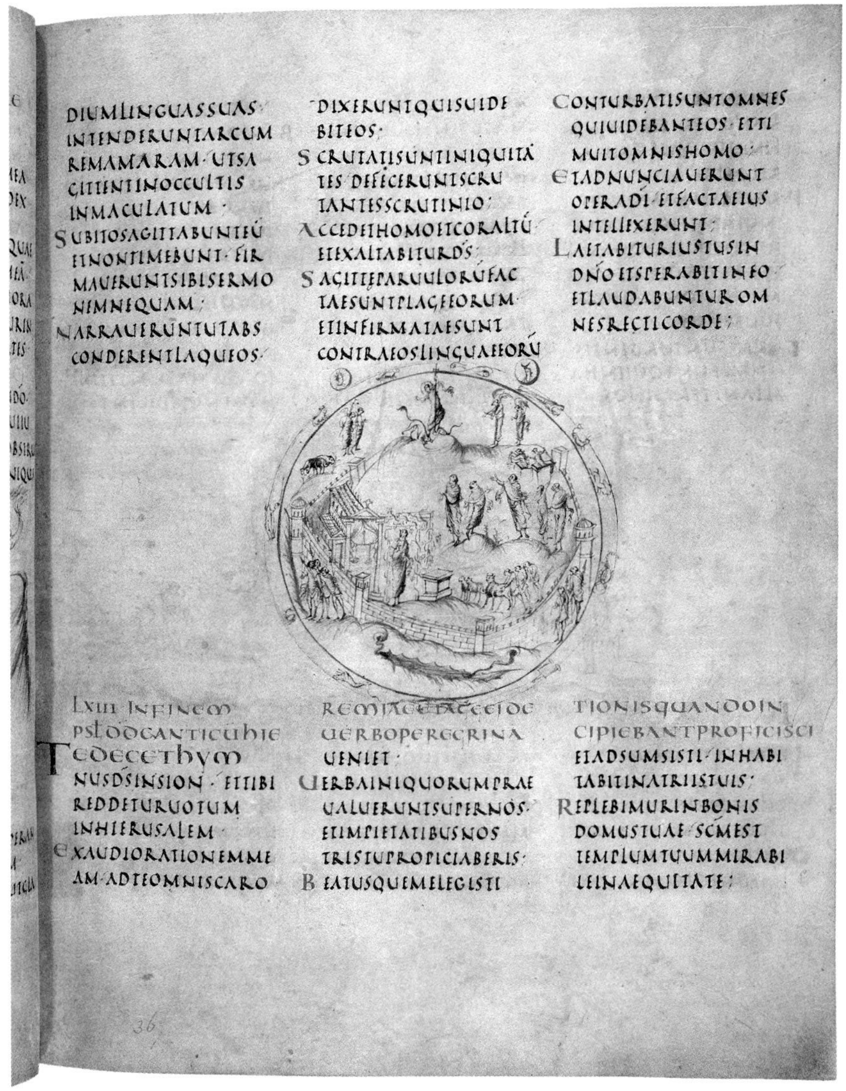

FIGURE 49

Zodiac for Psalm 64. Utrecht Psalter, Universiteitsbibliotheek Utrecht, MS 32, fol. 36r. Photo: Universiteitsbibliotheek Utrecht.

nuances of meaning. The representation of Pisces in Drogo's copy of the *Handbook of 809* exemplifies the conventional form associated with zodiac wheels. Arguably, this conscious decision underscores the placement of Pisces at the end of the season of death and dying. The image of Aries with its hoop referred to the renewal awaiting nature in the spring and summer months of the year. Symbolically, as in the image from the Utrecht Psalter, Aries ushered in the time of redemption and spiritual renewal, which is one reason Christ triumphs over the lion and serpent at 12:00 in the zodiac from the Utrecht Psalter (fig. 49). These iconographic subtleties would have been appropriate for inclusion within a Christian computistical compilation such as the *Handbook of 809*, emphasizing the links between the zodiac, their constellations, and Easter.

An assistant to the first master painter finished the three images on folio 59r of Perseus, Lyra, and Cygnus (plate 10). The profile image of Perseus resembles Serpentarius on folio 55v (plate 3), seen from the side and back. The musculature of Perseus recalls that of Hercules on folio 55r overall (plate 2), with ochre shadows modeling the human form. Perseus's bulging pelvis (and minuscule sex organs added by a later hand), however, fails to convey the confident, albeit neutered, appearances of the Gemini on folio 56v (plate 5) and of Aquarius on folio 59v (plate 11). The head of the Gorgon, Medusa, in Perseus's left hand resembles the figure of Virgo on folio 56r (plate 4). It appears that the master painter provided the harsh-brown contour lines and then permitted an assistant to finish the work on folio 59r:

> Perseus has bright stars in each shoulder, one bright one in the right hand, three stars in the left hand holding the head of the Gorgon, one star in the stomach, one bright star on the right side, one in the knee, one bright star in the foot, one bright star in the left thigh, two stars are in the shin, and three bright stars are around the head of the Gorgon. The head and the scimitar, however, are accepted as having one star each.[128]
>
> Lyra, also known as Fidis, has one star in each playing quill, one at the tip of both sides, one in both of the arm supports, one star in the base, and one star in the back. They total eight.[129]
>
> Cygnus, the swan, has one bright star in its head, five per wing, one star in the body, and one star in the tail. They total thirteen.[130]

The first painter resumed work on folio 59v. The form of Aquarius (plate 11) resembles the figural style seen in the depiction of the Gemini on folio 56v (plate 5). There is a particularly strong resemblance between the picture of Aquarius and the image of the left-hand musical twin, Amphion. The description of Aquarius notes the water flowing from his overturned pitcher, which was his identifying feature: "Aquarius has two dim stars in his head, one bright star per shoulder, one bright star in the left elbow, one star in the right elbow and one in that hand, one per pectoral muscle, one in the right leg, and one bright star per foot. They total twelve. The profusion of water was also designated by thirty stars, from which two in the set are brighter than the rest, while the others are dim."[131]

The representation of Aquarius in the Leiden *Aratea* on folio 48v portrays Ganymede, the cupbearer to the Olympian gods, with a Phrygian cap that recalls his Trojan heritage (fig. 50).[132] The related image of Ganymede-Aquarius in Drogo's copy of the *Handbook of 809* portrays the mantle of the figure as a full cape slung over his left arm, opposite the overturned pitcher raised in his right hand. It is interesting to conjecture that the removal of pants (or chausses) from the half-clad Aratean image in Leiden (fig. 50) resulted in, or contributed to, the sexless physiognomy of the miniature in Drogo's book. A similar disrobing of the clothed figure of Hercules in the Leiden *Aratea* also resulted in an arguably sexless heroic figure in Drogo's copy of the *Handbook of 809* (plate 2 and fig. 51). In the depictions of the Gemini found both in the Leiden *Aratea* and in Drogo's manuscript (Madrid 3307), however, the figures are effectively nude (plate 5 and fig. 46). The first master painter in Drogo's copy of the *Handbook of 809* nevertheless depicted the Gemini without any recognizable indications of sex. Perhaps this demonstrated a sacralization of the figures that undermines the pure anatomical beauty of the human form, resulting in asexual, and occasionally sterile, nonclassical depictions. The depiction of Aquarius from the *Recensio interpolata* in Saint-Gall 902 on page 92 displays the figure fully clothed with a mantle (fig. 40). Ironically, the reversal of the left-right and back-front orientation of the depiction of Perseus in Leiden on folio 40v results in another sexless figure for the handbook in Madrid (plate 10).

These changes argue in favor of the following conclusions:

1. The artists of Drogo's copy of the *Handbook of 809* censored their representations of figural forms in the pictorial cycle by removing sex organs from their paintings of the constellations.[133]

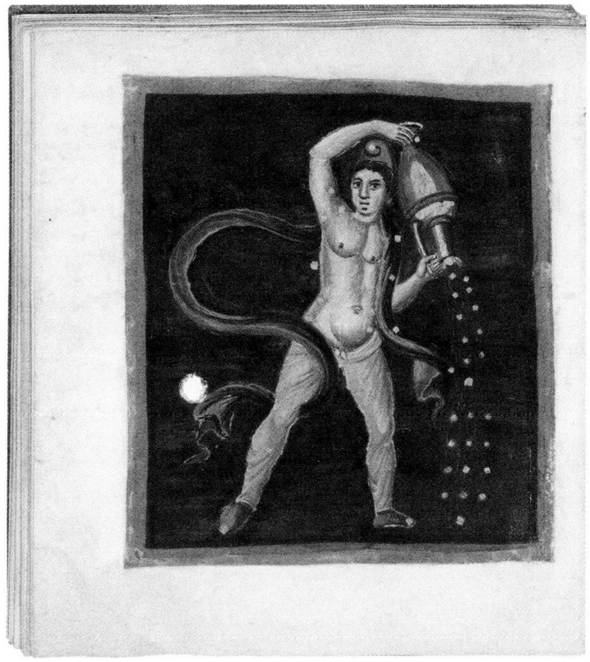

FIGURE 50

Aquarius. Leiden *Aratea,* Bibliotheek der Rijksuniversiteit, MS Voss. lat. Q. 79, fol. 48v. Photo courtesy of Universiteitsbibliotheek Leiden.

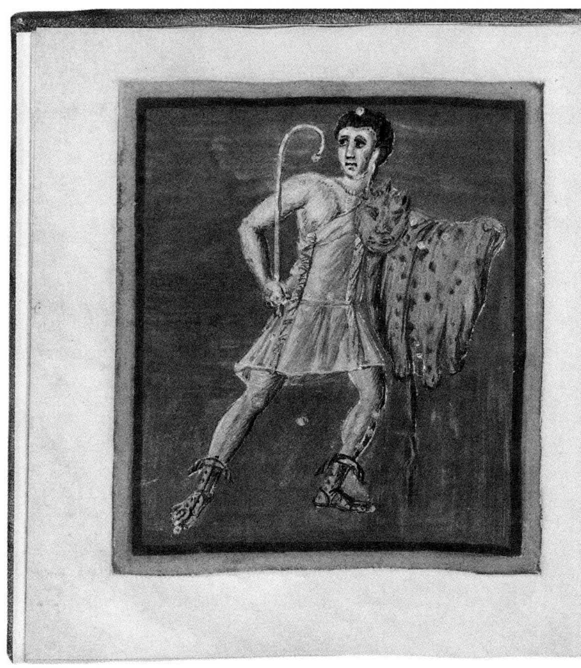

FIGURE 51

Hercules. Leiden *Aratea,* Bibliotheek der Rijksuniversiteit, MS Voss. lat. Q. 79, fol. 6v. Photo courtesy of Universiteitsbibliotheek Leiden.

2. The imagery in the lost *Handbook of 809* harks back to pictorial prototypes related to those that also inform the creation of the cycle of miniatures in the Leiden *Aratea* for some, but not all, of the images, revealing a polycyclic approach to the creation of an astronomical pictorial program.

It is interesting in this context that the images of feminine constellations such as Virgo, Cassiopeia, and Andromeda in Drogo's copy of the *Handbook of 809* (presumably representative of the original archetypal cycle) are all fully clothed, unlike the bare-breasted image of Andromeda from the Leiden *Aratea*. The neutered presentations of the male constellations preserve, arguably, a more direct link to the classical pictorial content than do the feminine star pictures. Certain details, such as gender, were deleted in keeping with a contemporary Carolingian understanding of what constituted an appropriate and normative depiction of the stellar form, which overlapped with concerns about moral issues for the local monastic readership of the *Handbook of 809*.

One of Charlemagne's favorite books was Augustine's *City of God*, according to his biographer, Einhard.[134] Augustine was quite clear about the importance of genitalia and human sex for the human body after the Resurrection, in fulfillment of the promise of eternal life upon which the early Christian Church had established its greatest hope. In *City of God*, Augustine argues not only for the presence of both male and female sex within resurrected and perfected bodies but also for the disappearance of "sexual lust, which is the cause of shame."[135] In fact, in defense of female genitalia, Augustine explained: "Now a woman's sex is not a defect; it is natural. And in the resurrection it will be free of the necessity of intercourse and childbirth. However, the female organs will not subserve their former use; they will be part of a new beauty, which will not excite the lust of the beholder—there will be no lust in that life—but will arouse the praises of God for his wisdom and compassion, in that he not only created out of nothing but freed from corruption that which he had created."[136] The perpetuation of sexual difference into the afterlife prolonged a celebration of God's creation. This belief about the glorification of human bodies in the heavens was not extended to the images of the constellations rendered in human form. For the early medieval world, these images of heavenly bodies came closest to capturing the classical content that the *Handbook of 809* sought to convey, and the androgyne constellations proclaim more than the necessity of astronomical study. Classicizing images of the constellations also emerged as the harbingers of Greco-Roman content in its purest form for later medieval and Renaissance reinterpretation. In the hands of Drogo and his pupils at Metz, his copy of the *Handbook of 809* continued to be read by mortal men who occupied a world in which "the lust of the beholder" could be elicited. This resulted in presentations of the constellations denuded of their sex, which nevertheless preserved aspects of their gendered sexual and virile identities for early medieval readers.

Interestingly, a zodiac wheel associated with the series of manuscripts of the *Recensio interpolata*—in other words, manuscripts of the *Revised Aratus*, such as the one on page 100 of Saint-Gall 902 (fig. 52)—displays Aries with a laurel at the head of the new year, the two Pisces joined by a cord at their mouths, and Aquarius-Ganymede cloaked solely in a mantle gathered about his shoulders. One more conclusion can be advanced at this time: the representations of the zodiac signs in Drogo's copy of the *Handbook of 809* appear to refer back to standard images derived from zodiac wheels, but they do not always conform to any known cycle in particular. Certainly, no such cycle has survived. The Gemini twins in the zodiac wheel from Saint-Gall, for example, are clothed and bear arms. The more typically truncated depiction of Taurus was rendered full-length in Drogo's copy of the *Handbook of 809*, as another counterexample to the archetypal *Handbook of 809*'s direct reliance on the zodiac linked to the *Recensio interpolata* in Saint-Gall 902 (fig. 52). In contrast, two of the remaining seven representations of the zodiac from Drogo's book, Virgo (plate 4) and Sagittarius (plate 12), shared significant iconographic details with the zodiac on page 100 in Saint-Gall 902 (fig. 52). The zodiac sign Libra was not included in the *Handbook of 809* or its catalogue of the constellations. The constellation, however, received its own depiction in the twelve-part zodiac wheel as a matter of course. This also could help explain, art-historically speaking, why the scales are separated from the angelic vision of Virgo/Dike in Drogo's book, as they also are in the zodiac wheel. Similarly, the artists of the zodiac wheel in Saint-Gall 902 represented Sagittarius in the less common form of a satyr, as it is portrayed in Drogo's handbook as well (and discussed below).

Reforming the Zodiac and Star Pictures South of the Ecliptic

The work of the first painter's team breaks off in Drogo's copy of the *Handbook of 809* in the middle of folio 59v. The next six miniatures are the craftwork of a superior early medieval artist who preferred a large contour line juxtaposed with fine textural details. This second master painter rendered the forms of the constellations through fine hatching and feathery touches with great textural variety. These miniatures, despite their sorry state of conservation at present in Drogo's copy of the *Handbook of 809*, are among the major accomplishments of the pictorial program. The style used to render these forms resulted from the artistic innovations of the team working in Metz. Their textual descriptions follow (plates 11–13):

> Capricorn has one star per horn, one bright star in its nose, two stars in the head, one star beneath the neck, two stars in the chest, one star in the first foot, one star at the tip of the foot, seven stars in the back, five stars in the stomach, and two bright stars in the tail. They total twenty-four.[137]
>
> Sagittarius has two stars in its head, two stars at the tip of the arrow, one star in the right elbow, one star in the hand, one bright star in the stomach, two in the back, one in the tail, one in the first knee, one star at the tip of the foot, one star at the back knee. They total thirteen.
>
> Aquila, the eagle, has four stars; of these the one in the middle is bright. The arrow has four stars: one in the tip, one in the middle, two at the other end.
>
> Delphinus, the dolphin, has one star in its ear, two in its horn, three in the little pectoral fin, one star in the back, two stars in the tail. Altogether they total nine.[138]

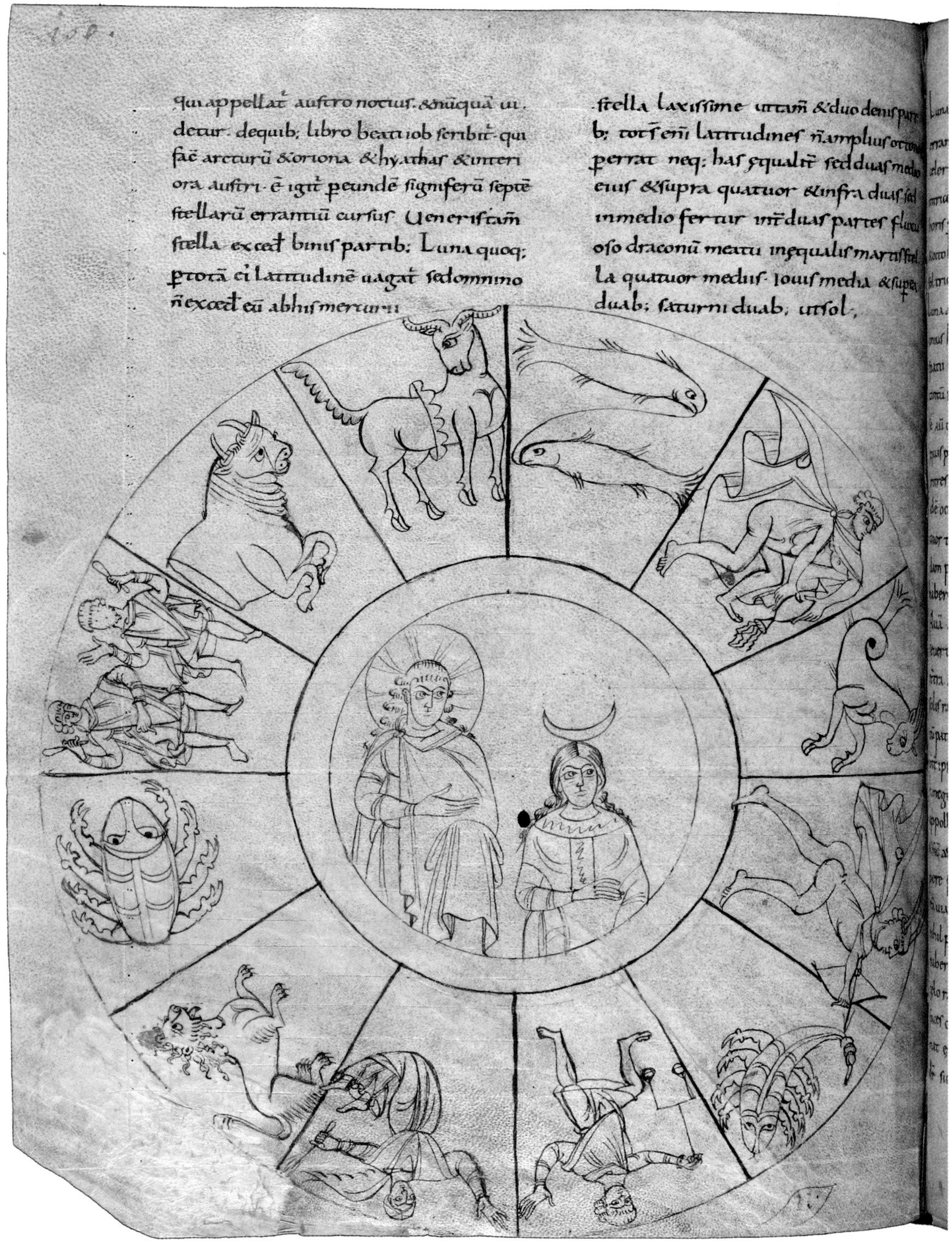

FIGURE 52
Zodiac. Stiftsbibliothek St. Gallen, Cod. Sang. 902, page 100. Photo: Stiftsbibliothek St. Gallen.

> Orion has three bright stars in his head, a bright star in each shoulder, one dim star in the right elbow, one star in the right hand, three stars in the belt, three bright stars in the dagger, one bright star per knee, one star in each foot. They total seventeen.[139]

> Canis Major, the dog, has one bright star in its tongue, which is called Sirius, the Dog Star, one dim star in each of its front legs, two stars in the chest, three stars in the front paw, two in the stomach, one star in the left thigh, one bright star in the paw farthest back, and four stars in the tail. They total sixteen.[140]

Beginning with Orion (plate 13)—which straddles the celestial equator, thereby rendering the constellation nearly universally observable[141]—and following the order established by Aratus, the list of the constellations shifts to descriptions of the constellations to the south of the ecliptic: "but many others rise below this, between the south and the track of the sun."[142]

Two important shifts take place in the dramatic presentations of these constellations. First, the more delicate technique of modeling the forms with three shades of the same color placed alongside one another was abandoned in favor of a dramatic contour and textural build-up of color with hatching, employed by a new master painter's team. Second, this technique complements virtuosic moments of freehand painting unparalleled by other Carolingian artists working in the first half of the ninth century. For example, the body of the mythical hybrid half-goat, half–sea monster Capricorn (plate 11) displays harsh blue-grey contour lines and an exceptionally refined and confident, sinuous watery line washed across the folio surface to define the other side of the creature. Strident forelegs emerge through the freehand colored brushstrokes, without any discernible contour line beyond the knee joints, in confident, rapid washes. The hind legs of Canis Major show a dramatic contrast (plate 13). Different shades create a sense of overlapping perspective in the dog's anatomy. If the first painter and his assistant employed a more controlled, quiet style motivated by economy and accuracy, then this second painter and his pupil were the dramatic expressionists at work within the small team employed at Metz.

Rapid brushstrokes and juxtaposed layers of color define the tortuous forms of Capricorn's horns. Similarly, the hirsute pelt of the satyr Sagittarius (plate 12) defies the controlled contour line of the inner thigh on the right leg, by freely overflowing that bestial boundary with dark brown and tan ticks of paint. This chaotic display, reminiscent of manuscript illumination from Reims (discussed in chapter 3), nevertheless permitted the Carolingian painter to model the thigh's cylindrical form.[143]

The fan-like tails of the dolphin (plate 12) and Capricorn (plate 11) required the juxtaposition of a lighter and darker color to create a feathery effect. The forms recall the style displayed by the fan-like tails found in carved ivories of the Carolingian period made in Metz. This second master created the plumage of Aquila the eagle (plate 12) with a tan underpainting accentuated by medium brown and dark brown tick marks. Like the other images by this painter, the contour line was used to define one side of the bird's body. The expressive potential of the layered hues (salmon-pink flesh tone for the talons, and brown hues ranging from ochre to chocolate) suggests the remainder of the fowl's plumage with a sense of arrested motion, as though the bird of prey has just alighted on its fletcher's perch, the *Sagitta* (or arrow). The talons themselves are fascinating from a codicological point of view, since the second master painter used the salmon-pink ink, at hand for rubrication in many Carolingian astronomical manuscripts, to render rapidly a gestural expression of their gnarled, knuckled form.

Although Orion (plate 13) has been badly damaged, the remaining paint best reveals the working technique of the second master painter. Light-blue washes define the left-hand side of the figure, opposite the darker contour lines on the right side of the legs. In some areas, such as the elevated right arm, light and thin contour lines were probably intended to define the form, yet would have been concealed by the introduced pigments. In other zones, such as the calf and shin of the hunter's right leg, the washes supply the graceful curves of Orion's musculature. Additional hatching identifies the elevated contour of the knee on the same leg. Applications of an even darker blue to the right forearm and Orion's underbelly supply deeper shadows. More saturated red-orange lines define the folds of the mantle. Lighter washes of the same hue define its form.

The image of Sagittarius (plate 12) in the form of a satyr derives from an iconography associated with ancient celestial globes (figs. 53–55) and Carolingian zodiac wheels (fig. 52). Nearly all the important Aratean images of Sagittarius depict the hunter as a centaur. Sagittarius is portrayed as a centaur in the Leiden *Aratea*, for example (fol. 52v; fig. 21).[144] Aratus also suggests that he intended his Archer to be a centaur, since Aratus notes the "forefeet" of the constellation and consequently draws attention to the form of a quadruped.[145] This form is arguably the most ancient, with precursors in the ancient Near East. The satyr type could also have two heads (in which case one resembles a sparrow hawk). Egyptian sources gave rise to this alternative. A fourth, rarer possibility is the presentation of Sagittarius as a childlike creature.[146]

As noted above, a consistent effort was made to provide information in the *De ordine*, which drew attention to the sorts of meteorological phenomena described in the *Phaenomena*. The south winds rose with violent force and bitter cold whenever "the sun meets up with Capricorn."[147] Before the trying month linked to Capricorn, Aratus adds that "when the sun inflames the bow and the Drawer of the bow, you should put ashore in the evening and not continue to trust the night. A sign of that season and that month will be the rising of the Scorpion at the end of night. The Archer actually draws his great bow near the sting."[148] This passage adequately describes the appearance of the lower, right-hand corner of the zodiac wheel from the Saint-Gall 902 copy of the *Recensio interpolata* on page 100 (fig. 52), mentioned above. As already noted, there was an apparent tendency for the artists of the *Handbook of 809* to adopt aspects of the image cycle linked to the traditions of the *Revised Aratus*, revealing a preference for those manuscripts which included the text of the *Recensio interpolata*, as revealed by Drogo's copy of the *Handbook of 809*. Whether the compilers of the imagery for the *Handbook of 809* adopted the satyr type for Sagittarius because of a pictorial tradition linked to certain redactions of the *Revised Aratus* is impossible to say for certain. The drawing of Sagittarius on page 93 of Saint-Gall 902 relied on the more ordinary form of a centaur instead. There was not even consistency within any one manuscript's pictorial program, which indicates just how tentative any hypotheses must remain about the archetypal pictorial program in the *Handbook of 809*.

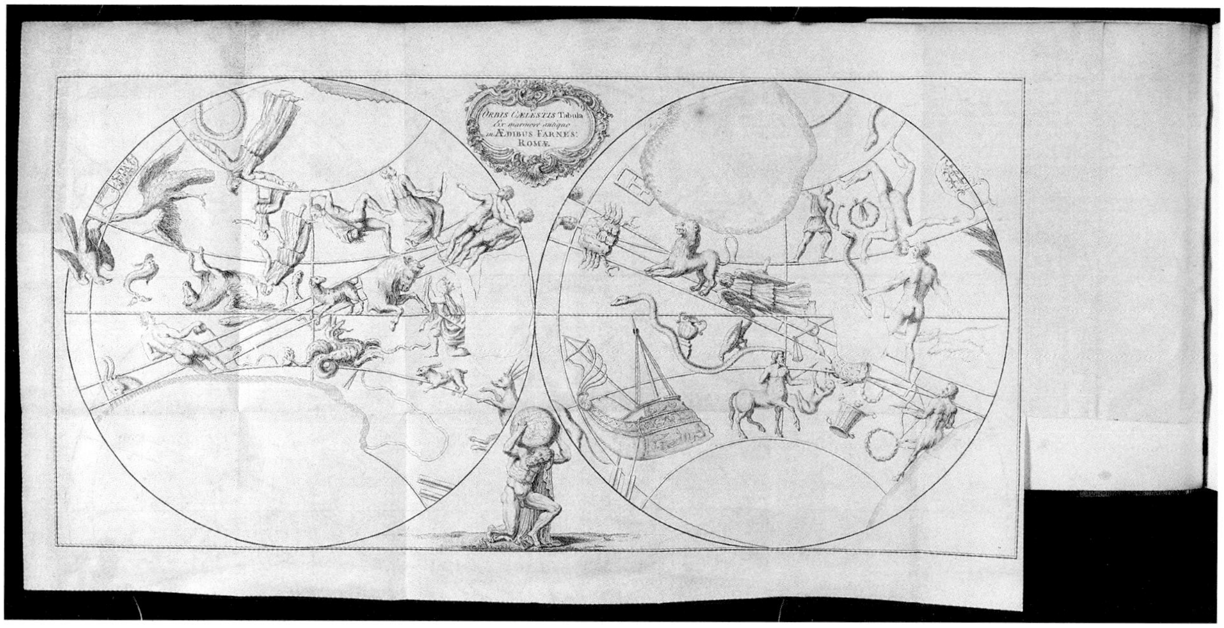

FIGURE 53

Martin Folkes, illustration of the Atlante Farnese for Bentley's Manilius of 1739, *M. Manilii Astronomicon*, PA6500 .M4 1739. Photo: Special Collections, University of Virginia, Charlottesville, Virginia.

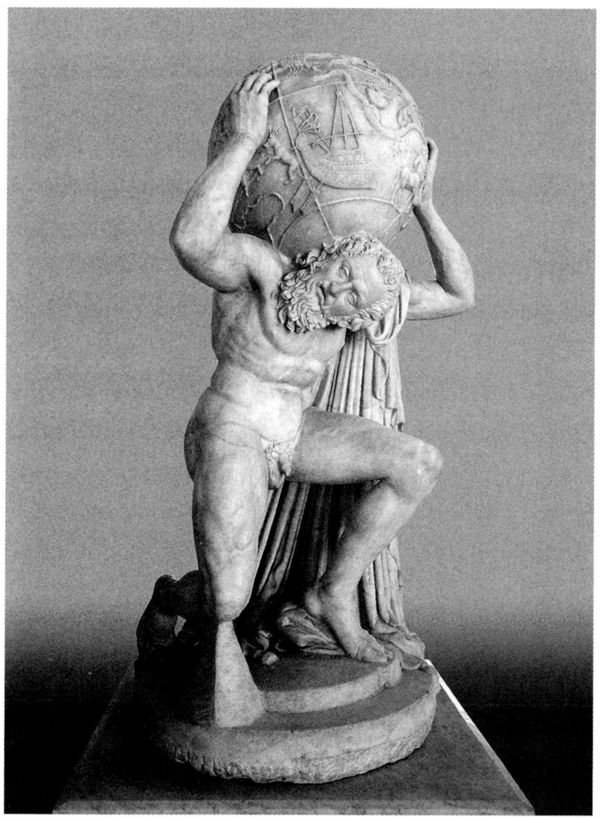

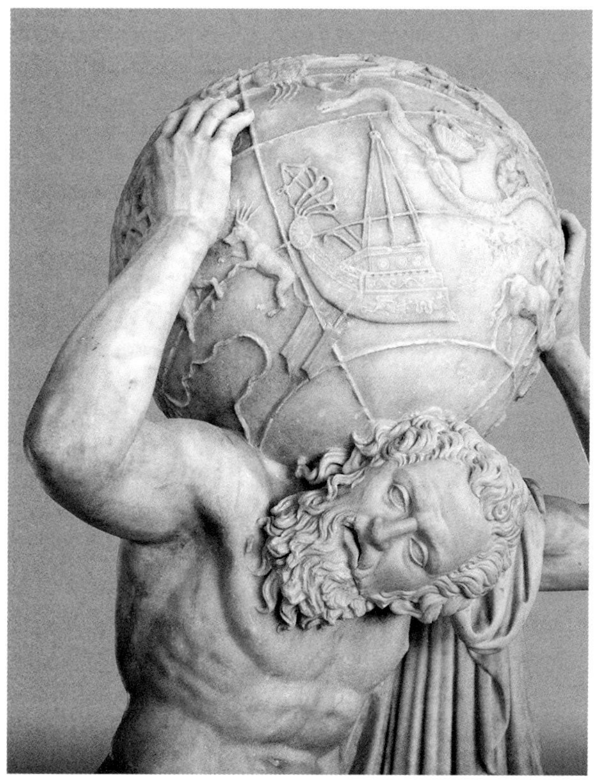

FIGURE 54

Farnese Atlas Globe. National Archaeological Museum of Naples, inv. 6374. Photo © Vanni Archive / Art Resource, New York.

FIGURE 55

Farnese Atlas Globe, detail. National Archaeological Museum of Naples, inv. 6374. Photo © Vanni Archive / Art Resource, New York.

Given the widespread iconographic diversity revealed by the manuscript record, it is reasonable to conclude that a typically Carolingian polycyclic strategy of image compilation was at work in these astronomical codices.[149] Decisions about the sorts of pictures to include in various pictorial programs required the consultation of multiple sources either because of resource scarcity or because of resource *availability*. There is no reason to discount the latter possibility. The manuscript evidence does not support the conclusion that monastic scribes and illuminators, though in relative isolation, engaged exclusively in a straightforward rote practice of copying individual exemplary codices without reflection, taking one classical exemplar and restoring its tarnished splendor. In the *Handbook of 809* and the traditions of the *Revised Aratus* manuscripts, the reoccurrence of iconographically related and clearly derivative forms of star pictures, discovered in tandem with illuminations displaying moments of aesthetic interpretation and painterly invention, suggests instead that a collaborative courtly workshop practice was fostered throughout the Frankish lands. Scribes and illuminators shared their visual presentations of the heavens with one another from Reims to Metz to Aachen, as the promulgated texts revealed the mysteries to be interpreted through artistry and were captured anew, time and again, on the parchment preserve of the foliated text. Both pictures and texts disseminated slippery traditions that migrated from one context to another without excessive discretion, so long as the image type remained, along with its text, an acceptable record of period *scientia*. For this reason it is best to recognize the general point that the satyr type of Sagittarius arose more commonly in conjunction with zodiac wheels or images related to the scant evidence pertaining to celestial globes (a point discussed in detail in chapter 3).[150] Either the satyr type or the centaur type of Sagittarius picture may have reminded readers of the Aratean admonishment to bring vessels into harbor each night during late fall and winter voyages. Incidentally, Aquila is also linked to navigational difficulties, whenever the eagle "rises from the sea at the departure of night."[151]

The satyr form of Sagittarius in Drogo's copy of the *Handbook of 809* vied with the more ancient representational form of the constellation as a centaur. This relatively innovative break with other pictorial cycles containing Carolingian star pictures has suggested several important conclusions. First, the representations of the constellations within zodiac wheels or pictures related to star globes provided alternative forms of images for artists to choose to include in their visual representations of the heavens. Material objects, and not just manuscripts, were probably consulted when creating manuscript cycles of illustration. In the case of Carolingian star pictures, the material objects of choice were precisely those whose use facilitated mastery of the pedagogical aspects of the *Handbook of 809*, such as a celestial globe.[152]

Second, the selection of a less common iconographic form, such as the satyr for Sagittarius, was a significant and purposeful decision made by a medieval painter. Such atypical forms nevertheless remained connected to the normative, iconographic meanings associated with the more common symbols.

Third, it is possible that the form changed for practical reasons. In fact, the *Catasterisms* 28 derived from Eratosthenes, as explained above, describes Sagittarius as a satyr. Douglas Kidd has argued that the satyr form of Sagittarius could more readily be differentiated from a suitably rendered star picture of the Centaur in astronomical texts and extended pictorial cycles.[153] In any case, Georg Thiele has noted that the satyr form of Sagittarius was also linked to Krotos of Sositheos in the texts of Hyginus.[154]

The next three miniatures bear the mark of a style that is related to that of the second master painter but inferior in its expressive force. The images on folio 61r (plate 14) are the work of the assistant to the second master. Like the depiction of Canis Major (plate 13), the haunches and rear legs of the hare (plate 14) contrast dramatically with each other by the juxtaposition of the farther haunch cast in shadow and a lighter rear leg that comes forward in a form of overlapping perspective from the viewer's vantage point. The graceful and sinuous form of the dog, which the second master rendered in a bold and confident profile, contrasts with the sketchy and mottled hues, which were intended to model the hare's pelt. The colors of the hare's coat resemble the basic color scheme of Leo on folio 57r (plate 6). This demonstrates that the first and second painters and their assistants all occasionally shared certain workshop practices. In fact, the vibrant hues rendered in gentle washes belong to a color scheme and style of painting associated with the manuscripts in Metz, such as the Drogo Sacramentary (fig. 33). This is an important result, for it permits a synchronic discussion in chapter 3 of the style of Carolingian manuscript painting in Metz.

There is no need in Drogo's copy of the *Handbook of 809* to posit two discrete campaigns of illustration or competing aesthetic impulses. The art-historical evidence argues instead in favor of four talented painters of sophisticated specialized training who plied their trade in medieval Metz for and alongside Bishop Drogo:

> The hare (Lepus) has stars in each eye, two stars in its body, one star at the end of the tail, and one star per paw in the back legs. They total seven.[155]
>
> The ship, which is called Argo among the Greeks, has four stars in the stern, five on the side, three at the apex of the mast, and five stars below in the keel. They total seventeen, because the whole ship in the sky is not illustrated, but it is fashioned from the rudder to the mast.
>
> Coetus has two bright stars in its tail, and five stars from the tail to the bend in its body; six stars are beneath its stomach. They total thirteen.[156]

Interestingly, the assistant to the second painter created an innovative depiction of the vessel *Argo* (plate 14), which transported Jason and the Argonauts, according to legend. He chose to paint the ship in its entirety rather than follow the text and portray only the stern and midsection in Drogo's book. Other Aratean examples, such as the miniature in the Leiden *Aratea* on folio 64v, adhere more accurately to the textual tradition, offering a bifurcated nautical vessel. When half of the *Argo* was dashed upon the rocks (the Symplegades), it rose like a fallen hero from the cold emptiness of the sea to assume its commemorative place in the celestial sphere. The image in Drogo's copy of the *Handbook of 809* is an original composition derived from other half-length boats of the Aratus recensions and a competing tradition of seaworthy nautical vessels.[157] The Carolingian assistant to the second painter completed the vessel with three bowsprits.

This assistant used the same technique as his master to supply a minimally modeled form, by placing a lighter tone next to a darker shade of the same or a related hue. For example, the bold contour line along the bottom of the boat dramatically defines its keel, using the same technique as the second painter in his stark rendering of Capricorn's underbelly (plate 11). The lightblue and cinnabar of the sails recall the garments of Orion (plate 13), painted by his master. In palette and design, the link between master and pupil remains apparent, though the unrefined quality of the assistant's work ensured that his participation was restricted to merely two folios.

The same holds true for the image of the sea monster Coetus (plate 14). According to the myth, Perseus rescued the sacrificial daughter Andromeda by slaying this sea monster.[158] As noted above, the tails of the dolphin (plate 12) and of Capricorn (plate 11) by the second painter display a sophisticated, feathery texture achieved through the overlapping of pigments and small, quick brushstrokes. His assistant painted the tail of Coetus with a geometric sterility instead. The feathery texture of the fins on the head of the sea monster resembles the work by the second master, and may have been used as an example to train his student. In any case, the assistant could not replicate the technique in the large wing-like fins below the belly of Coetus. The stark contrast between the evergreen contour and the lime-green body of the serpent visually separates into clumsy green stripes that do not coalesce into a presentation of rounded form. The technique was inspired by his teacher's use of a stark contour line in an image such as that of Capricorn. The second painter's assistant attempted to duplicate the pointillist technique that the first painter used to great advantage in his representation of the scaly flesh of Serpens on folio 54v (plate 1). In the image of Coetus, however, the dots merely provide a decorative pattern along the winding back and belly of the sea monster.

The painting of the second master displays a greater array of colors than those of the first artist. He also delicately renders the forms, which are painted freehand in certain areas but always delineated with a confident draughtsman's sense of line, or ductus. The next four miniatures are also by the second master painter:

> The river, which is called Heridanus, has three stars at the first bend, three at the second bend, and seven at the third bend, which are called collectively the Mouths of the Nile; these total thirteen. A star nearby is called Canopus.
>
> Piscis Magnus, the Great Fish, has twelve stars placed in order from head to tail.
>
> The sacrificial altar (Ara), or the shrine, has four stars: two stars are in the fire which burns where it was placed on the altar and two stars are in the pedestal (plate 15).[159]
>
> The Centaur has one star in each shoulder, two stars per elbow, four stars in the chest, ten stars in the rest of the body, two stars in the feet, two in the garment, two stars in the [Bacchic] thyrsus, which he carries, ten stars in the animal, which he holds in his hand, two stars in that hand, and five stars in the head. They total forty-three.[160]

These four images on folios 61v–62r (plates 15–16) are among the most accomplished of the pictorial program, which indicates that the second master painter received the honor of carrying out the more complicated compositions in the manuscript.

It is important to recall that in Drogo's copy of the *Handbook of 809*, Sagittarius and Capricorn were also painted by the second master painter. On these grounds, it is fair to propose that the second master painter was the most important master working in Metz. The second master painter was probably also the head of the artistic team of four painters encountered throughout this detailed analysis of the pictorial program. There was one assistant under the direction of each of the master painters, whose styles have already been discussed. The very fact that this team was working at an episcopal center on a scientific manuscript suggests that the artists may not have been permanently situated in Metz. Apparently, an astronomical manuscript was a specialized commission made at a particular site, such as Metz, for a specific recipient, such as Drogo. There is no evidence to suggest instead that a specialized scientific scriptorium arose in the ninth century to produce on speculation luxurious or functional schoolbooks, akin to the way that Touronian pandects emerged as a lucrative venture linked to the renowned monastic scriptorium of Saint Martin's, beginning with the abbacy of Alcuin (d. 804) and continuing well into the tenure of lay abbot Vivian (d. 851).[161] And the diversity of styles manifested by the individual copies of the *Handbook of 809* under consideration confirms this result.

Given all of the evidence considered in part 1 of this volume, from the *Annales Prumienses* to this review of the pictorial program, it is highly likely that Louis the Pious was the royal patron who commissioned Drogo's book and had the order carried out in medieval Metz, where his half-brother could oversee the project. How long a band of four painters was assembled in Metz is difficult to determine, given the limited number of illustrations in a copy of the *Handbook of 809* and the specialized nature of the manuscript. It is not difficult to imagine a situation in which one or both of the two master painters were brought in from elsewhere for a unique commission. Following the creation of Drogo's book of star pictures, they could have remained in Metz or, as itinerant painters, moved on to subsequent commissions throughout the Frankish lands. This theme is taken up again in chapter 3.

For now, it is important to underscore that these painters need never have worked together as a team on any other book. Related activity can be tracked during the ninth century. A group of four painters, however, coalesced around the specific project of creating Drogo's book primarily because Drogo was a prelate with royal kinship ties to Charlemagne. But, as time would reveal, on the witness of his sacramentary and the gospel books from early medieval Metz (figs. 30–34), Drogo was also enamored with luxury codices. If the book is treated as an important fact, as argued in the introduction—that is, if the manuscript becomes a nodal point of creative activity in a Carolingian network of scribal composition and painterly creation—then the fluidity of forms and artistic styles entering into its creation can better be appreciated. In other words, against standard historiographic models that taxonomically cluster stylistic currents into schools (in the work of Koehler and Mütherich, for example), the evidence available in Drogo's copy of the *Handbook of 809* argues instead for a more dynamic scenario. Painters such as the second master found themselves in novel temporary environments that influenced the production of specific books. Those working conditions established limitations that informed the iconographic models available, and that resulted in diverse dynamic relationships among masters, assistants, and patrons. Pragmatic considerations of economy, the successful completion of the commission, and sites where standing libraries, such as Corbie, were available for consultation set limits on the degree of fluid exchange. To treat these four painters in Metz as a mobile atelier is, however, to repeat the methodological error identified here. On the contrary, the work of the painters in Metz on Drogo's book offers insights into creative activity circa 830 in the Austrasian heartland, as four discrete early medieval lives converged. The book's relationship to other manuscripts documenting eighth- and ninth-century creative activity is the focus of chapter 3, including a discussion of the innovative diagrams in the *Handbook of 809* and the relationship of Drogo's copy to other illuminated manuscripts.

Bold, stark contour lines separate the drapery of Heridanus into well-lit and shaded areas (plate 15). The second painter also used this technique in his attempt at modeling the form of Capricorn (fol. 59v; plate 11). The personification of the god for the River Po, otherwise known as Heridanus,[162] is damaged and demonstrates the second painter's lack of skill at rendering human faces. Large round eyes suggest sockets sketchily. The second master painter employed a large shadowy swath for visual effect. This vied, however, with his effort to render a reasonably accurate human form. In the profile of Sagittarius the satyr (fol. 60r; plate 12), and in the animal forms such as Canis Major (fol. 60v; plate 13), the second master painter used this technique to more satisfactory effect. Harsh contour lines divide out the regions of pectoral and abdominal musculature in the depictions of Sagittarius (plate 12), Heridanus (plate 15), and the Centaur (plate 16). The chubby fingers and small hands in these three miniatures also suggest that they are by the same artist who painted Orion (plate 13). These stumpy hands differ from the long outstretched fingers associated with the first master painter in his depictions of Cepheus, Cassiopeia, and Andromeda (plates 7–8) on folios 57v and 58r. The harsh contour line beneath the dorsal fins of Piscis Magnus (plate 15) delineates the upper boundary of the fish's body and recalls how the same painter painted the Capricorn picture (fol. 59v; plate 11).

The signature textural studies of the second master painter supply an expert method for painting the transition from equine fur to human flesh in the image of the Centaur. A noteworthy novelty was the introduction of the white paint, added for highlights in both the image of the Sagittarius (around the arm drawing the bow in plate 12) and that of the Centaur (along its hindquarters in plate 16). Finally, many of the animal or hybrid forms painted by this second master display a ferocious or earnest eye: Capricorn (plate 11), Aquila and even Delphinus (plate 12), Canis Major (plate 13), and Piscis Magnus (plate 15).

The odd inclusion of a *thyrsus*, or shaft of pine cones, in the image of the Centaur (plate 16) appears to reinforce the distinction between the working techniques and philosophies of the two master painters in Metz. The *thyrsus* was a symbol of Dionysus carried by his followers.[163] But the *thyrsus* is mentioned in the text. A change in the artistic vision of the second compared to the first master painter fails to explain its inclusion. The second master painter created an acceptable interpretation of the hegemonic text of the *De ordine*.

Rather than dismiss the reference to the *thyrsus* as an insignificant detail or an odd authorial choice, it would be better to propose an art-historical rationale for its inclusion in the text. Any proposal should respect the prevailing tendency in the handbook to attribute oddities to scientific rather than mythopoetic origins. The textual source for the passage on the Centaur from the *De ordine* is the Basel *Scholia*. The term "veste" from the description of the constellation in Drogo's copy of the *Handbook of 809* on folio 62r appears in the other important Carolingian copies of the handbook as well (i.e., Reg. lat. 309, folio 98v, and Vat. lat. 645, folio 65r). This emendation was intended to clarify the confusing text of the Basel *Scholia* in Corbie. The Basel *Scholia* includes a list of the stars for the constellation named Chiron the Centaur. The passage on folio 38r–v of the Basel *Scholia* states that "in vestis ex [*sic*] in thyrso iiii" (in the garment [slung over the arm] and in the *thyrsus* are four stars).

The Carolingian editor of the handbook preferred to emend the words "in vestis" to the singular ablative form "in veste." This would have been fine had the draughtsman who worked on the archetypal *Handbook of 809* also included the garment slung over Chiron's left arm in the image from the Basel *Scholia* on folio 38v. The copies of the *Handbook of 809*, such as Monza f–9/176, also leave any sort of garment out of the picture on folio 69r.[164] The emendation in Corbie was a sensible change to the text accompanying the sorts of pictures seen in Aratean manuscripts, because the Centaur typically did have only one animal slung over his shoulder (typically a panther) that in certain manuscripts resembled a free-flowing pelt suggestive of a mantle. That motif also resembled a skin or pelt in the depiction of the Centaur once found in the Leiden *Aratea*, if a later copy from Boulogne-sur-Mer of the missing miniature can be trusted (Bibliothèque municipale, MS 188, folio 29r, ca. 1000).[165]

Given the textual traces that lead back from the *De ordine* to the Basel *Scholia*, it is reasonable to identify the Centaur depicted in Drogo's book as Chiron, the pious teacher of Achilles. This centaur was reportedly an expert on medicine and astronomy. In *Etymologies* IV.ix.12, however, Isidore of Seville extended the Centaur's résumé to include expertise in veterinary science.[166] According to Isidore, Chiron's treatment of animals resulted in his representational form as a hybrid creature. In the image from Drogo's copy of the *Handbook of 809* (plate 16), the rabbit (perhaps a hare) replaces the sacrificial panther and the *thyrsus* replaces the club found in the image from the Boulogne-sur-Mer *Aratea*. In other words, the choice of a centaur holding a *thyrsus* and rabbit for display was intentional. The image of Chiron paid homage to classical medicine, which was intrinsically linked to astronomy and even astrology in both the classical world and in the Middle Ages. Unlike the mantle, the cone-tipped rod of the *thyrsus* remained in the pictorial tradition because of the demands placed upon the picture, which differed from the demands placed upon the text. The *thyrsus* drew attention to the mythological physician, who is pictured similarly in the image accompanying the copy of the Basel *Scholia*. The second master painter, like the first, apparently felt compelled to accentuate these details, which permitted the iconographic connections to medieval science; the pagan, spiritual role of the rod as a reference to Dionysian bacchanals had been suppressed. Indeed, some might say that the second master painter in Metz painted the *thyrsus* in Drogo's book in such a way that it resembled a hunting spear used for slaying the rabbit, even though the rod was referred to explicitly as a *thyrsus* in the text.[167] Hunting was much more palatable to ninth-century royals than were pagan rituals.

The Centaur's name was also removed from the text of the *De ordine*. The erasure shows the effort of the editors in Corbie to accentuate scientific points while suppressing *certain* mythological details. In this instance, the silences are as informative as the ideas expressed in the text. Details pertaining to the form of the constellation in space needed to be retained. Further artistic decisions were motivated by a spirit of scientific pragmatism, provided such iconographic decisions conformed to, or at least did not vie with, Christian orthodoxy.

The image of the Centaur in the Cologne *Aratus* on folio 166v is probably related to the one in the archetypal *Handbook of 809* and was probably its immediate temporal precursor within the Frankish lands (ca. 805 as opposed to ca. 809). As suggested above, the *Revised Aratus* manuscript from Cologne provided a model for the creators of the original *Handbook of 809* and the painters of Drogo's copy to bear in mind. The Centaur from the copy of the Basel *Scholia* is also related to the form that probably harks back to a prototypical depiction of Chiron. It was the text of the Basel *Scholia*, however, that designated the Centaur as Chiron. This detailed analysis of the text and image of the Centaur in Drogo's copy of the *Handbook of 809* provides art-historical evidence for the direct reliance of the *Handbook of 809* text about the constellations, the *De ordine*, on the text of the Basel *Scholia*.

The last four depictions in the handbook were all done by the assistant to the second master painter:

> Serpens, the Serpent, which they call the Hydra, has three bright stars in its head, six stars where it first bends, three stars at the second bend, two stars in the fourth bend, and eight bright stars from the fifth bend through the tail. They total twenty-six.[168]
>
> Corvus, the raven, is on the tail [of the Hydra] looking out over a depression [in the serpent's body]; the raven has one star in its eye, two stars in its wings, two in its tail feathers, and one star per foot. They total seven.[169]
>
> The Crater, or the urn, which was placed after the first bend of the serpent's body, has two dim stars at its lip, three stars in the midsection, and two stars in the foot. They total seven.
>
> Anticanis, Canis Minor, has three stars.[170]

The assistant to the second master revealed his true expertise in the second representation of the Raven on folio 62v (plate 17). The image is first displayed on the back of the Hydra on folio 62r (plate 16). In both, the attention to detail is limited to the heavy contours likewise encountered in the images on folio 61r, for example (plate 14). The assistant to the second master painter created the solid form of the second raven freehand and without underdrawing, carefully reticulating the gnarled appearance of the bird's talons. The image of the raven is an impressive example of medieval pen-and-ink drawing.

The Hydra on folio 62r is a lackluster attempt at modeling the physiognomy of the creature (plate 16). The assistant to the second master similarly exaggerated the juxtaposed layers of hues in the image of Coetus on folio 61r (plate 14). Neither image captures the textured modeling seen in the image of Serpens on folio 54v (plate 1), attributed to the first master painter. The image of the Crater (plate 16) on the back of the Hydra resembles the second image on folio 62v (plate 17). Both have a nearly hemispherical tonal arrangement, such that the right side displays an exaggerated shaded zone typical of this painter's style. In the individual image of the Crater, the assistant to the second master painter mimicked his teacher's preferred technique of hatch marks texturing quasi-modeled volumes. The hatching, however, became a disappointing series of banded striations in the hands of the pupil.

The image of Canis Minor (plate 17) clearly shows the contrast in style between the assistant to the second painter and his master. Whereas the depiction of Canis Major on folio 60v (plate 13) shows the stern gaze typical of the second hand, the assistant inserted a beady pupil into the eye socket of Canis Minor on folio 62v (plate 17). The second master painter painted the far back leg of Canis Major darker than the one closest to the observer, whereas the assistant misinterpreted the style and divided both rear legs of Canis Minor into striped zones. In Canis Minor the left rear leg has the dark shaded strip on the bottom, whereas the right rear leg closest to the viewer has the shaded region on top, along the dog's haunches, demonstrating the second assistant's confusion. This assistant was clearly learning from his teacher but had yet to master the logic of his master's techniques.

As noted already, the repetition of the images of the Raven and the Crater was a conscious decision motivated by the preference for visual parataxis in the *Handbook of 809*. This form of parataxis was a departure from earlier Aratean pictorial programs and a creative presentation of the pictorial program attached to the *De ordine* in Drogo's copy of the *Handbook of 809*. As a result, the Carolingian astronomers and artists of the handbook, as well as their later copyists, demonstrated a value placed upon the individual representations of celestial forms for scientific understanding and also for the aesthetic pleasure of their learned patrons.[171]

Greco-Roman Visions of the Stars and Their Christian Avatars: Panofsky, Saxl, and Carolingian Creativity

This iconographic analysis has revealed that the creative process in Metz was much more than a straightforward copying or reclamation effort. The tendency to render the individual images of the constellations according to an early medieval pictorial strategy of visual parataxis indicates that even long-standing forms could be recombined in new ways. The probable reliance of aspects of the archetypal *Handbook of 809* on textual and visual components of the *Revised Aratus* also places the cycle of illustration in Drogo's personal copy within an interpretative framework that emphasizes active processes of organization (*ordinare*) and critique (*emendare*).

Adopting a compositional strategy appropriate to encyclopedic literature, the *Revised Aratus* of the late eighth century included excerpts of texts that recast the traditional significance of Aratus's *Phaenomena*. The selection of texts from authors such as Isidore of Seville did more than pay homage to the legacy of classical antiquity, as Le Bourdellès would suggest.[172] Isidore was not only an encyclopedic exegete of classically inspired natural history and the liberal arts, including astronomy, but imbued the rich rhetorical, literary, and scientific legacies of his pagan and Christian *auctores* (such as Varro, Pliny, Macrobius, Martianus Capella, Donatus, Solinus, Servius, Vergil, Lactantius, Cassiodorus, Jerome, and Augustine)[173] with a novel early medieval Christian *auctoritas*. Nowhere was this point made more explicit or with more visual force than in the fifth wheel diagram (Munich, Bayerische Staatsbibliothek, Clm 16128, folio 16r; fig. 1) depicting the *mundus-annus-homo*, or the microcosmic-macrocosmic harmony, in Isidore's *De natura rerum*.[174]

As addressed above in the discussion of the image of Hercules on folio 55r, the first master painter included pictorial references to the Nemean lion and club (plate 2). But what is *missing* from the image of Hercules is arguably the most important factor to appreciate. Unlike the image of Hercules in the Cologne *Aratus* on folio 156v (fig. 8), in the image of Hercules in Drogo's book, the mythological reference to the Tree of the Hesperides has been ignored. This reflects a point of view advanced by Isidore in book III of his *Etymologies*: "Therefore, observations of the stars, or horoscopes, or other superstitions that attach themselves to the study of the stars, that is, for the sake of knowing the fates—these are undoubtedly contrary to our faith, and ought to be so completely ignored by Christians that it seems that they have not been written about."[175]

As Isidore had prescribed, one of the express responsibilities of a compiler or early medieval author was to relay what could be helpfully gleaned from the classical authors while packaging the astronomical information in terms that did not border on heterodoxy and while effacing questionable content. Such deliberate selection to guide the reader through a proper mode of interpretation consists of a creative process of textual composition and image collection that is referred to here as *exegetical emendation*. The methodological consequences of this design process are discussed in greater detail in chapter 4. For now, it suffices to note that the interpretative or exegetical strategy of compilation appropriate to the creation of the *Handbook of 809*—which derived in part from the same pedagogical impulse that Isidore had manifested in his *Etymologies*—recognized that the ultimate utility of training in the liberal arts lay in the application of such knowledge by professional clerics. As Arno Borst has argued, the *Handbook of 809* benefited from this kind of interdisciplinary spiritual and tactical focus (see the introduction).[176]

The constellation Hercules from Drogo's copy of the *Handbook of 809* provides an example of this focus. The image of Hercules himself was essential to the star catalogue contained within the *De ordine*. The mythological content pertaining to Hercules's attack on the serpent that guarded the golden apples in the Tree of the Hesperides was omitted, even though this larger, elaborate image was linked to the *Revised Aratus* and depicted in the Cologne *Aratus* (plate 2 and fig. 8). Similarly, on folio 14r, the Basel *Scholia* reports, "Hic est Hercules sup- anguem incumbens clava" (Here is Hercules, attacking a serpent with a club), before the corresponding image of the constellation follows on folio 14v (fig. 7).[177] The image from the copy of the *Revised Aratus* in Cologne was created in 805, and the text associated with the Basel *Scholia* shows Hercules attacking the serpent, which makes it all the more remarkable that the image in Drogo's copy lacks the same pictorial references.[178] This shows that early medieval processes of image selection for large natural-historical cycles of illustration involved *exegetical* strategies of discovery. Choosing certain iconographic types and disregarding others when culling through extant prototypes is a meaningful rather than rote activity, and deviating from standard options to advance a non-mythological point of view in the Carolingian *Handbook of 809* was an act of purposeful *emendation* on the part of Adalhard's team.

This level of creative involvement in the production of new manuscript anthologies has not traditionally been associated with aspects of the Carolingian renewal. It is easier to focus on the Carolingian star pictures that remain and use them as evidence of both the extant and the lost pictorial traditions that would have been passed down from late antiquity to Frankish artists. This was the important art-historical project of Fritz Saxl and Erwin Panofsky, as mentioned in the introduction.

Saxl and Panofsky's "Classical Mythology in Mediaeval Art" appeared in 1933 in *Metropolitan Museum Studies*.[179] In this essay, the coauthors argued that the Franks had played a vital role in linking the High Renaissance celebration of classical form and proportion to their ancient forebears. Along the way, however, Carolingian artists necessarily undermined the pure expression of classical *Latinitas* to advance a period-specific agenda and Holy Christian imperial ideology. However, Panofsky and Saxl attest to an unnecessary putative divide, in which they claim that the effort to promote a nascent yet already hegemonic Christian interpretation of the star pictures obviates the possibility for a reinsertion or continuity of classical content into the Carolingian images of the heavens. In their words, "The Carolingian Renaissance differed from the 'Rinascimento' of the fifteenth century in many respects. Where the latter was based on the irresistible feeling of the whole people and was brought forth in popular political and spiritual excitement, the earlier was the result of the deliberate efforts of a few distinguished men, and thus was not so much a 'revival' as a series of improvements in art, literature, calligraphy, administration, etc."[180]

Panofsky and Saxl are here willing to recognize that Carolingian star pictures were closer to the classical models than other, later images of the constellations "in style and technique," thus adhering to long-standing "representational traditions" offering "integral unities of subject matter and form."[181] This measured praise for the ability of Frankish artists and scribes to copy with rote precision came bundled with a disparaging dismissal of Carolingian creativity, disregarding the possibility of exegetical emendation or the role of conscious decisions in the formation of the text and image cycles contained within manuscripts such as the *Handbook of 809*. It is not truly fair to say that the Frankish artists and scribes working for Adalhard copied either the Aratean texts or aspects of their corresponding cycles of illustrations when compiling the original *Handbook of 809*. On the contrary, original strategies of selection and composition resulted in the organic creation of a novel Carolingian presentation of the cosmos that laid the foundation for similar decisions in Drogo's later copy.

The more definitive statement of Panofsky's belief in the exceptional renewal of classical *Latinitas* during the High Renaissance appeared in his 1960 *Renaissance and Renascences in Western Art*.[182] Here Panofsky altered the emphasis from an inquiry into the transmogrification of celestial forms over the *longue durée* to a heartfelt apology for the celebrated uniqueness of the classical renewal experienced in fifteenth-century Florence and Rome.[183] Carolingian star pictures, in Panofsky's view, could only represent aspects of classical form and content through the sedulous copying efforts exerted by the Frankish painters.

It is important to underscore here a vital point and thereby to correct an occasional misconception. Panofsky celebrates the Carolingian painters who created classically inspired images of the zodiac and constellations, especially when this required novel pictorial research reflecting upon examples of cycles in mosaics, on wall paintings, on celestial globes, or in extant manuscripts that were believed to be classical and not just early Christian precedents. Carolingian artists saved from oblivion these classical prototypes and breathed new life into these "figures (or groups of figures) classical not only in form—this, needless to say, applies to numberless motifs handed down by the Antique to Early Christian art and thus held in readiness for the Carolingian revival—but also in significance."[184] Panofsky develops his argument: "Carolingian art, we remember, revived and re-employed scores of 'images' in which classical form was happily united with classical content. . . . The iconographical significance of these images, however, remained unchanged."[185] And again: "Nowhere in Carolingian art, however, do we seem to encounter an effort to infuse into a given classical image a meaning other than that with which it had been invested from the outset. . . . We are confronted with quotations or paraphrases, however skillful and spirited, rather than with reinterpretations."[186] It is in this context that Panofsky moots his "Principle of Disjunction": if a medieval artist adopts a classical form for a composition, then its corresponding content is typically recast in compliance with Christian orthodoxy; alternatively, if a medieval artist opts to illustrate a narrative with its roots in pagan antiquity, the corresponding visualization of the content will reflect spatio-temporally contingent medieval mores and courtly life.[187]

Panofsky could not envision, however, the precise situation in which the creators of the *Handbook of 809* found themselves. Under Adalhard, an official delegation recombined and revised myriad fragments of classical text and images, assembling the truthful record of Carolingian astronomy one sanctioned piece at a time in the years that followed the synod of 809. The text of the *De ordine* was illustrated with classical forms and advanced classical content, but did so within a programmatic Christian platform calling for religious reforms.[188]

The agenda pragmatically set the parameters on the semiotics of interpretation and established the framework in which the creation of the *Handbook of 809* could take place. In other words, the exegetical emendation of the text-image pair pertaining to Hercules in the *De ordine* preserved the classical Aratean form and content of the constellation within the star catalogue, while it simultaneously excluded the pagan mythological narrative alluded to by the Tree of the Hesperides from the folios of Drogo's copy of the *Handbook of 809*.

Modern readers and viewers may at times glimpse the late antique classical past through the shadowy reserve of the choices made by the Carolingian scribes and artists working for Adalhard in Aachen. But to focus exclusively on the images in Drogo's copy of the *Handbook of 809* as records of lost pictorial precedents (or as records at one level of remove from the archetypal *Handbook of 809*) is to miss a vital insight about Frankish life—namely, that the artistic and creative decisions which resulted in the extant pictorial programs in the handbook supply *firsthand* information about Carolingian decision-making processes. The extant star pictures do not simply inform the art historian about iconographic precedent. Rather, these pictures (such as the one of Hercules) declaim aspects of the Frankish identity of their creators, revealing facts about Frankish priorities, values, preferences, and prejudices, without the need for a *Weltanschauung*.

Plates

PLATE 1

Haelice/Arcturus Maior, Cinosura/Arcturus Minor, and Serpens. Madrid, Biblioteca Nacional de España, MS 3307, fol. 54v. Photo © Biblioteca Nacional de España.

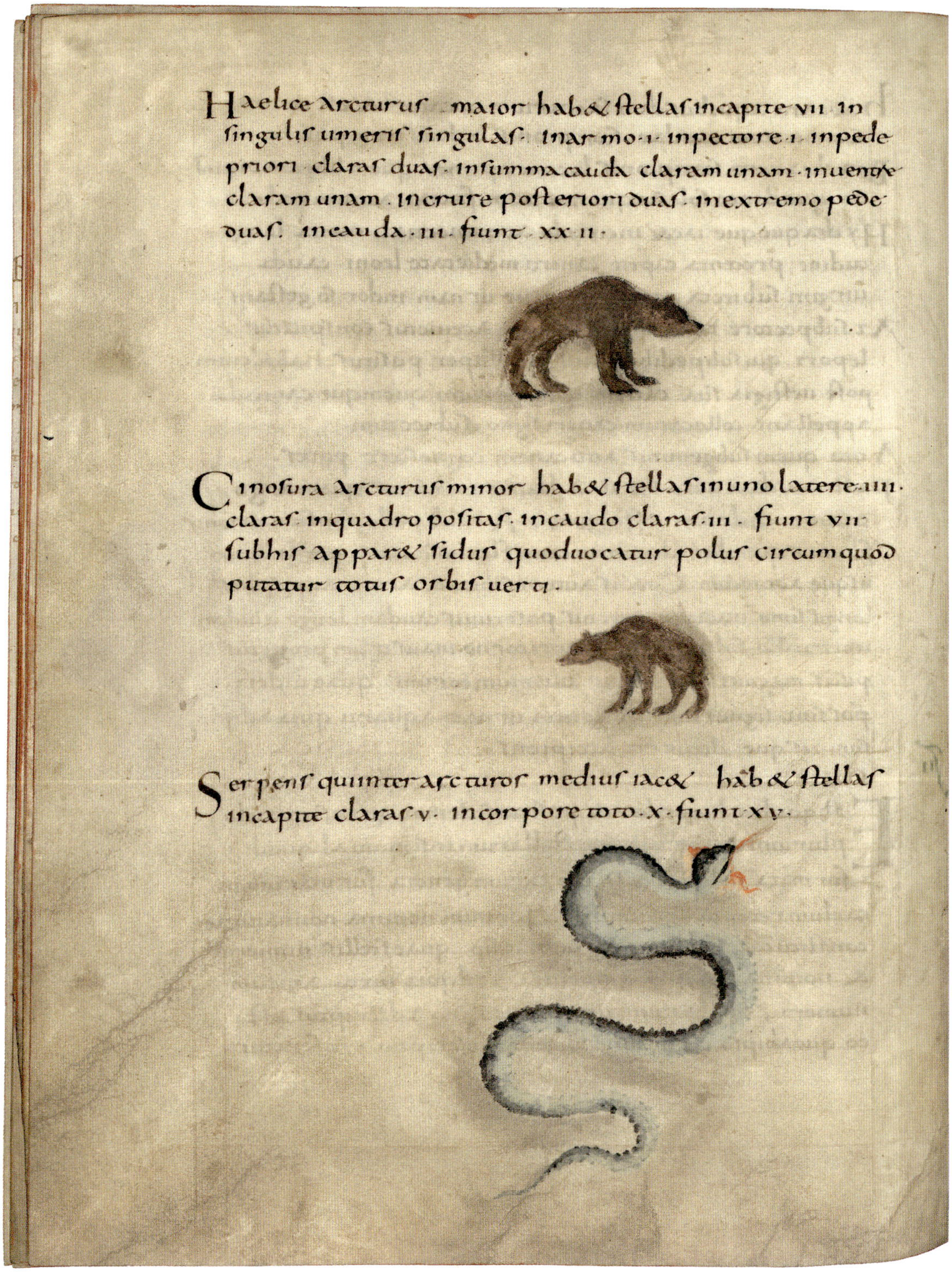
Haelice arcturus maior hab& stellas incapite vii in
singulis umeris singulas. inarmo .i. inpectore .i. inpede
priori claras duas. insumma cauda claram unam. inuentre
claram unam. incrure posteriori duas. inextremo pede
duas. incauda .iii. fiunt xxii.

Cinosura arcturus minor hab& stellas inuno latere .iiii.
claras. inquadro positas. incaudo claras .iii. fiunt vii
subhis appar& sidus quoduocatur polus circumquod
putatur totus orbis uerti.

Serpens quinter arcturos medius iac& hab& stellas
incapite claras .v. incorpore toto .x. fiunt xv.

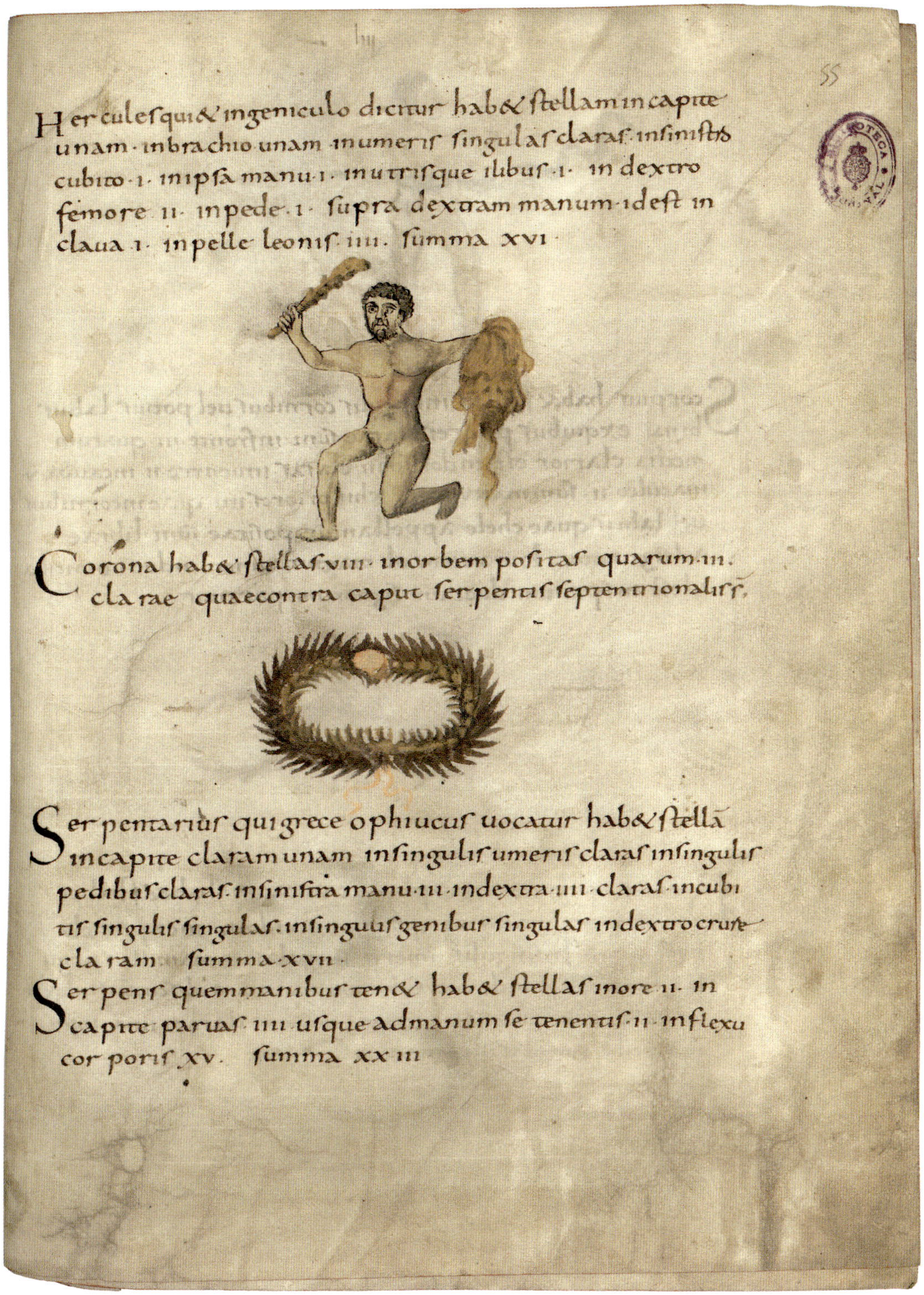
Hercules qui & ingeniculo dicitur hab& stellam in capite
unam · in brachio unam · in umeris singulas claras · in sinistro
cubito · i · in ipsa manu · i · in utrisque ilibus · i · in dextro
femore · ii · in pede · i · supra dextram manum · id est in
claua · i · in pelle leonis · iiii · summa xvi ·

Corona hab& stellas viii · in orbem positas quarum · iii ·
clarae quae contra caput serpentis septentrionalis.

Serpentarius qui grece ophiucus uocatur hab& stellā
in capite claram unam in singulis umeris claras in singulis
pedibus claras · in sinistra manu · iiii · in dextra · iiii · claras · in cubi
tis singulis singulas · in singulis genibus singulas in dextro crure
claram summa · xvii ·

Serpens quem manibus ten& hab& stellas in ore · ii · in
capite paruas · iiii usque ad manum se tenentis · ii · in flexu
corporis xv · summa xxiii

PLATE 2

Hercules and Corona. Madrid, Biblioteca Nacional de España, MS 3307, fol. 55r. Photo © Biblioteca Nacional de España.

PLATE 3

Serpentarius/Ophiuchus and Scorpius/Scorpio. Madrid, Biblioteca Nacional de España, MS 3307, fol. 55v. Photo © Biblioteca Nacional de España.

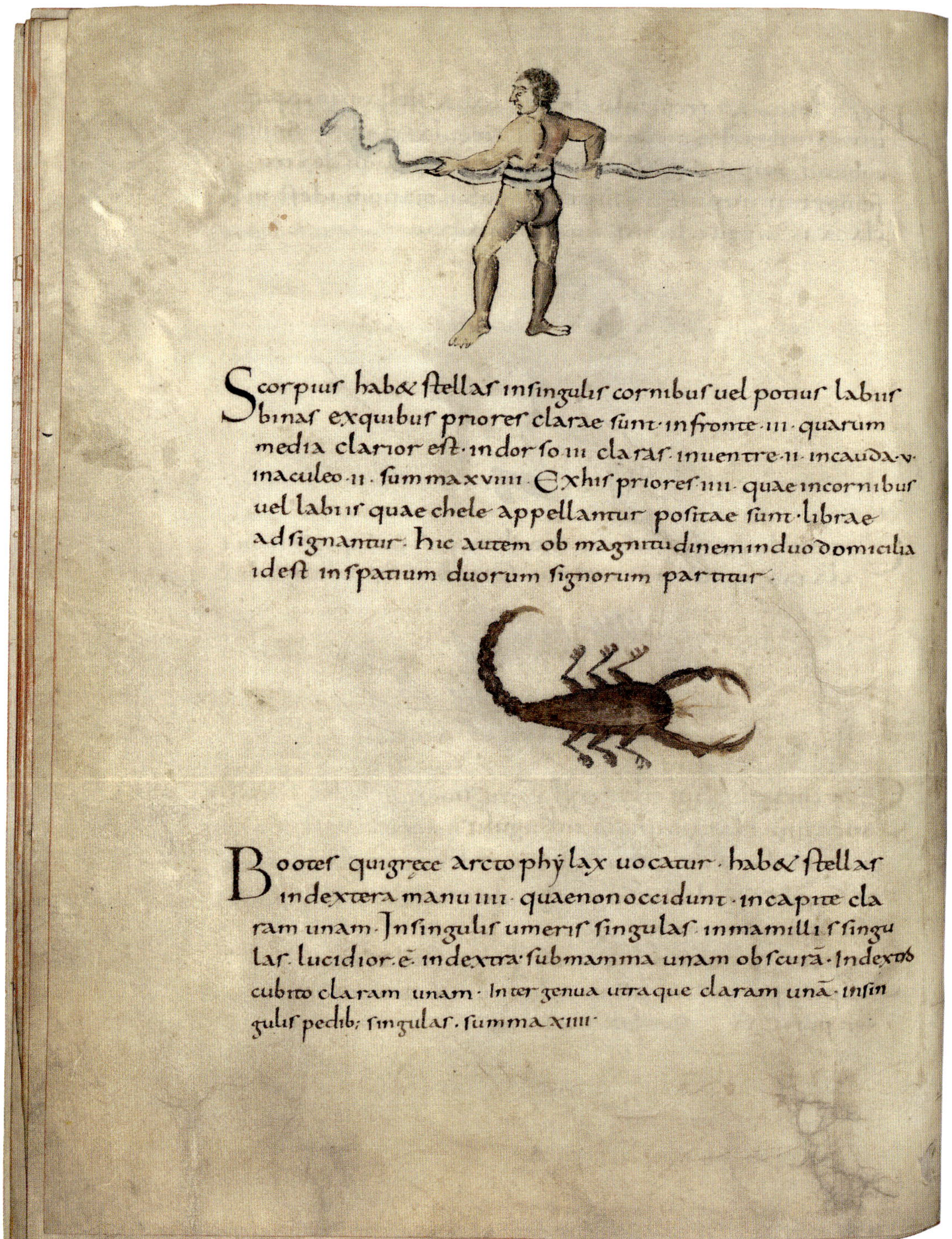

Scorpius hab& stellas insingulis cornibus uel potius labius
binas exquibus priores clarae sunt. infronte. III. quarum
media clarior est. indorso III claras. inuentre. II. incauda. V.
inaculeo. II. summa XVIII. Exhis priores IIII. quae incornibus
uel labiis quae chele appellantur positae sunt. librae
adsignantur. hic autem ob magnitudinem induo domicilia
idest inspatium duorum signorum partitur.

Bootes quigrece arctophylax uocatur. hab& stellas
indextera manu IIII. quaenon occidunt. incapite cla
ram unam. Insingulis umeris singulas. inmamillis singu
las. lucidior ē. indextra submamma unam obscurā. Indextro
cubito claram unam. Inter genua utraque claram unā. insin
gulis pedib; singulas. summa XIII.

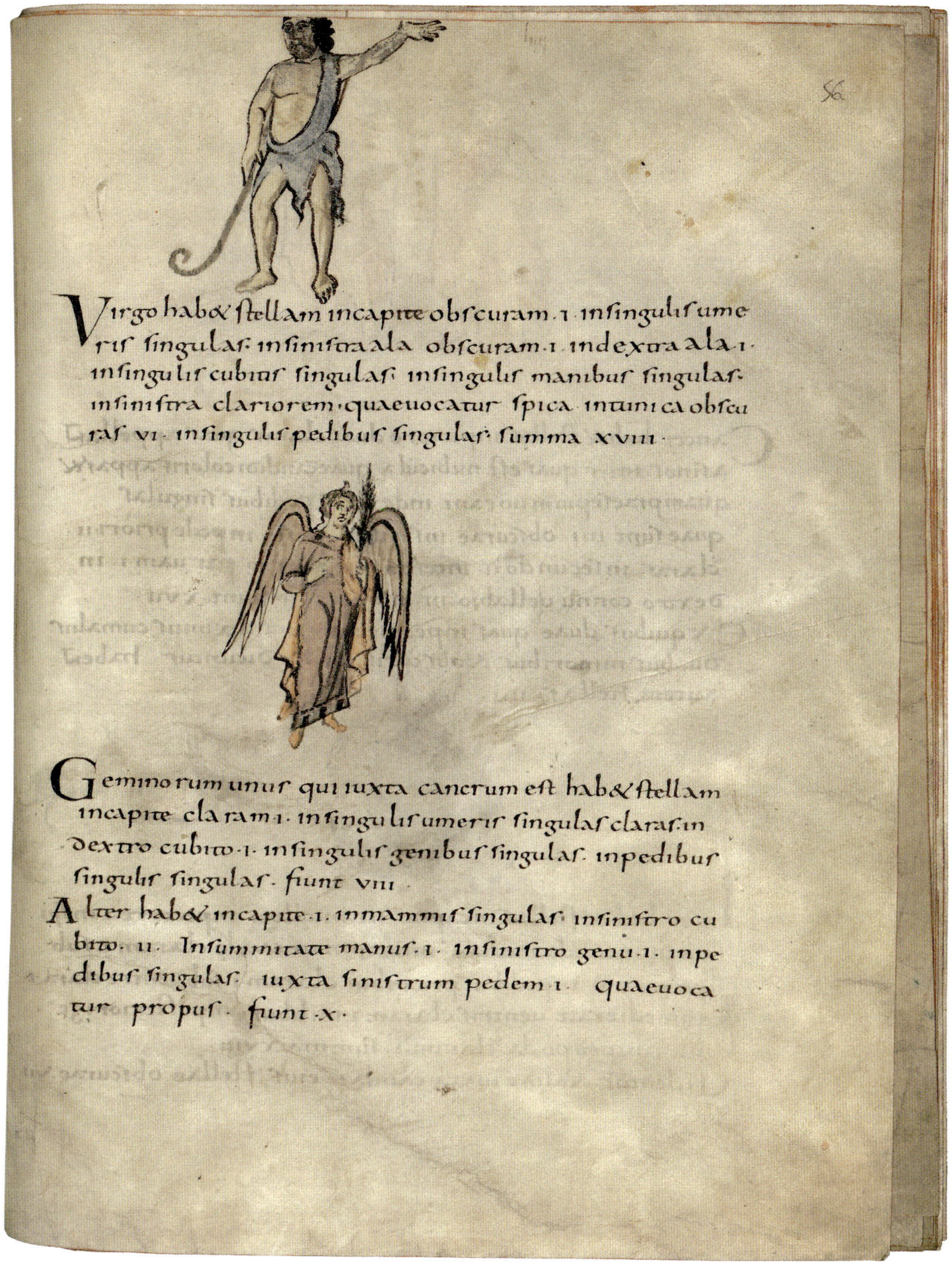

PLATE 4

Boötes/Arctophylax and Virgo. Madrid, Biblioteca Nacional de España, MS 3307, fol. 56r. Photo © Biblioteca Nacional de España.

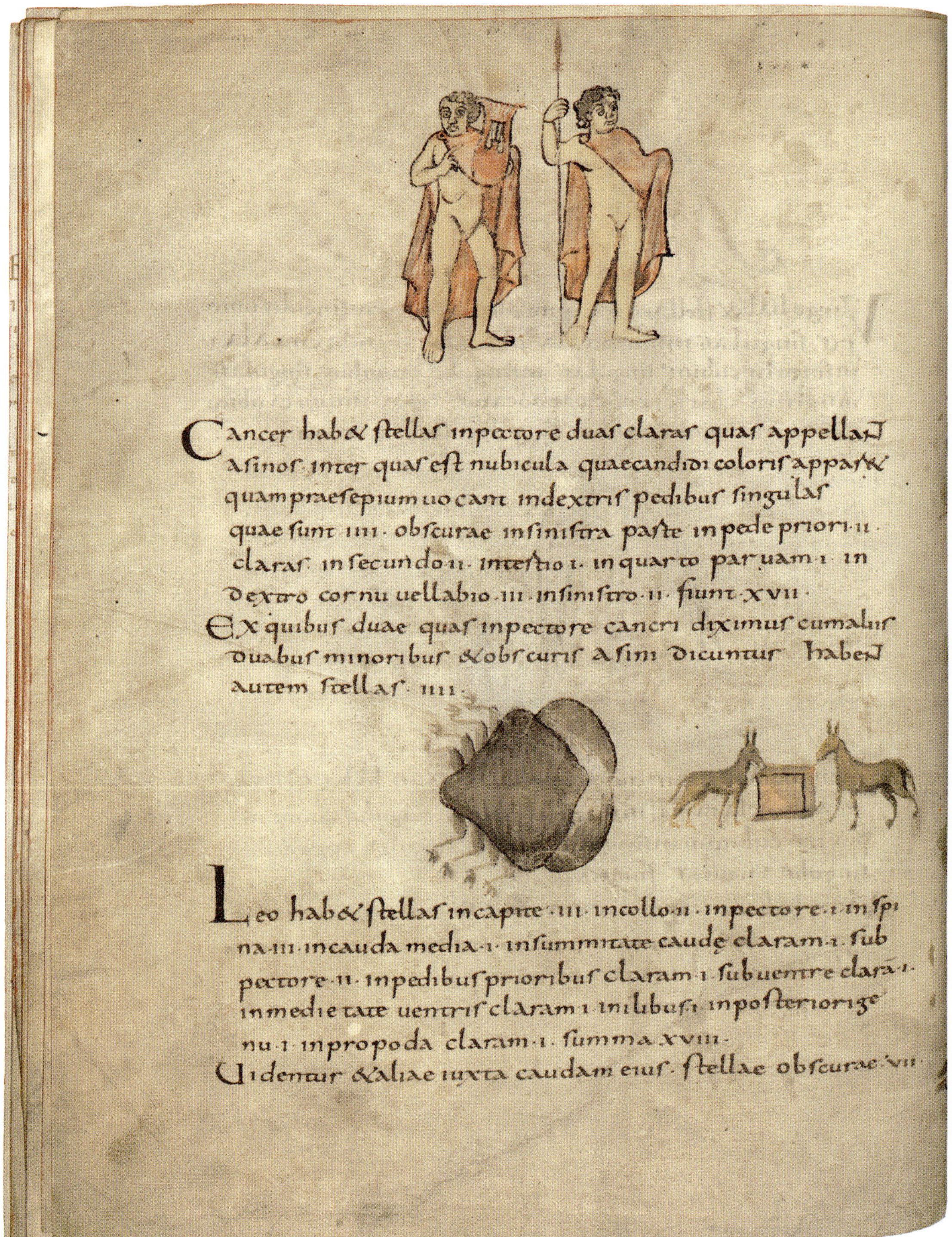

Cancer hab& stellas inpectore duas claras quas appellant
asinos. inter quas est nubicula quae candidi coloris appar&
quam praesepium uocant. indextris pedibus singulas
quae sunt IIII. obscurae. insinistra parte inpede priori II.
claras. insecundo II. intertio I. inquarto paruam I. in
dextro cornu uel labio III. insinistro II. fiunt XVII.
Ex quibus duae quas inpectore cancri diximus cum aliis
duabus minoribus & obscuris asini dicuntur. habent
autem stellas IIII.

Leo hab& stellas incapite III. incollo II. inpectore I. inspi
na III. incauda media I. insummitate caudę claram I. sub
pectore II. inpedibus prioribus claram I. subuentre claram I.
inmedietate uentris claram I. inlibus I. inposteriori ge
nu I. inpropoda claram I. summa XVIII.
Uidentur & aliae iuxta caudam eius stellae obscurae VII

PLATE 5

Gemini and Cancer. Madrid, Biblioteca Nacional de España, MS 3307, fol. 56v. Photo © Biblioteca Nacional de España.

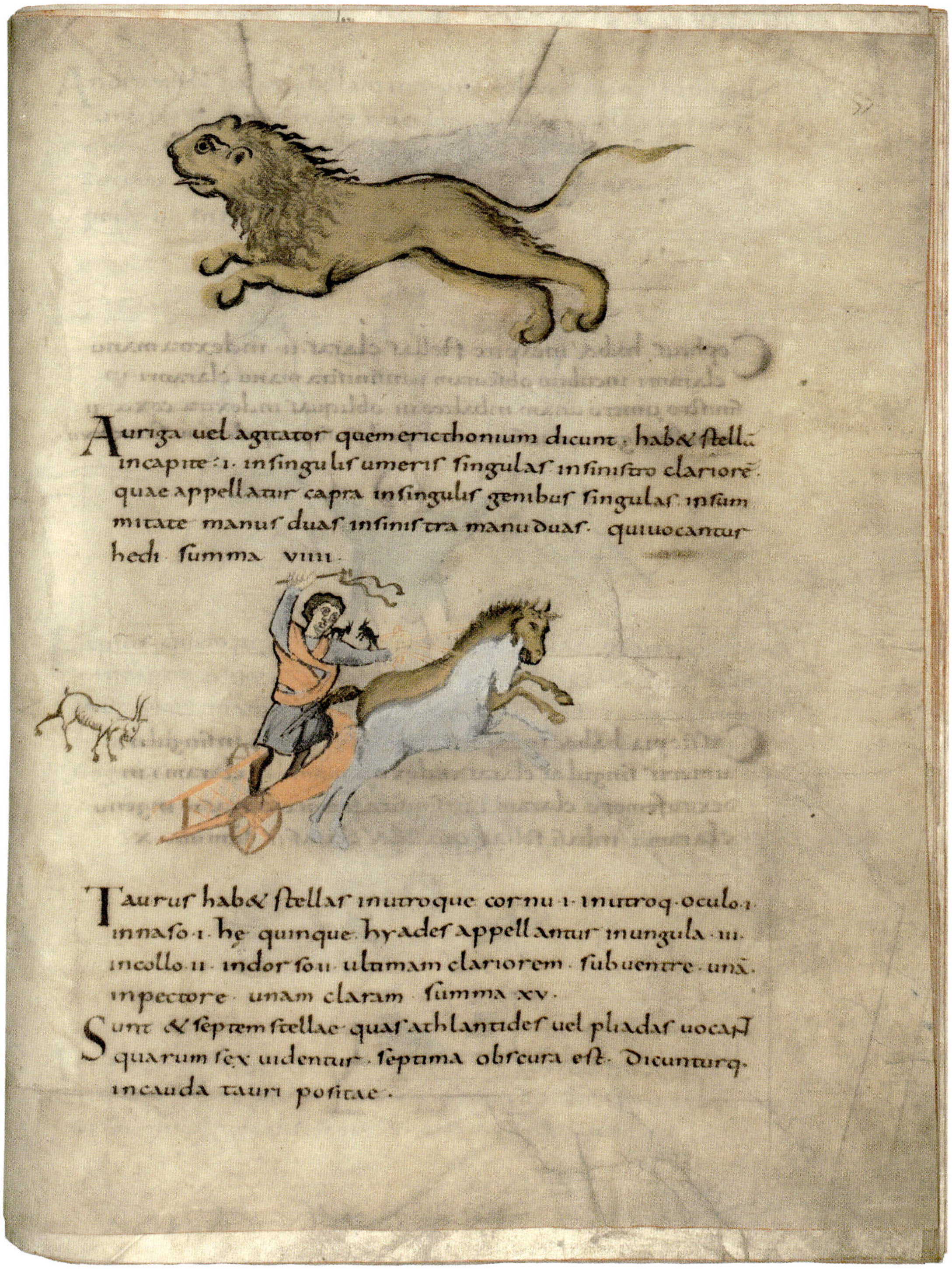
Auriga uel agitator quem ericthonium dicunt. hab& stellā
incapite .i. insingulis umeris singulas insinistro clariorē
quae appellatur capra insingulis genibus singulas. insum
mitate manus duas insinistra manu duas. qui uocantur
hedi. summa VIIII.

Taurus hab& stellas inutroque cornu .i. inutroq. oculo .i.
innaso .i. hę quinque hyades appellantur inungula .II.
incollo .II. indorso .II. ultimam clariorem. subuentre unā
inpectore unam claram. summa XV.
Sunt & septem stellae quas athlantides uel pliadas uocant
quarum sex uidentur. septima obscura est. Dicunturq.
incauda tauri positae.

PLATE 6

Leo and Auriga. Madrid, Biblioteca Nacional de España, MS 3307, fol. 57r. Photo © Biblioteca Nacional de España.

PLATE 7
Taurus, Cepheus, and Cassiopeia. Madrid, Biblioteca Nacional de España, MS 3307, fol. 57v. Photo © Biblioteca Nacional de España.

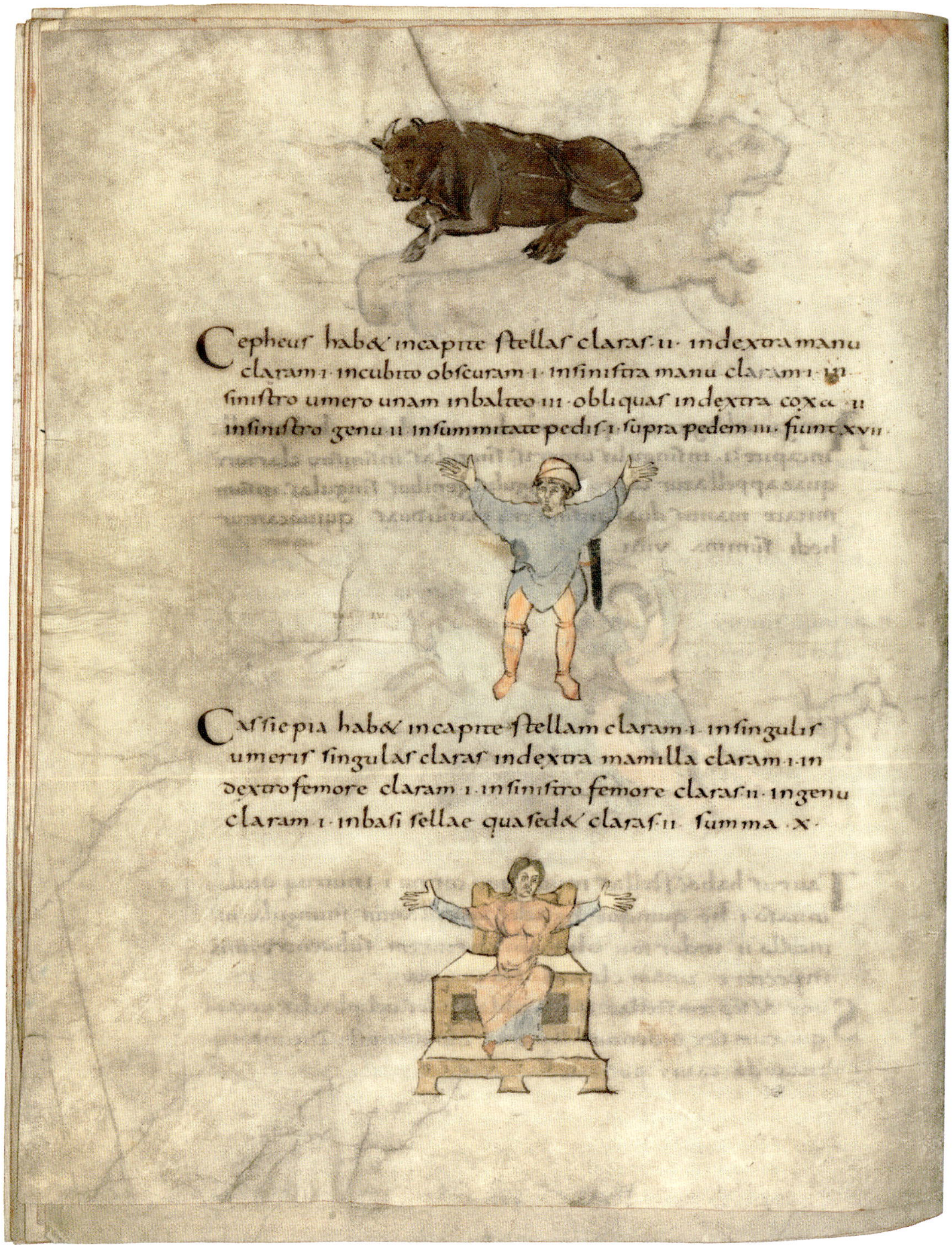

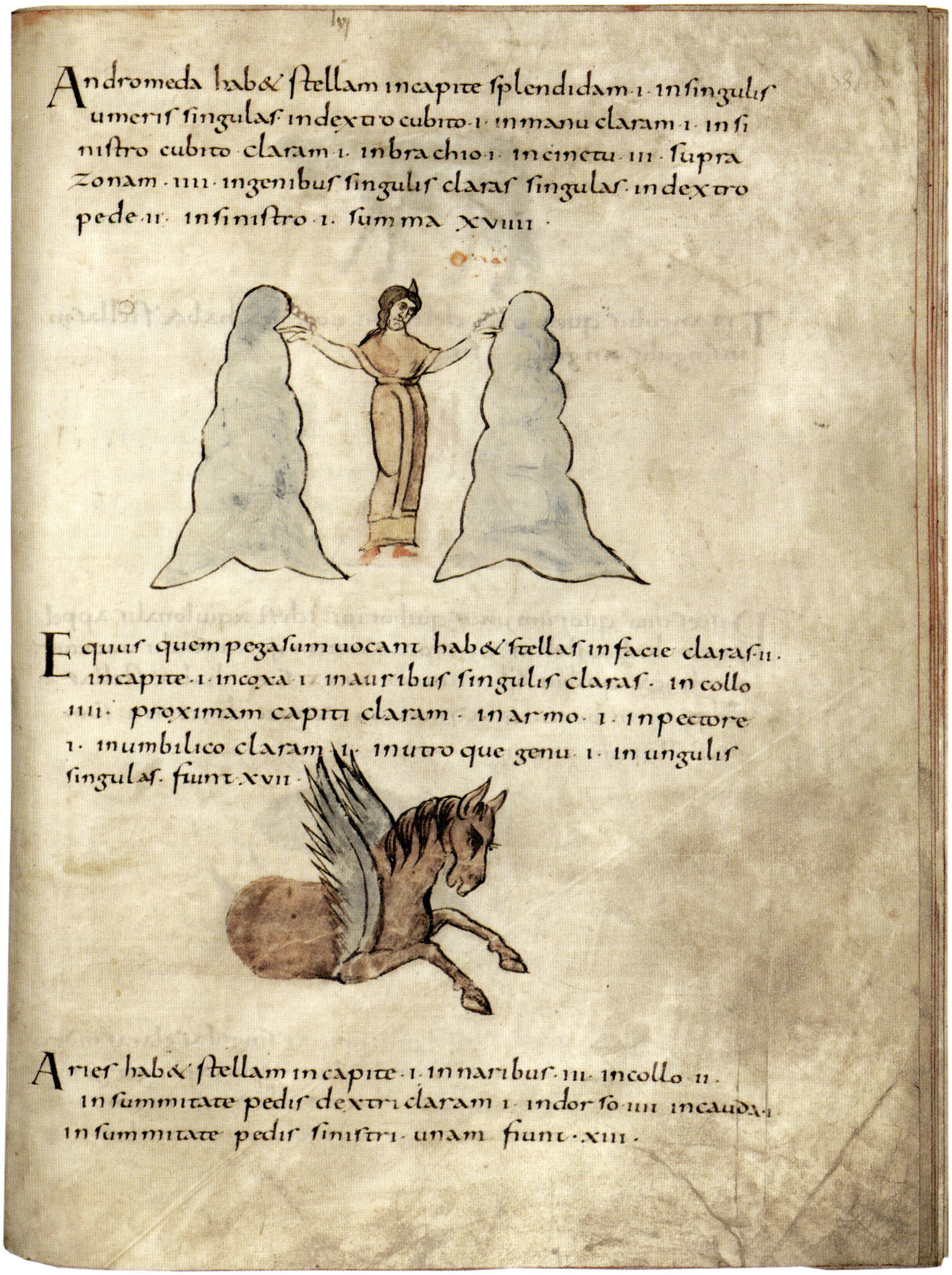

Andromeda hab& stellam incapite splendidam .i. insingulis umeris singulas. indextro cubito .i. inmanu claram .i. in sinistro cubito claram .i. inbrachio .i. incinctu .iii. supra zonam .iiii. ingenibus singulis claras singulas. indextro pede .ii. insinistro .i. summa .xviiii.

Equus quem pegasum uocant hab& stellas infacie claras .ii. incapite .i. incoxa .i. inauribus singulis claras. in collo .iiii. proximam capiti claram. inarmo .i. inpectore .i. inumbilico claram .i. inutroque genu .i. in ungulis singulas. fiunt .xvii.

Aries hab& stellam incapite .i. innaribus .iii. incollo .ii. insummitate pedis dextri claram .i. indorso .iiii. incauda .i. insummitate pedis sinistri unam fiunt .xiii.

PLATE 8

Andromeda and Pegasus. Madrid, Biblioteca Nacional de España, MS 3307, fol. 58r. Photo © Biblioteca Nacional de España.

PLATE 9

Aries, Deltoton, and Pisces (Aquilonalis and Australis). Madrid, Biblioteca Nacional de España, MS 3307, fol. 58v. Photo © Biblioteca Nacional de España.

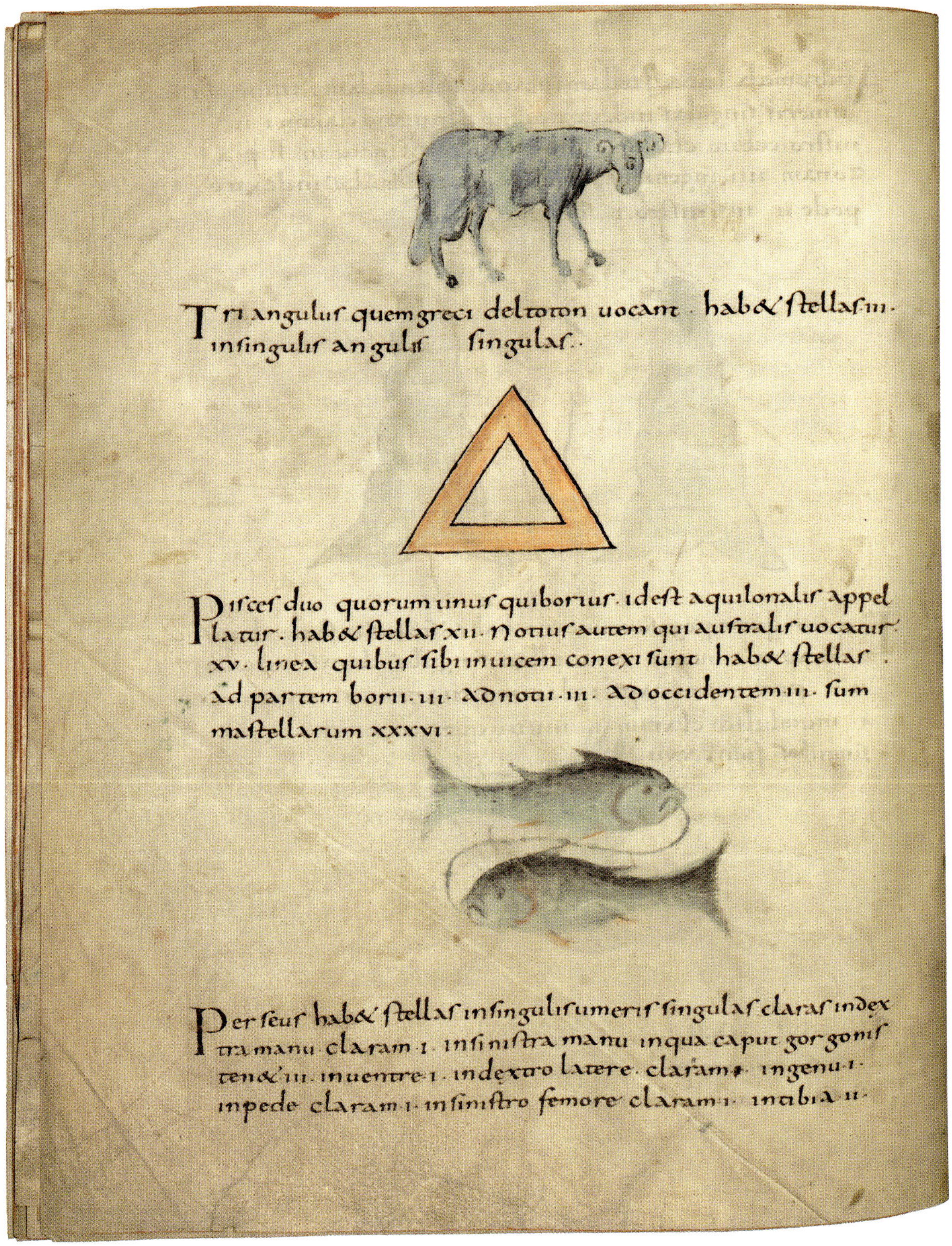
Triangulus quem greci deltoton uocant . hab& stellas iii .
in singulis angulis singulas .

Pisces duo quorum unus quiborius . id est aquilonalis appel
latur . hab& stellas xii . Notius autem qui australis uocatur
xv . linea quibus sibi inuicem conexi sunt hab& stellas .
ad partem borii . iii . ad notii . iii . ad occidentem iii . sunt
in stellarum xxxvi .

Perseus hab& stellas in singulis umeris singulas claras in dex
tra manu claram i . in sinistra manu in qua caput gorgonis
ten& iii . in uentre i . in dextro latere . claram i . in genu i .
in pede claram i . in sinistro femore claram i . in tibia ii .

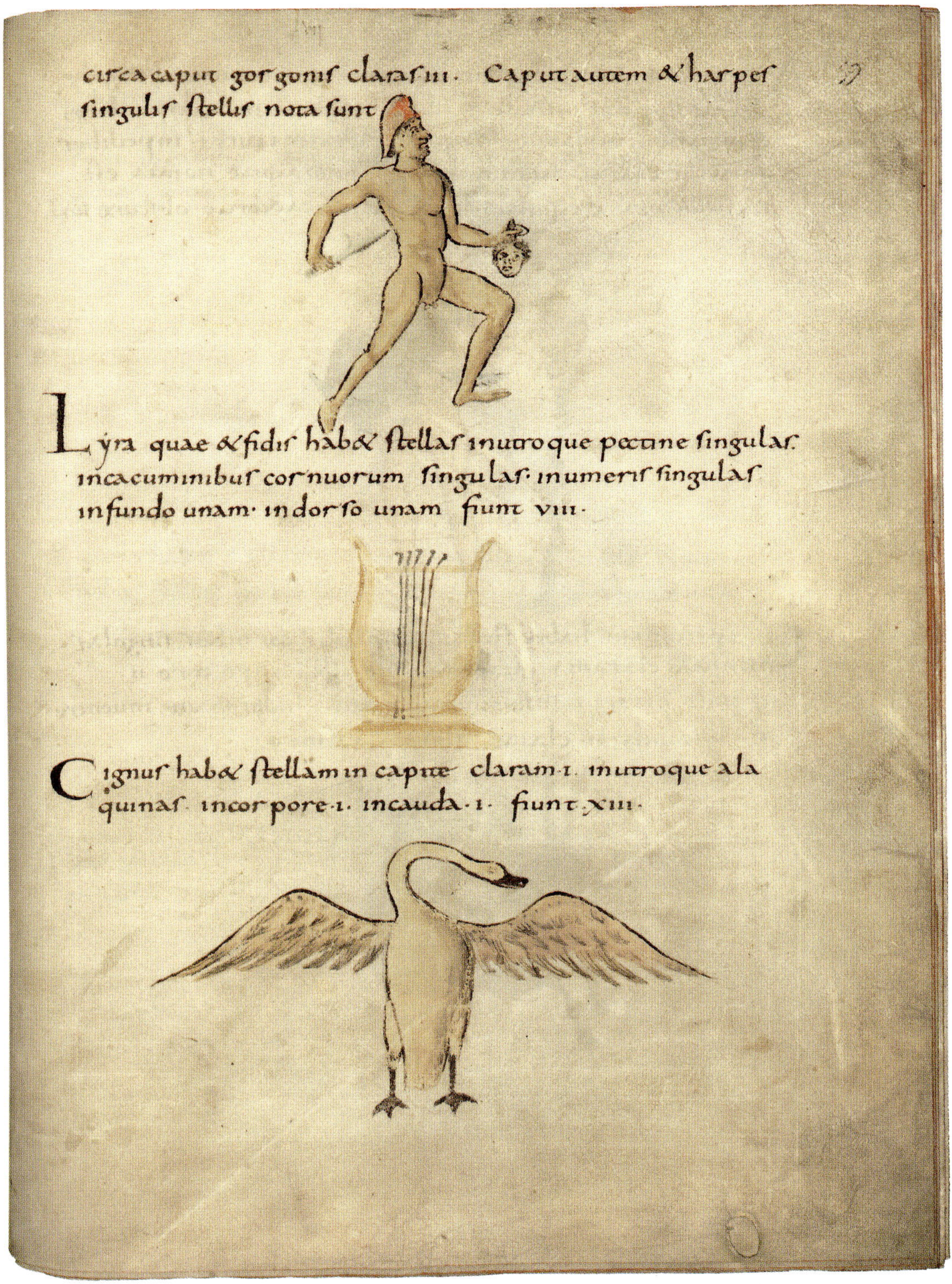

circa caput gorgonis claras iii. Caput autem & harpes
singulis stellis nota sunt

Lyra quae & fidis hab& stellas inutroque pectine singulas.
incacuminibus cornuorum singulas. inumeris singulas
infundo unam. indorso unam fiunt viii.

Cignus hab& stellam in capite claram .i. inutroque ala
quinas. incorpore .i. incauda .i. fiunt xiii.

PLATE 10

Perseus, Lyra, and Cygnus. Madrid, Biblioteca Nacional de España, MS 3307, fol. 59r. Photo © Biblioteca Nacional de España.

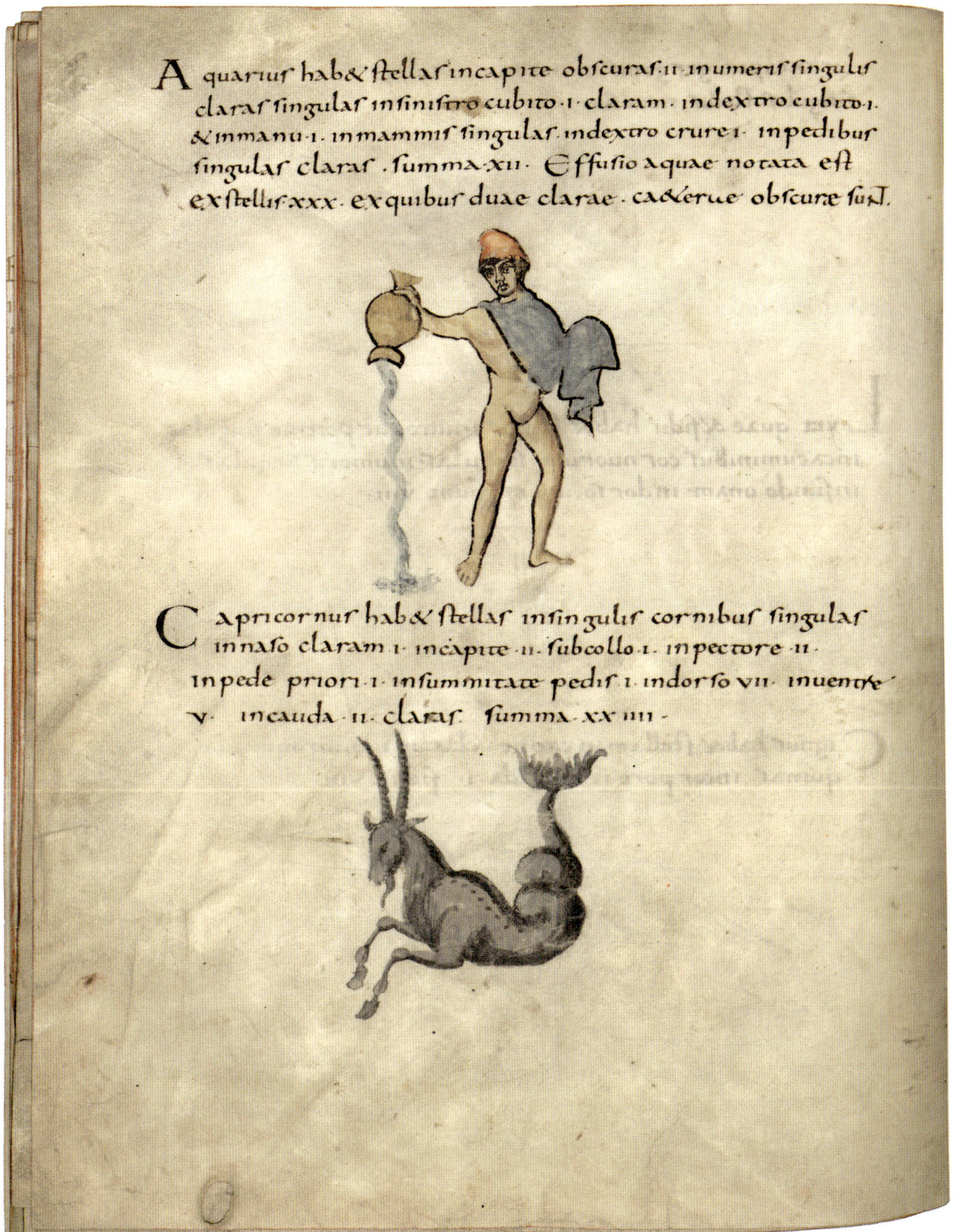

PLATE 11

Aquarius and Capricorn. Madrid, Biblioteca Nacional de España, MS 3307, fol. 59v. Photo © Biblioteca Nacional de España.

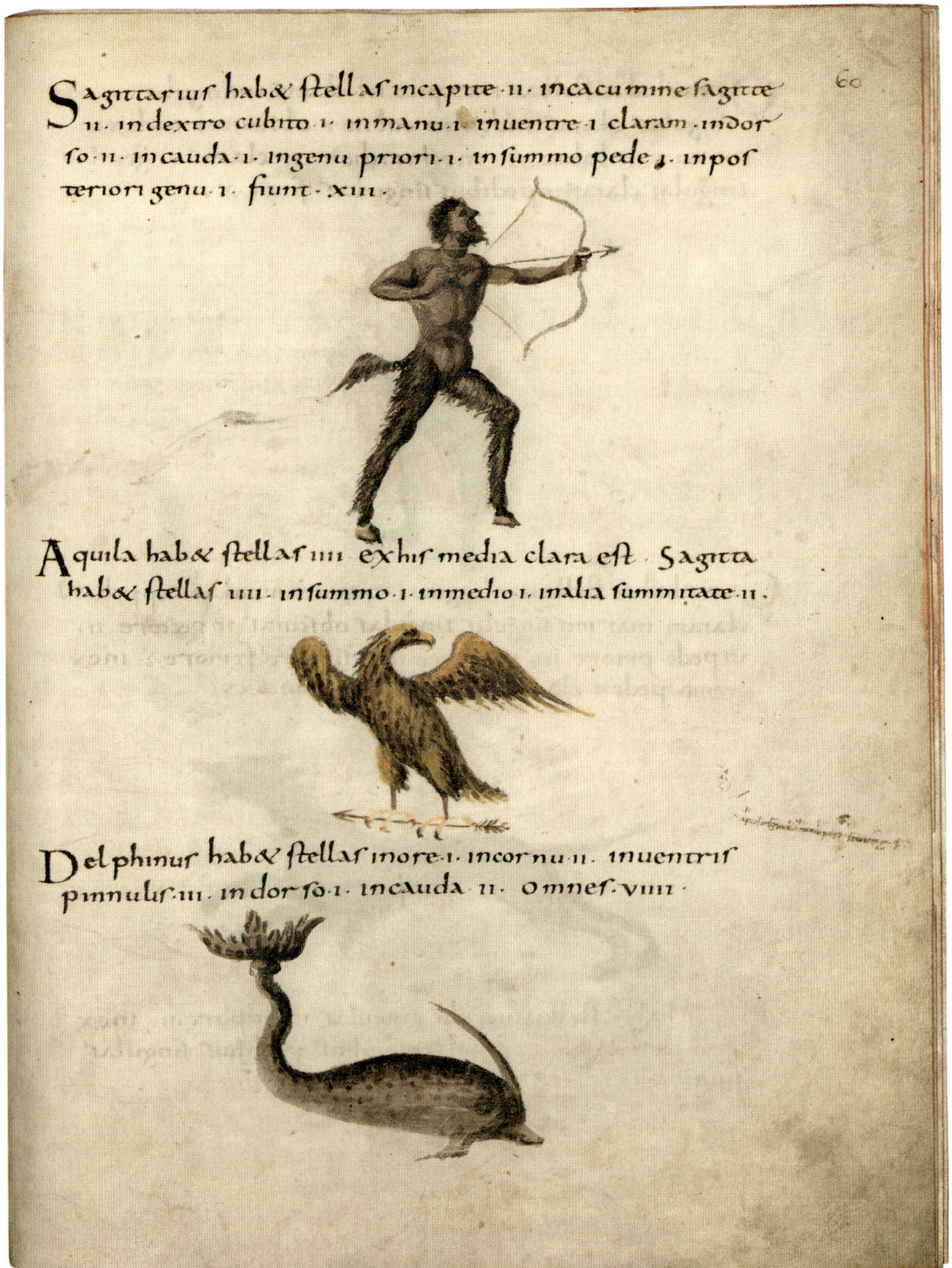

PLATE 12

Sagittarius, Aquila, and Delphinus. Madrid, Biblioteca Nacional de España, MS 3307, fol. 60r. Photo © Biblioteca Nacional de España.

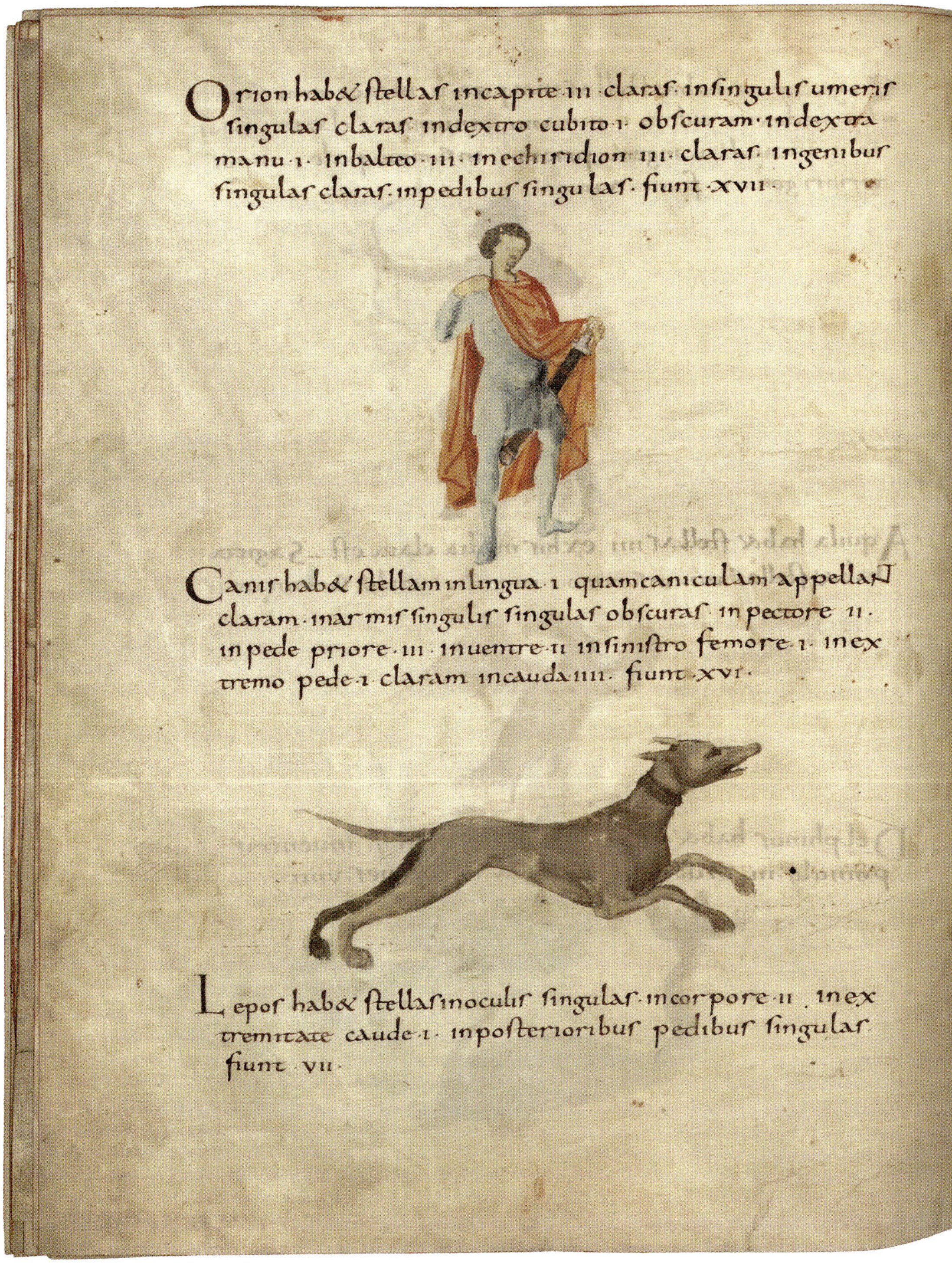
Orion hab& stellas incapite III claras. insingulis umeris singulas claras. indextro cubito I. obscuram. indextra manu I. inbalteo III. inechiridion III. claras. ingenibus singulas claras. inpedibus singulas. fiunt XVII.

Canis hab& stellam inlingua I quamcaniculam appellant claram. inarmis singulis singulas obscuras. inpectore II. inpede priore III. inuentre II. insinistro femore I. inextremo pede I claram. incauda IIII. fiunt XVI.

Lepus hab& stellas inoculis singulas. incorpore II. inextremitate caude I. inposterioribus pedibus singulas. fiunt VII.

PLATE 13
Orion and Canis Major. Madrid, Biblioteca Nacional de España, MS 3307, fol. 60v. Photo © Biblioteca Nacional de España.

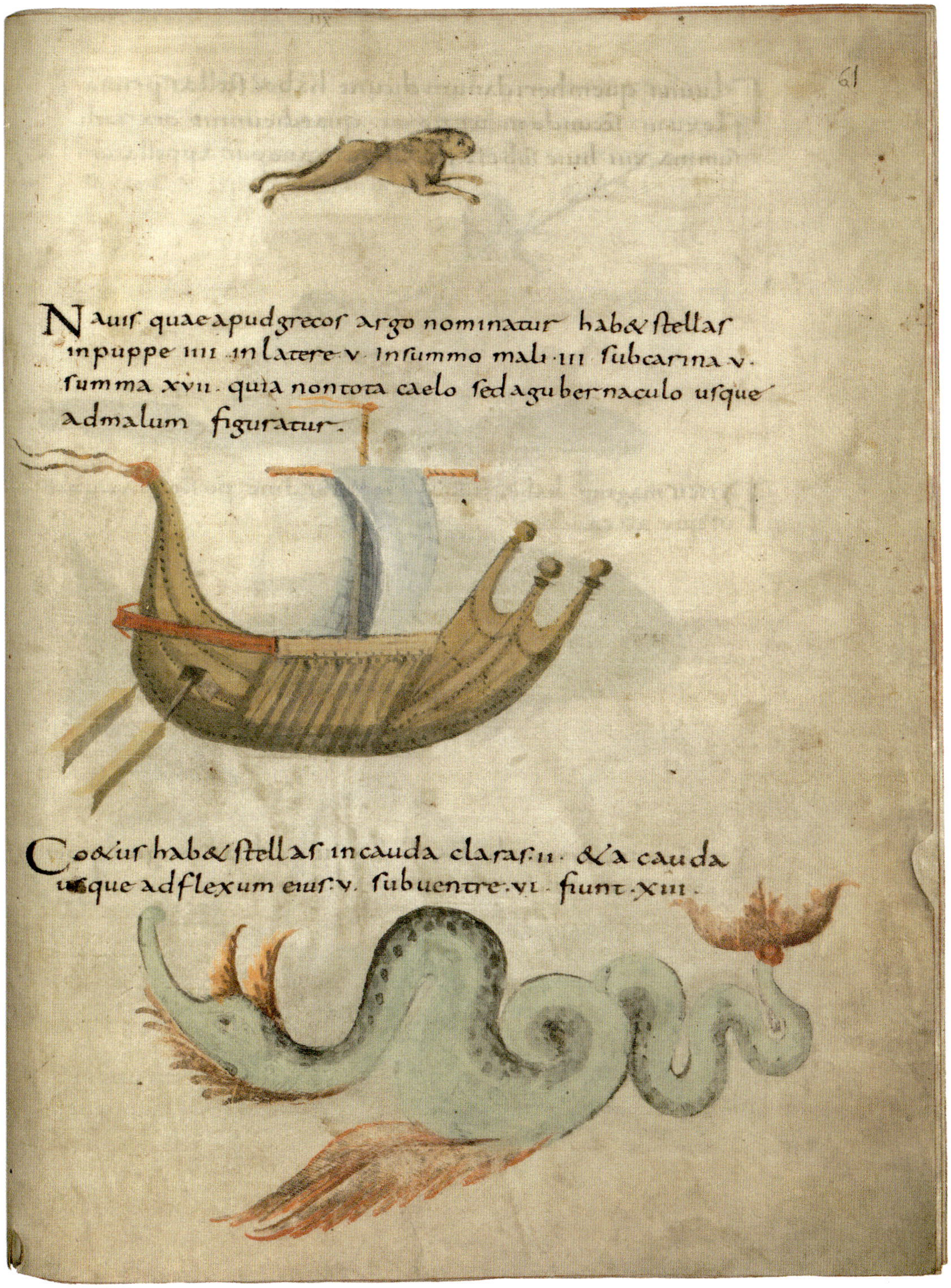
61

Nauis quaeapudgrecos argo nominatur hab& stellas
inpuppe iiii inlatere v insummo mali iii subcarina v.
summa xvii. quia nontota caelo sedagubernaculo usque
admalum figuratur.

Coetus hab& stellas incauda claras ii. &a cauda
usque adflexum eius v. subuentre vi. fiunt xiii.

PLATE 14

Hare, Argo, and Coetus. Madrid, Biblioteca Nacional de España, MS 3307, fol. 61r. Photo © Biblioteca Nacional de España.

PLATE 15

Heridanus/River Po, Piscis Magnus, and Ara/The Shrine. Madrid, Biblioteca Nacional de España, MS 3307, fol. 61v. Photo © Biblioteca Nacional de España.

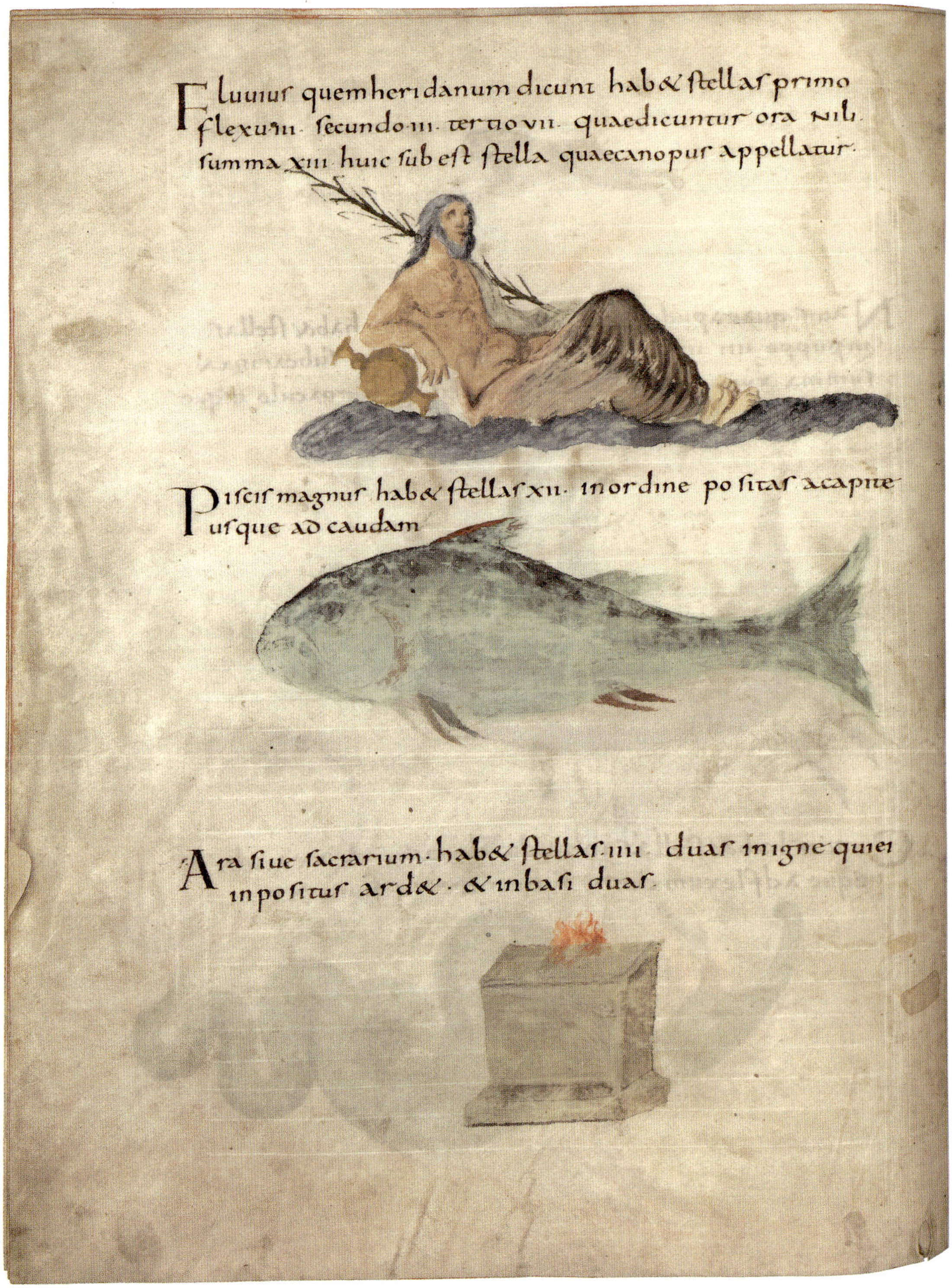
Fluuius quem heridanum dicunt hab& stellas primo flexu iii· secundo iii· tertio vii· quae dicuntur ora nili· summa xiii· huic sub est stella quae canopus appellatur·

Piscis magnus hab& stellas xii· in ordine positas a capite usque ad caudam

Ara siue sacrarium· hab& stellas iiii· duas in igne quiei inpositus ard&· & in basi duas·

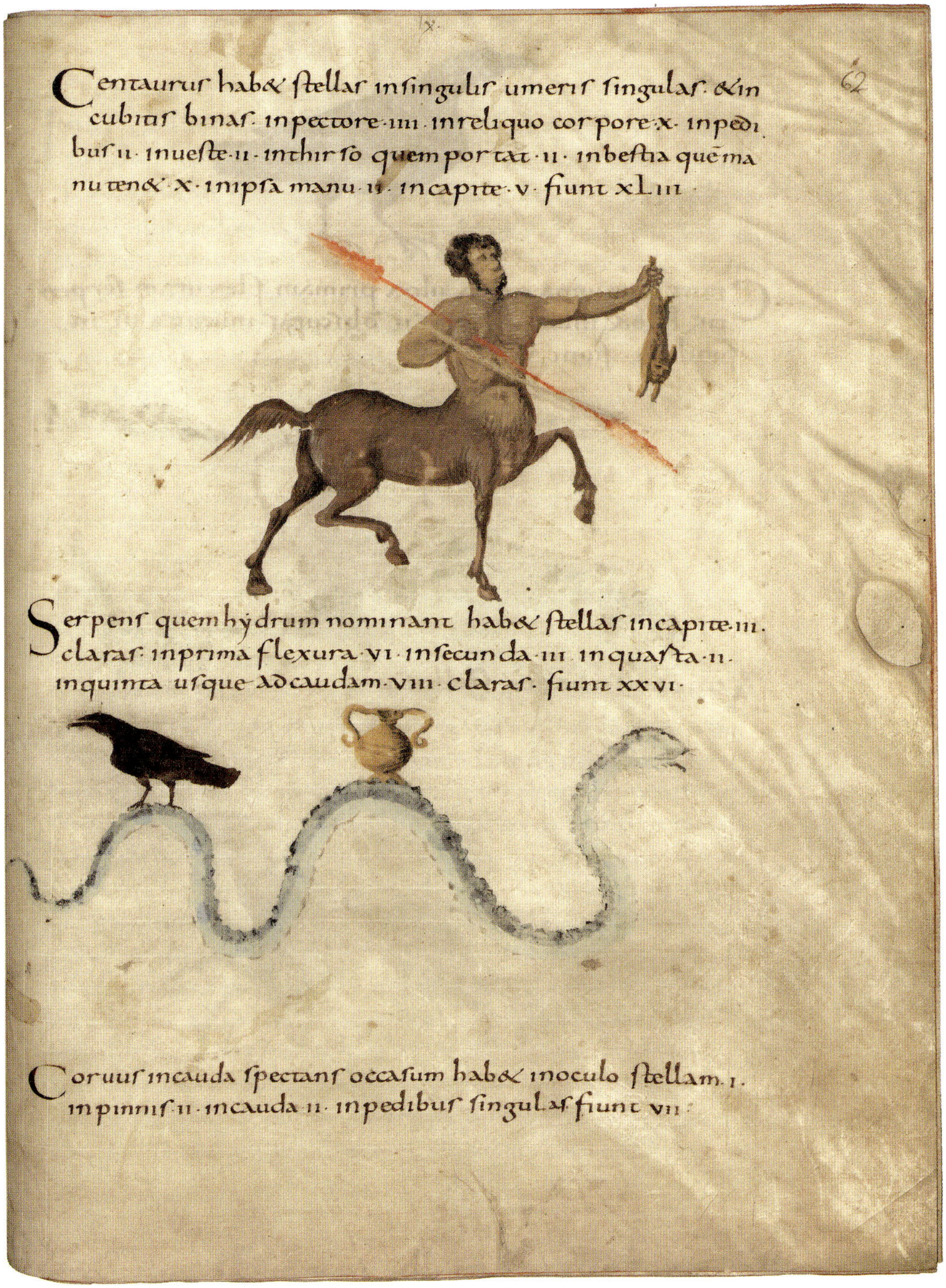
Centaurus hab& stellas in singulis umeris singulas. & in
cubitis binas. in pectore iiii. in reliquo corpore x. in pedi
bus ii. in veste ii. in thirso quem portat ii. in bestia quē ma
nu ten& x. in ipsa manu ii. in capite v. fiunt xliii.

Serpens quem hydrum nominant hab& stellas in capite iii.
claras. in prima flexura vi. in secunda iii. in quarta ii.
in quinta usque ad caudam viii. claras. fiunt xxvi.

Coruus in cauda spectans occasum hab& in oculo stellam i.
in pinnis ii. in cauda ii. in pedibus singulas fiunt vii.

PLATE 16

Centaur and Serpens/Hydra. Madrid, Biblioteca Nacional de España, MS 3307, fol. 62r. Photo © Biblioteca Nacional de España.

PLATE 17

Corvus/The Raven (or Crow), Crater/The Urn, and Anticanis/Canis Minor. Madrid, Biblioteca Nacional de España, MS 3307, fol. 62v. Photo © Biblioteca Nacional de España.

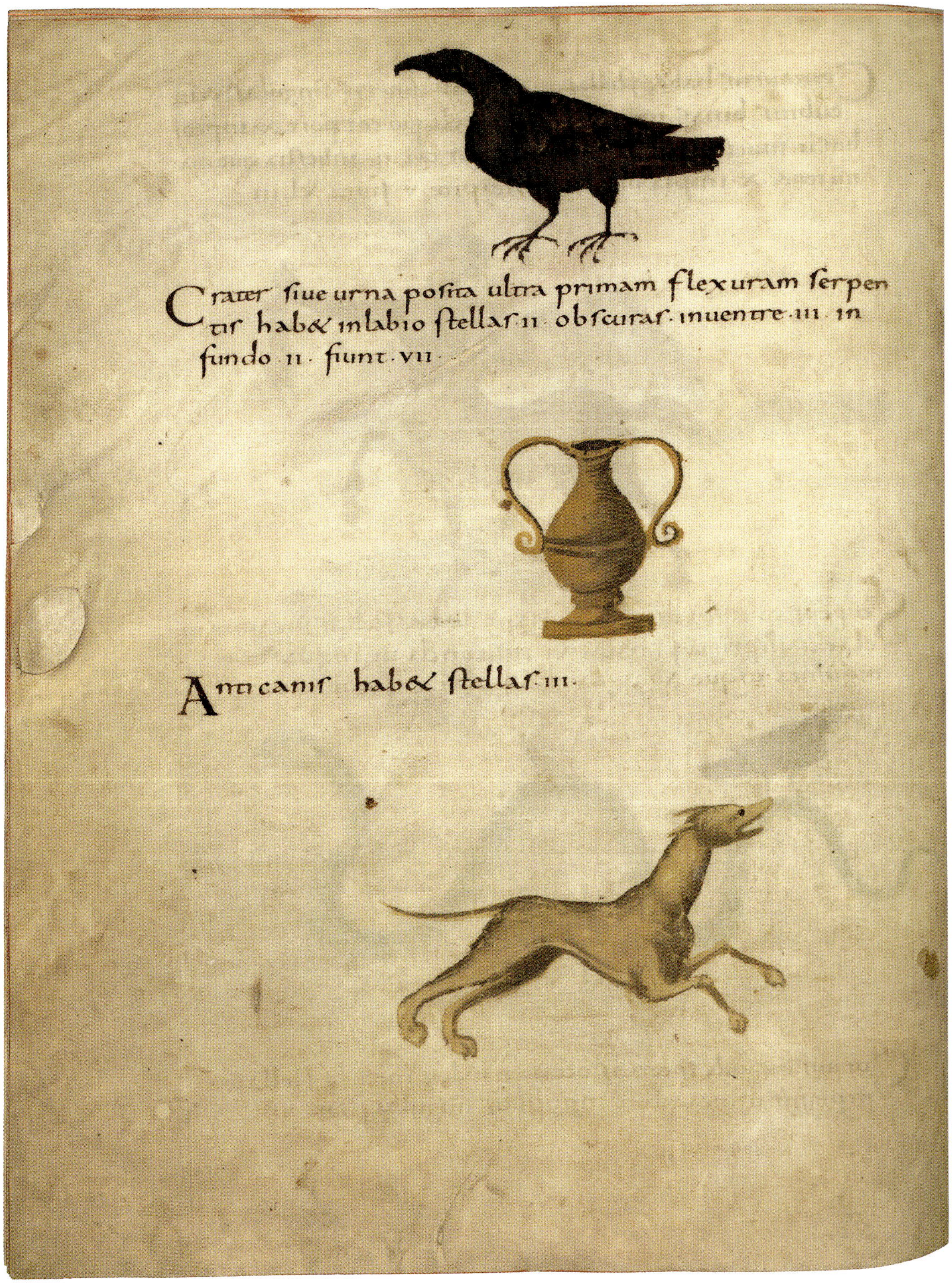

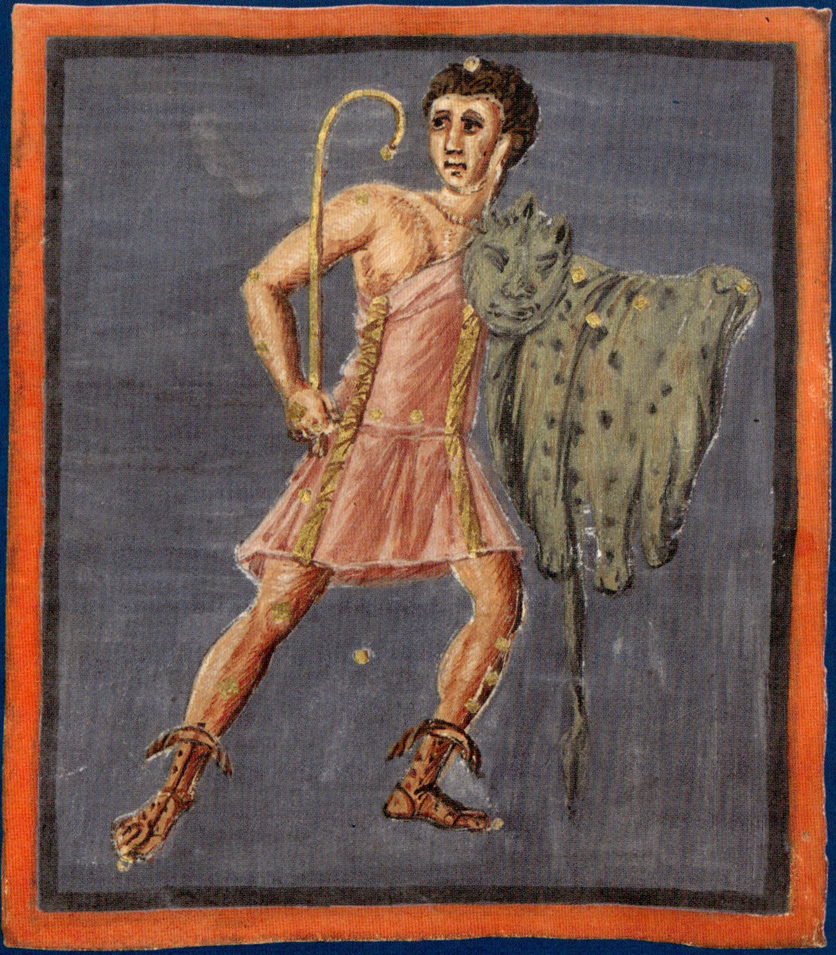

PART TWO

Representing the Cosmos for Carolingian Hearts and Minds

CHAPTER THREE

REVEALING ASTRONOMY

Itinerant Painters and Shifting Signs

A full appreciation of Carolingian aesthetics has often suffered from relatively terse and pedantically epistemological descriptions in Frankish texts. A Carolingian theological or pedagogical tractate is far more likely to be concerned with questions of belief or truth than modalities of sensory engagement with crafted things, so modern readers often fall back upon interpretative aspects of the Frankish image debate.[1] Although the semantic content expressed by Carolingian signs is addressed in chapter 4, the physical properties of paint also bore historic semantic import for Carolingian viewers and readers of the *Handbook of 809*. Considerations of style likewise emerge as an indication of the itinerant lives of some Carolingian scribes and painters, revealing the interconnected scriptoria of Austrasia as a longstanding transgenerational network of occasionally competitive but mostly collaborative skilled craftsmen primarily interested in the beautification of Ecclesia rather than individual foundations.

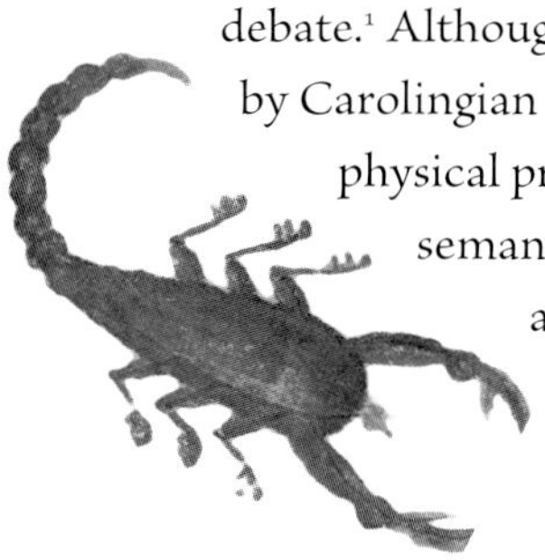

In part 1 of this book, a thorough examination of Bishop Drogo of Metz's relationship to his copy of the *Handbook of 809* (Madrid 3307) reveals that the complex history of his manuscript is intertwined with an emphasis on kinship in Carolingian rule and ecclesiastical responsibility.[2] The review of the visual and textual contents of Drogo's copy of the *Handbook of 809* in chapters 1 and 2 also offers a detailed record of the history of Madrid 3307 and its significance for Bishop Drogo. These chapters show that the cycle of images in Drogo's copy was derived from various Aratean traditions, with particular attention to zodiac wheels in the polycyclic process of image selection. The iconographic themes strongly point to Carolingian creativity and exegetical emendation at work in the pictorial cycle.

Here in part 2, the discussion turns from the significance of Madrid 3307 for Bishop Drogo to larger themes of import for Frankish painters and students of the liberal art of astronomy—themes suggested by this examination of Drogo's personal copy of the handbook. In the present chapter, I examine both the cycle of diagrams and the pictorial program illustrating the constellations to identify the aspects of Frankish creativity on display in Drogo's book. To contextualize this creative achievement, I provide brief introductory remarks to assess Carolingian ideas about pictorial imaging and the significance of painting for Frankish culture. Drogo's copy of the *Handbook of 809* is above all the achievement of four painters working in Metz. Emphasizing the itinerant careers of these artists requires an investigation into the connections between the images in Drogo's handbook and those in other Carolingian manuscripts or on celestial globes, resulting in a tentative reconstruction of the portions of the archetypal *Handbook of 809* about which it is possible to speculate.

Taking raw material and rendering it meaningful had profound implications for a theologically astute Frankish understanding of the rapport between the individual and the heavens. The cosmological aspects of this important topic, discussed in chapter 4, presuppose a material substrate underpinning the intrinsic harmonies of the celestial spheres. Mediated material models of these relationships resulted, therefore, in novel Plinian diagrams (discussed below) linked to the archetypal original *Handbook of 809* and also to subsequent Carolingian copies of the handbook. The physical act of medieval book manufacture, transmuting the skin of a calf into pristine vellum ready for the written word, was already a metamorphosis that took the fecundity of the Creation and, through human intervention, stored portions of discerned truth for safekeeping in the potent magical imprint of script.[3] In a related way, the revelation of Christ in the gospel text of John 1:14 as the "Word [that was] made flesh" established a metonymic connection among the Incarnation, a taut parchment folio, and a crystalline ideal of inscribed textual belief for medieval readers.[4] Flesh was the vehicle for truth, serving the apostolic mission and supplying a ready membrane ruggedly suitable to the dissemination of learned texts as well as to repeated consultation.[5]

Drogo's elevation to his episcopal see made possible the development of a local painterly and (as already discussed) scribal style at Metz beginning in 823.[6] Whether this painterly activity was centered at the Carolingian royal necropolis associated with the monastery of Saint Arnulf or linked directly to the cathedral is a matter of controversy, although perhaps Occam's razor would slice in favor of Saint-Étienne.[7] It is important in this context not to lose sight of Drogo's significance as a patron and recipient of illuminated manuscripts. In his early treatment of the manuscripts from Metz, Wilhelm Koehler assesses the dearth of luxury codices prior to Drogo's elevation.[8] This event was recorded in the *Annales regni Francorum* (*Royal Frankish Annals*) for 823: "[Louis] made his brother Drogo, who was leading the life of a monk, head of the church of Metz upon the consent and election of the clergy of that city and believed that he should be made bishop."[9] Although losses during World War II cloud the picture of early manuscript production in Metz, the scant evidence that does remain suggests the lack of a thriving center rather than a stable scriptorium. There is the likelihood of manuscript activity during the tenure of Bishop Angilram (served 769–91), but that may be because he was also linked to the court chapel as *apocrisarius* (chaplain) as early as 784.[10] The presence of books in the early collections of the monastery or the cathedral does not say anything about their site of origin, even though it does indicate diverse library activity. The cathedral library laid the foundation on which Bishop Drogo could build in the performance of his liturgical duty to his see, cultivating a school for instruction in the liberal arts in early medieval Metz.[11]

At the end of the records for the year 823 in the royal Frankish annals, omens and portents were adduced as occurrences worthy of remembrance. Alongside reports of hailstones large enough to be confused with rocks, an extraordinary excess of lightning, and a twelve-year-old from Commercy who fasted for ten months, there is the narrative description of the apotheosis of Gravedona:[12]

> Near the Italian city of Como, in the village of Gravedona, there was a picture painted in the apse of the church of St. John the Baptist of Holy Mary holding the infant Jesus in her lap and of the Magi offering presents that was dimmed and almost wiped out with age. This picture shone for two days with such clarity it seemed to viewers that its ancient beauty almost surpassed the splendor of a new picture. But the same clarity did not brighten the images of the Magi except for the presents which they offered.[13]

Granted, the miraculous nature of the revelation at Gravedona renders the example somewhat inapplicable to the quotidian aspects of Carolingian viewing experiences. The emphasis on ocular experience that was integral to the record, however, contributes to our understanding of how human beings engaged with painting and painted forms in the ninth century.[14] Several aspects of this recorded event help illuminate key features of the ways in which Carolingians understood their viewing experiences. Worshippers in Como (and apparently Aachen) recognized the Adoration scene's disrepair. In addition, the omen manifested metaphysical confirmation of a supernatural hierarchy, juxtaposing the refulgent Virgin and Child with the shadowy figures of the Magi, withdrawn into the obscure reserve of memory. Only the Magi's gifts emerged from the ground into the light through the transformative properties of paint. The gold, frankincense, and myrrh had become one with the Holy Family they were intended to serve, and a portion of their blessing had therefore effected its magic along a reverse path to these presents (a blessing that was not extended to their bearers or the long-forgotten makers or retailers of such items in the East).

This view of Carolingian aesthetic experience fits nicely with the terse description of painting in Hrabanus Maurus's encyclopedia *De universo* (or *De rerum naturis*, ca. 842–46): "Pictura est imago exprimens speciem rei alicujus, quae dum visa fuerit, ad recordationem mentem reduxit" (A picture is an image expressing the appearance of a thing, such that when seen restores the mind to [a state of] recollection).[15] The image of the Madonna and her child, effaced over time at Gravedona, received the momentary renewal that was possible when a special light activated its form, restored the status of imagery to the holy figures, and permitted awareness to resurface from the recesses of forgetfulness. Recollection, reverie, and restoration were all intertwined components of Carolingian image theory, and they play a vital role in a proper understanding of the ways the Franks understood their star pictures.

As was common practice for Hrabanus, he took his definition of the picture almost verbatim from Isidore of Seville's *Etymologies* of circa 636.[16]

Curiously, Isidore (XIX.xvi.1–2) immediately proceeds to evaluate the utility of the imagery vis-à-vis pervasive epistemological considerations, ever intertwining the efficacy of painting with concerns about objectivity and warranted intellectual assent. Mary Carruthers is right to draw here a fine distinction between mimetic worries about beds that function as or look like beds, on the one hand, and an intellectual appreciation of relatively trustworthy bed-pictures, on the other. For Isidore, only the latter are significant. In the Isidorean account, a good image is a well-constructed summary, operating according to its own visual cues and principles, laying a visual pathway to recollection: "the 'good' of a picture, its underlying aesthetic principle, is thus understood in terms of its role in cognitive function."[17]

None of this enters into the account offered by Hrabanus, who moves on in his next section to discuss the symbolic significance of discrete hues, expanding on Isidore's introductory remarks and personalizing them as an exegete interested in the spiritual development of a specifically Carolingian audience. Hrabanus's text on colors (*De coloribus*, book XXI.10) follows Isidore through book XIX.xvii.1–2 of the *Etymologies*, recounting a division between man-made pigments and those found throughout the world. The geographic localization of pigments and their hues concerns Isidore for the remainder of book XIX.xvii.[18] Hrabanus opts instead to identify the symbolic properties of colors. These should be taken as his personal comments on the topic, even if they perhaps derive in part from other, unidentifiable sources:[19]

> But, the diverse colors signify different aspects of virtue, by which the image of God is adorned in the likeness of the Creator, as the redness of red ochre or of minium may reveal in the ardor of charity, and purple pigment reveals in the name of Christ the things suffered with regard to the martyrdom and the Passion, in orpiment is revealed the splendor of wisdom, in cerulean sky blue [*venetum*] and sapphire are revealed the heavenly realm, in ceruse [or white lead, *cerus(s)a*] is revealed the purity of chastity, and all the other kinds of virtue, as each is to be conveyed at an appropriate moment of time, and as the rule of order requires.[20]

Colors themselves had the power to convey symbolic content, making them semiotically charged. Through Christian traditional schemes of association rather than exemplification, the colors of the rainbow revealed the manifold work of God in the Creation. For Hrabanus, this was an important aspect of Frankish painting. In addition to the debate over the proper use of imagery, pictures were charged with an illustrative value that was linked to their homiletic potential. Just as a color like sapphire blue could indicate the celestial realm frequented by God—a point also made by Bede (d. 735)[21]—so the astronomical pictures in the *Handbook of 809* could participate, through education and exegesis, in larger conceptual networks aimed at the recollection of the divine aspects of astronomy. The colors and forms of the diagrams and star pictures in Drogo's copy of the *Handbook of 809* are also highly informative about the archetypal nature of its imagery and the stylistic import of the paintings for the development of a regional center of artistic renown at Metz. These stylistic conventions became ways to infuse images with rich layers of aesthetic meaning beyond straightforward considerations of content.

Princely Pictures: Charting the Heavens in the *Handbook*

As seen in the preceding remarks, the ardent Frankish effort exerted to visualize the heavens relied on classical tradition, ninth-century innovation, and painterly acumen. The joint pursuits of scientific accuracy and pleasing artistic creation in the archetypal *Handbook of 809* resulted in the inclusion of four additional illustrated Plinian texts clustered in book V. These diagrams should be considered illuminated corollaries of the star pictures that precede them. Taken together, they offered the ninth-century reader a comprehensive presentation of the aspects of astronomical and cosmological theory that were considered accurate and beneficial for young Christian clerics in training. The fact that the most important copies of the *Handbook of 809* have been linked to Carolingian prelates who had kinship ties to the Frankish royal family but were elevated to clerical dignities only underscores the joint devotional and intellectual import of the schematics accompanying book V.3–6. The Plinian excerpt from book XVIII of the *Naturalis historia*, which became book V.12 (the *De presagiis tempestatum* in the handbook), had contributed to the Carolingian innovation of the novel form of Cancer the Crab in Drogo's copy of the *Handbook of 809*. Four additional excerpts from book II of Pliny's *Naturalis historia* similarly received novel Carolingian diagrams:[22]

Book V.3, *De positione et cursu vii planetarum* with the diagram of "The Order of the Planets" (fol. 63v; fig. 3)

Book V.4, *De intervallis earum* with the diagram of "The Pythagorean Theory of the Harmonic Intervals" (fol. 64r; fig. 4)

Book V.5, *De absidibus earum* with the diagram of "The Apsides of the Planetary Orbits Within the Zodiac" (fol. 65v; fig. 5)[23]

Book V.6, *De cursu earum per zodiacum circulum* with the diagram of "The Latitudes of the Planets Through the Zodiac" (fol. 66r; fig. 6)

These diagrams, which were included in the original *Handbook of 809*, given the record of the table of contents provided in the introduction to the present volume, attest to the scientific acumen of the prelates in attendance at the synod of 809 and to the artistic creativity of those trained in the liberal art of astronomy at Charlemagne's court. The second master painter, who was responsible for the best miniatures in Drogo's copy of the *Handbook of 809*, such as those of Capricorn (plate 11) and Sagittarius (plate 12), also painted the copies of the diagrams in the manuscript. Stylistic comparison reveals that the personifications of the planets in the diagram of the geocentric model of the solar system on folio 63v (fig. 3) resemble the faces of the Centaur (plate 16), Heridanus (plate 15), and Orion (plate 13) in the pictorial cycle of miniatures. The pouty lips, bulbous noses, and heavy eyebrows were all stylistic traits of the second master painter, who accomplished splendid miniatures but was actually less skilled at rendering faces than his colleague, the first master painter in Drogo's book. The representations of Chiron, the Centaur (plate 16), and Orion (plate 13) share these characteristics with the personifications of Sol (the sun) and Venus in the diagram (fig. 3), while the heavy, slanted eyebrows of Heridanus (plate 15) resemble those of Mars in the diagram (fig. 3). These considerations are highly significant. The attribution of the best paintings and diagrams to the same master painter demonstrates that the diagrams were not ancillary additions; rather, they were considered valuable artistic contributions that enhanced the overall quality of Drogo's copy of the *Handbook of 809* and elevated it to *de luxe* status.

According to Bruce Eastwood's careful analysis, the excerpts from Pliny in book V of the *Handbook of 809* treat planetary theory as a complement to the descriptions of the constellations in the *Excerptum* and the *De ordine*. As a result, the handbook moves from the relative placement of the constellations in the *Excerptum* (book V.1) to the list of constellations proper in the *De ordine* (book V.2). Then, book V proceeds to examine the order of the planets between the earth and the fixed stars in V.3, *De positione et cursu vii planetarum*.[24] For this reason, the editors of the handbook omitted an independent Aratean image of the planets, which ordinarily followed the discussion of the constellations in other Aratean manuscript recensions.[25] This removal was already a creative editorial decision resting with Adalhard of Corbie and the other Carolingian redactors of the original *Handbook of 809*. In so doing, they demonstrated a preference for Pliny's geocentric study of planetary theory as they compiled the passages and added informative diagrams to illustrate the texts.[26]

In "The Order of the Planets" diagram in book V.3, the order of the planets proceeds from the earth at the center outward toward the sphere of fixed stars: Earth, Moon, Mercury, Venus, Sun, Mars, Jupiter, Saturn (fig. 3). In the geocentric model of the solar system, the moon and the sun were counted among the wandering stars, as opposed to the fixed stars, and therefore labeled the *errantia*, or the planets. Pliny noted, however, that "none wanders less than these," affirming the predictability of their motion.[27] The planets were adduced according to the Chaldean system derived from the classical authority of the *auctores* Archimedes and Cicero, according to the formulation presented in Macrobius's widely diffused *Commentarii in somnium Scipionis* (*Commentary on the Dream of Scipio*, ca. 430–40).[28] During the Carolingian period, in the years 811–62, Dungal, Lupus of Ferrières, and Heiric of Auxerre all supplied additional notes and qualifications to the presentation of early Christian astronomy and cosmology contained within Macrobius's development of the relatively brief vatic comments from the end of book VI of the *De re publica* (52 B.C.E.) by Cicero (d. 43 B.C.E.).[29] Macrobius himself preferred an alternative arrangement attributed to Egyptian precedent: the moon, the sun, Mercury, Venus.[30]

The much-discussed planetary configuration on folio 93v of the Leiden *Aratea* provided an alternative and revised Carolingian view of the solar system, setting Mercury and Venus in concentric, heliocentric orbit about the sun (fig. 56). This view had also been advanced by Martianus Capella (fl. ca. 475) in book VIII.857 of the *De nuptiis Philologiae et Mercurii* (On the marriage of Philology and Mercury), although, as Eastwood has argued, the insertion of Plinian passages concerning the exaltations of the planets into the diagram on folio 93v makes it an interesting hybrid document, attesting to the Carolingian predilection for Plinian rather than Capellan astronomy prior to midcentury.[31] By the third quarter of the ninth century, there were competing views not only about the order of the planets in the cosmos but also about their deployment and arrangement. The geocentric diagram in the *Handbook of 809* offered the simplest and most straightforward initial foray into planetary astronomy for Frankish students (fig. 3). This ideal, pristine geometry reflected the prevailing view in 809 about the embedded spheres that constituted the cosmos. As Eastwood argues, the initial diagram in the *Handbook of 809* was a heuristic introduction of limited ongoing pedagogical relevance, used and discarded as an ideal abstraction when the standard Plinian planetary order had been memorized. As called for by Hrabanus's definition of *pictura* above, the diagram reminded the reader of an authoritative, privileged point of view, but it also invoked understanding, supplying a convenient and easily recalled structural presentation of discernibly accurate information.[32] The well-constructed schematic encouraged recollection of the planetary order and fulfilled Frankish ideas about both astronomy and appropriately manufactured diagrams.

The diagram of the embedded circular orbits was created for the Plinian excerpt by Carolingian scribes and painters, but in the ninth century the circular structure had also achieved a certain dominance as an appropriate formulation of early medieval Christian scientific information. The concentric circles agreed in form with the wheel diagram, which harked back to Aristotelian precedent yet for Frankish artists was associated with the example of the *rotae* in Isidore of Seville's *De natura rerum* (see, for example, fig. 1).[33] As a book for teaching princes, the *Handbook of 809* was embellished with authoritative presentations of the heavens for readers such as Drogo.

In addition to the presentations of planetary order in the *Handbook of 809* and the Leiden *Aratea*, alternative models of planetary order came to be associated with a diagram found in Capellan manuscripts, such as a ninth-century diagram now in Paris (Bibliothèque nationale de France, MS lat. 8671, fol. 84r).[34] Although the creation of both the archetypal *Handbook of 809* and Drogo's individual copy predate the halcyon days of Capellan influence on Carolingian astronomy, it is important to note that later ninth-century exegeses of partial circumsolar

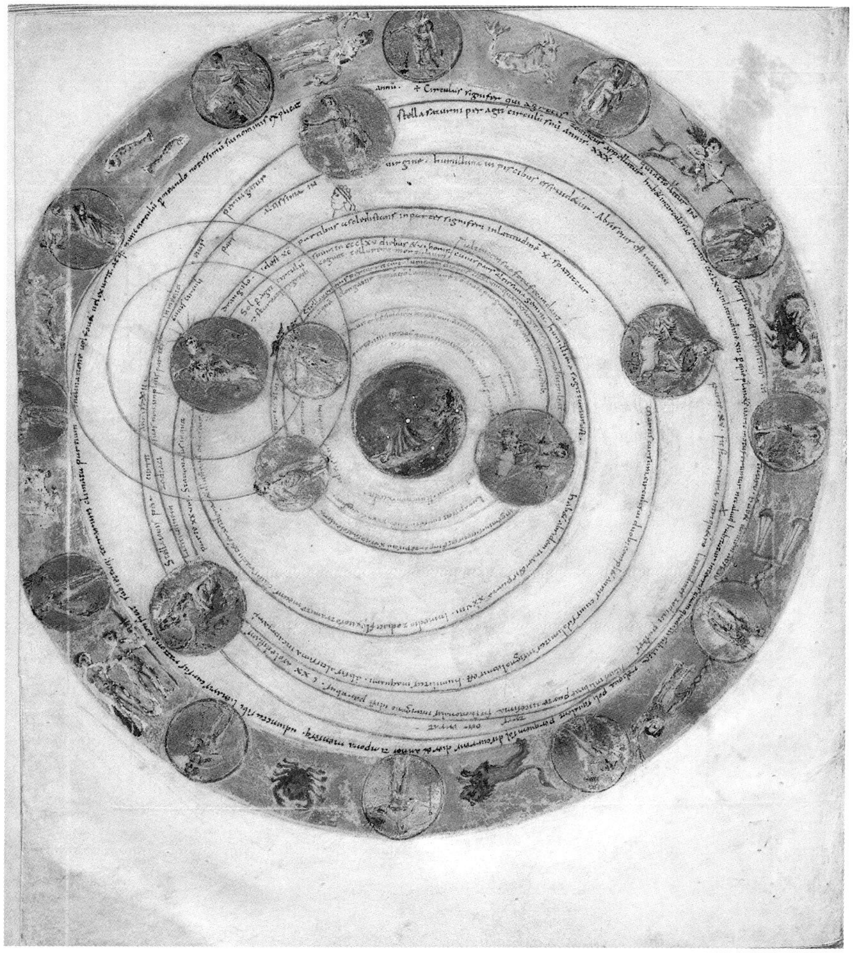

FIGURE 56

Planetary configuration. Leiden *Aratea*, Bibliotheek der Rijksuniversiteit, MS Voss. lat. Q. 79, fol. 93v. Photo courtesy of Universiteitsbibliotheek Leiden.

heliocentrism were derived from Martianus Capella's *De nuptiis*: "[N]ow Venus and Mercury, although they have daily risings and settings, do not travel about the earth at all; rather they encircle the sun in wider revolutions."[35] Interpretations of this straightforward assertion varied among Frankish scholars, however. There was a view attributed to Capella that called for twice-crossing, overlapping heliocentric orbits for both Mercury and Venus. Another view, alternatively attributed to Plato (via Calcidius) or Bede, claimed that the circumsolar orbits of Mercury and Venus were concentric and intersected nowhere (as in the Leiden *Aratea*; fig. 56). Finally, Pliny was pinpointed as the proponent of a view in which the orbits of Mercury and Venus crossed twice but always remained between the earth and the sun, like a pair of embedded soap bubbles surrounding the hot surface of Sol, as Eastwood explains.

As can be seen from this brief discussion, the ninth century experienced a renewed interest in creative and interpretative classical planetary theory that required exegetical approaches to textual analysis.[36] The astronomical texts at hand were pored over in search of novel interpretative frameworks that could clarify confusing issues about planetary theory for *astronomia* or computistics. The first diagram in Drogo's copy of the *Handbook of 809* recalls the beginning of this sophisticated enterprise but was quickly surpassed by further pedagogical refinements throughout the course of the ninth century.[37] The very fact that the Carolingian manuscript painters emended their diagrams and offered competing circumsolar heliocentric and geocentric views reveals that such supplements participated in a strategic unveiling of increasingly complex pictorializations of the cosmos. These diagrams fulfilled a straightforward "illustrative" role, to be sure, but they also drew upon the established visual tradition of the *rota*. In the *Handbook of 809*, the diagrams were thus a visual mark of accuracy, authenticity, and authority that resulted from the disciplined efforts of Carolingian natural philosophers and exegetes who attempted to understand their world better through focused and prolonged inquiry, supporting widespread belief in the cultural benefits of education (as argued in the introduction and in the work of John Contreni).[38]

The second diagram on folio 64r of Drogo's handbook supplies a straightforward illustration of the Pythagorean intervals described in book V.4, *De intervallis earum* (fig. 4).[39] (Like the pictures of the constellations, the diagrams follow the texts they illustrate, which speaks not only to the integrated design of the archetypal *Handbook of 809* but also to the creative painterly practice employed by the second master painter in Drogo's personal copy.[40]) The second diagram portrays the Pythagorean theory of tonal relationships corresponding to celestial distances (fig. 4). In the diagram, the earth stands at the center and the zodiac lies in the last of eight concentric circles. Whereas the planets were designated by personifications when the artist sought to underscore an abstraction of spatial order, in the diagram emphasizing celestial periodic distance, the planets' role as wandering stars, or *errantia*, is highlighted through the use of symbols resembling stars. In both "The Order of the Planets" and "The Pythagorean Theory of the Harmonic Intervals," the most important information appears along the vertical axis. The concentric circles draw attention to the spherical model of the geocentric universe, presupposed by the Aratean globe described in the *Excerptum*. The diagram offers a planar representation of the bifurcated sphere of the heavens. For a heavenly observer, this model would resemble a series of concentric rings. The planets proceed centrifugally in the same order as in the first diagram. The corresponding tonal intervals established by Pythagoras were based on a base unit of measurement of one tone. A *tone* was equal to the distance between the moon and the earth, or 125,000 stadia, roughly 14,370 miles using an Attic stadium. In total there were seven tones, establishing seven intervals with dual terminals and therefore an eight-part diapason or octave, distributed as follows: to proceed from the earth to the moon required the distance equivalent to one tone. Travel lasted for the measure identical to half a tone, or a semitone, when moving from the moon to Mercury. A trip from Mercury to Venus was just as long. To go from Venus to the sun required one and a half tones. From the sun to Mars required one more tone. A trip from Mars to Jupiter lasted a semitone. The distance from Jupiter to Saturn was also a semitone. Finally, to proceed from Saturn to the zodiac required one and a half more tones, or roughly 21,555 miles.[41]

By setting the earth in spatial relation to the realm of the fixed stars and the ecliptic band of the zodiac, with its corresponding offset array of constellations (as a result of precession), through a series of harmonic intervals, the diagram revealed to the Carolingian reader the intrinsic aesthetic integration of the universe. An important aspect of this diagram within the *Handbook of 809* was its role as a mediated presentation of the sublime, proportionate order of all Creation. A Frankish student would have recognized his terrestrial home as the locus of origin for this celestial exploration from the vantage point of the earth, facing outward toward the embedded celestial planetary spheres and the great revolving sphere of the fixed stars. The Carolingian pupil and his teacher faced an established map to Paradise and to the Heavenly City of God, with fixed distances and a spiritual bridge, or melodic stepping stones, for attaining the rest assured to those who undertook life's journey to the New Jerusalem. As Bianca Kühnel has argued, such meaningful deployments of visual signs in diagrammatic structures emphasized iconographic symbolic associations with theological conceits and not just astronomical information presented as verifiable and factual data.[42] If a reader takes a standard *T* shape and rotates it 90 degrees clockwise, the form is identical to the layout of the diagram (fig. 4). The wandering stars are arrayed within the diagram along a horizontal axis. If the planets become the vertical axis, then the *T* shape is readily apparent, and it is highly plausible that the planetary information was arranged meaningfully in the form of a tau cross, one variety of a symbolic cruciform structure. The intersection of spiritual or eschatological symbolism and cosmological harmonies was a commonplace in Carolingian diagrams.[43]

The second master painter inserted the third diagram (fig. 5) on folio 65v rather than in the minimal space remaining at the bottom of 65r, because the diagram was clearly considered an integral component of Drogo's copy of the *Handbook of 809*. The diagram of the apsides accompanied book V.5, the *De absidibus earum*.[44] When Pliny's *Naturalis historia*, book II.12, paragraphs 59–70, was introduced into the *Handbook of 809*, the Carolingian editor ended book V.5 with the phrase "as the figure below shows."[45] This aside is extremely significant. First, this addition to the text provides the best proof that the images were indeed intended to *follow* their textual descriptions throughout the archetypal *Handbook of 809*, and it permits their prescribed inclusion within the contents of that lost codex. Second, the Carolingian editors, who revised and rearranged the Plinian excerpts from book II of the *Naturalis historia*, thereby noted their esteem for the diagrams they had both created and inserted.[46]

The diagram depicted the apsidal orbits of the planets. At the farthest distance from the earth on any apsidal orbit is the nodal point called the apogee; the closest point is called the perigee. The earth is set at the center of the circular composition in Drogo's copy of the *Handbook of 809* and is circumscribed by the orbit of the moon, considered the only wandering star, or planet, without an apogee (although later medieval thought assigned Taurus as the sign for its apogee anyway). The other eccentric circles in the diagram could be used to identify the apogees for the various planets in the text: "Saturn in Scorpio, Jupiter in Virgo, Mars in Leo, Sun in Gemini, Venus in Sagittarius, Mercury in Capricorn, at the midpoints of these signs, and on the contrary in the opposite signs being the lowest and closest to the center of the Earth."[47] This diagram of the apsides was intended to identify abstractly and without requisite longitudinal precision the association between two important orbital positions relative to the earth and their corresponding zodiac signs, as well as the effects of solar radiation on planetary motion. Rate and speed entered as corollary considerations, since the planets that were farthest away were naturally slower and less capable of completing their forward motion, thereby "producing the least motion."[48]

Not only, therefore, are placement and distribution considered by the diagram, but movement and fluctuation become a topic of discussion as well. Such abstract analytic Ptolemaic and pre-geometric spatial models of planetary motion have shifted the focus from the pristine elegance of the well-distributed models of planetary order offered in the first two diagrams in the handbook to a proposed visualization of displacement and seemingly arrested motion (vis-à-vis the stations; see below) in the diagram of the apsides.[49]

The extensions at right angles to some of the eccentric orbits, such as Mars in its apogee Leo, designated the so-called station, or a house of the zodiac in which the planet can be seen for a prolonged period of time. The zigzag patterns, like that of Jupiter in Virgo, arguably were intended to designate "retrograde motion," or the apparent backward movement of the planet. Such phenomena should take place at the apogee and perigee, but as Eastwood explains, they are hopelessly muddled in many copies of the diagram from the *Handbook of 809*. The sun was the alleged planetary impetus for these phenomena. As a result, the Carolingian artist ought not to have depicted either a putative station of the sun in Sagittarius nor the retrograde motion in its apogee Gemini, indications of apogee in the Plinian text notwithstanding.[50]

The original contribution to the archetypal *Handbook of 809* in Aachen was probably also lamentably flawed. The second master painter in Metz likely copied these errors and imported them into his precise and confidently rendered diagram. Originally, the creators of the *Handbook of 809* considered the diagram an appropriate supplement to the Plinian excerpts. The diagram more clearly revealed the apogees discussed in the text, in order to "clarify" (in Eastwood's terms) the spatial arrangement of the planets along their eccentric orbits relative to the earth. This would be nearly impossible to describe with text alone. Although it shared the schematic form of the *rota* with the previous two diagrams, as Eastwood argues, the innovative presentation of the apsides united the counterclockwise arrangement of zodiac signs with their relevant planetary apogees.[51]

In this instance, the diagram was a truly novel Carolingian complement to the Plinian text. This presentation of asymmetrical discs identifying orbital apogees, retrograde motion, and "stations" destabilized the pristine abstraction of order proffered by the first two diagrams, revealing a subtle tension that is also suggested by the discussion of the countervailing Carolingian models of planetary order alternating between circumsolar orbits for Mercury and Venus and a strict regularity witnessed by the geocentric diagram in the *Handbook of 809*.[52] The Carolingian artists who painted these mediated models of astronomical ideas intentionally permitted the unstable consequences of their science to enter into the diagrams. Unanswered questions and unverifiable assumptions established the limits of their understanding. Rather than an assertive dogmatism, these Frankish diagrams for Plinian texts actually affirm a creative intellectual culture grappling with real astronomical problems and conflicting textual records that the artists were attempting to synthesize and understand. The study of astronomy was one way to fill out the Frankish picture of the Carolingian Creation, and the Plinian diagrams in the *Handbook of 809*—at the intersection of thought experiment and spiritual presupposition—attest to the exegetical role that visualizations of science performed for Frankish readers.

After book V.6, *De cursu earum per zodiacum circulum*, folio 66r of Drogo's copy of the *Handbook of 809* depicts a diagram called "The Latitudes of the Planets Through the Zodiac" (fig. 6).[53] In classical planetary theory, the zodiac band was divided into twelve degrees of celestial space bifurcated by the ecliptic. The thirteen nested circles in the diagram define twelve equidistant concentric rings, with one ring for each of the zodiacal degrees, as Eastwood has explained. According to the geocentric model of the universe, the eccentric orbits of the planets stand at varying angles of incidence to the diagonal incline of the zodiac belt (otherwise known as inclination), and in the diagram the southernmost circular limit of the belt establishes the outside perimeter ring of the entire *rota*. If the observer treated the ecliptic as a planar median disc occupying the topological center of the tilted zodiac belt, then the eccentric orbits of the planets would still appear to move higher and lower through the twelve-degree spatial swath representing the zodiac (although Venus traveled two additional degrees of space beyond the limits therein established). When viewed from the top down—or, according to the diagram, from the inner ring to the outside ring—the eccentric orbits of the planets could be mapped according to their angles of inclination relative to the ecliptic located in the middle of the zodiac, thereby yielding the latitudes. The ecliptic is circle seven of the thirteen concentric circles, nested between rings six and seven. This extraordinarily creative Frankish version of a stereographic projection, which displayed vertical spatial relationships in a horizontal format, visually adduced a list of the varying eccentricities of planetary orbits (and is not to be confused with a presentation of either astronomical altitude or azimuth).[54]

This mediated presentation of a recondite spatial conceptualization of the cosmos required the creative artistic interpretation of perpetual planetary motion along divergent vectors. It self-consciously rendered a three-dimensional model for relative planetary placement in a convenient and elegant planar form. This not only connoted a coherent systematic presentation of ideal order through finely crafted geometric clarity, executed by Frankish hands, but also revealed the internal complexity and discord of the otherwise harmonious music of the spheres treated in text V.4 of the *Handbook of 809* (fig. 4). Differences among the orbits of the wanderers were integral to the structural fabric of the Creation. In the planetary diagrams for Plinian texts in the *Handbook of 809*, a perfect model of Carolingian creativity comes to the fore. Painters remained steadfastly within the scientific traditions of the *auctores* but rendered their interpretative models with novel aplomb.

The eccentric planetary orbits in Drogo's copy of the *Handbook of 809* therefore conform to normative and relatively archetypal arrangement and display, as indicated by Eastwood's analysis. But since these diagrams are geometric abstractions and derive from less-than-precise traditions, they allow a good deal of creative latitude. Mercury, labeled on the right, runs more or less from rings 4 to 11, because its orbit traverses 8 degrees of vertical space during its journey through the zodiac. The moon on the left in Drogo's book runs from ring 1 inside to ring 12 outside, traversing the entire zodiac belt. Venus above transgresses the inner and outer boundaries of the zodiac by one degree each, so that its orbit runs from rings 1 to 13. The eccentric orbit of Mars is drawn lightly on the diagram, but according to Pliny occupies "four degrees in the middle,"[55] and often in similar diagrams runs from rings 5 to 8, or circles 5 to 9. Although difficult to discern, Jupiter's orbit in Drogo's copy of the *Handbook of 809* accurately traverses rings 7, 8, and 9 between circles 7 and 10, or, in Pliny's words, "the middle degree and two above it."[56] Saturn's orbit travels the 2-degree swath of space along the ecliptic occupied likewise by the wavy solar orbit's "serpentine path."[57]

As a result, this diagram is a third and "explanatory" type of diagram, according to Eastwood.[58] Pliny sought to identify why "[the planets] change their sizes and colors and approach to the north and depart to the south."[59] Unlike the text, which at best presents the number of degrees for a single planetary latitude at a time, the diagram breaks with such restraints: it presents the entire eccentric orbit of any given planet and represents all the orbits united in one radial representation of planetary motion. This instantaneous and omnipresent view of the solar system, capable of conceptualizing all of the combined planetary movements and reorganizing them into a coherent visual presentation, is a complement to the text. In this respect, the second master painter of the handbook supplied the exegetical diagrams that helpfully explained Plinian planetary theory. These diagrams were useful pedagogical tools that encouraged group study, as master and pupil joined together to examine the text and its complementary diagram, which together became a meaningful text and image pair within the larger *Handbook of 809*. Contreni has described this form of close-knit textual study as a beloved enterprise shared by Alcuin and his pupils.[60]

Precursors to the Style of Painting in Drogo's Copy of the *Handbook of 809*

As discussed in chapter 2, a small team of four highly skilled painters participated in the production of Drogo's copy of the *Handbook of 809*. The paintings and diagrams were attributed to two master painters, each of whom had one assistant. The best work in Drogo's book belongs to the second master painter, who illuminated the final portion of the cycle of illustration in the *De ordine* along with the highly imaginative diagrams just described, which bring the cosmos down from the supraelemental regions of regulated space into the controlled and ordered presentation of the *Handbook of 809* itself.

To better appreciate the nature of this band of Carolingian painters, it is useful to review the various artistic precursors that could have played a role in the development of the style manifest in Drogo's copy of the *Handbook of 809*. A few points should be stated explicitly from the outset. In what follows, many of the proposed dates for manuscripts derive from methodologically suspect relative chronologies, which are intrinsically problematic even when they are occasionally plausible and perhaps even correct. In fact, one of the goals of this chapter is to supply some helpful insights about the connections between books that have long been associated with Metz and Reims. Drogo's copy of the *Handbook of 809* offers an excellent opportunity to do so. The art-historical appraisal of these cities and their presumed scriptoria benefits from the light shed by pictures of the planets, stars, and constellations in Drogo's handbook. Since illuminated manuscripts treating early medieval sacred science have not been suitably appreciated by art historians in the past, the good information and correctives that their cycles of illustration can offer have been undervalued. As a result, scholars positing connections between books associated with ecclesiastical centers, such as Reims and Metz, have often remained wedded to the outmoded idea of ateliers with singular house styles.

If this research paradigm is turned on its head and the manuscripts are given primary place as pieces of cultural evidence, with informative texts and pictorial programs treated equally, then the nature of exchange is necessarily altered. The art historian examining Drogo's copy of the *Handbook of 809* loses some of the foundational support of Wilhelm Koehler and Florentine Mütherich's monumental *Die karolingischen Miniaturen*[61] but gains the advantage of nonhierarchical comparisons among books that can be evaluated on their own terms. Treating each illuminated manuscript as one node in an interlocking network of social value, cultural exchange, and period science permits newfound connections to surface.

The presence of artistic work on the folios of early medieval manuscripts is the record of lived artistic identities, and the miniatures are the fundamental revelations of both their art and their science. In this chapter, the artistic and aesthetic significance of the pictorial program in Drogo's copy of the *Handbook of 809* reveals the connections to Italy, Corbie, and cultivated indigenous traditions that gave rise to an efflorescence of beautifully crafted confessions of sacred science during the reign of Louis the Pious.[62] Then, in chapter 4, the spiritual and intellectual significance of illustrated books of astronomy is shown to emphasize the sacralization of science, which was one long-standing Carolingian motivation for the liberal arts. Connections between Metz and Reims were partly established through the same Frankish kinship ties that gave rise to luxurious copies of the *Handbook of 809*, as argued in chapter 1. However, there would be no copies of the *Handbook of 809* or the *Physiologus* (discussed below) unless they had been valued for their content and deemed worthy of manufacture from the outset. The books themselves, not the putative scriptoria, are the nodal points from which an art history of early medieval exchange emerges. Prioritizing centers of production offers tendentious artificial anchors that limit the researcher's ability to discern fluid connections between illuminated manuscripts. The books studied here reveal a good deal about the working teams that traveled, learned, and stored experiences through their craft in books such as Drogo's copy of the *Handbook of 809*.

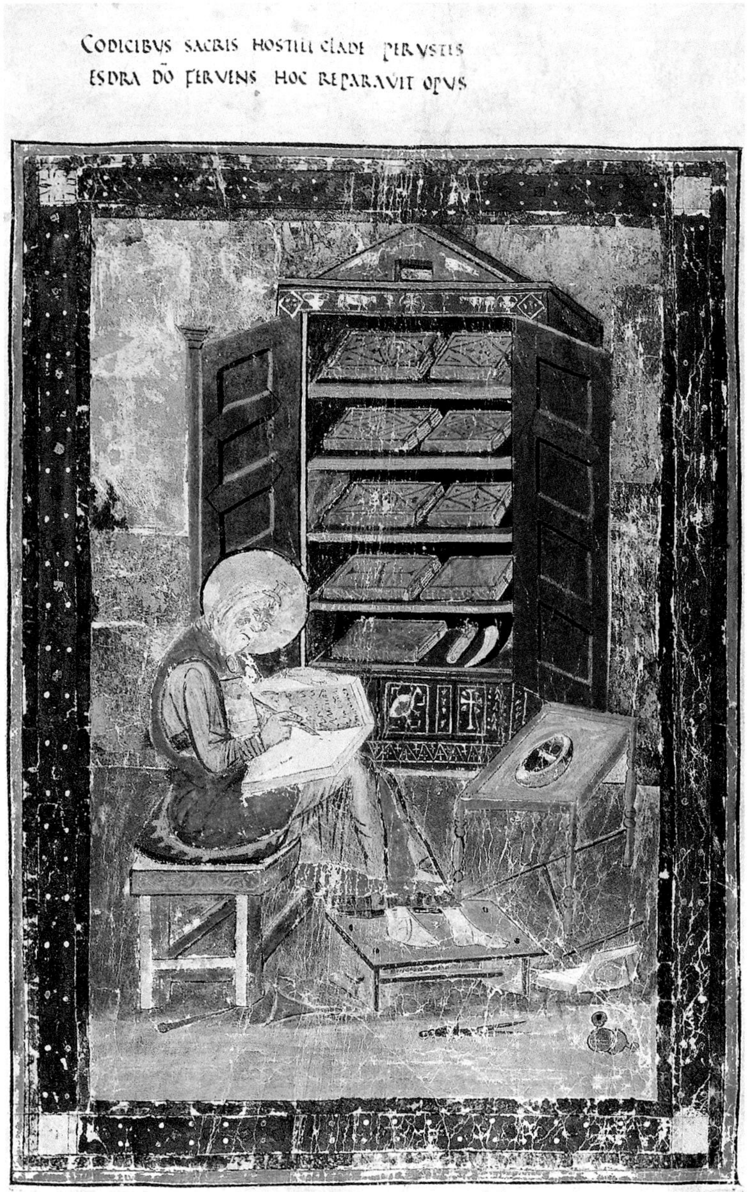

FIGURE 57
Ezra reconstructing the sacred texts from memory. Florence, Biblioteca Medicea Laurenziana, MS Amiatino I, fol. Vr. Photo © DeA Picture Library / Art Resource, New York.

It is important to underscore some of the connections among the arts of the Apennine peninsula and the development of the Carolingian book arts. The northern Italian lands were under the Carolingian "control" of Charlemagne's son Carlmann, otherwise known as Pippin, king of Italy (who ruled 781–810) at the time of the Aachen synod of 809. The same Adalhard who participated at the synod and played a prominent part in the creation of the *Handbook of 809* had also served as regent of the young King Pippin before returning to courtly life in Aachen. This relationship cemented the ties that, along with Adalhard's travels, resulted in the importation of books from Italy to the library at Corbie. In addition, it was probably at this time that Adalhard negotiated the marriage between his sister, Theodrada, and Pippin.[63] There are other records of cross-cultural diffusion and symbiotic collaboration during the early medieval period. As Lawrence Nees has noted concerning the exquisite interaction between line and painted forms in the image of Christ and the Evangelists from the Echternach Gospel Book in Trier, a Frankish artist collaborated with the Insular craftsman named Thomas on the painting (Trier, Domschatz, cod. 61, fol. 1v). Already in the second quarter of the eighth century, the Insular artists, who had accompanied the early Irish zealots or followed in the wake of their evangelistic peregrinations on the Continent, began to mix their understanding of acceptable compositional forms with that of their new neighbors.[64] Insular work ordinarily calls to mind the sublime labyrinthine carpet pages of the Lindisfarne Gospels, made in the first third of the eighth century (London, British Library, Cotton MS Nero D.IV).[65] Alongside this meditative linear tradition, however, an ongoing attention to figural proportion continued. For example, the Codex Amiatinus (made before 716) includes the exquisite framed portrait of Ezra, who is shown sitting at work reconstructing the sacred texts from memory, according to the inscription attributed to Bede that describes the picture (Florence, Biblioteca Medicea Laurenziana, MS Amiatino I, fol. Vr; fig. 57). The codex was the product of Anglo-Saxon scribes and painters working at Monkwearmouth or Jarrow.[66] The image of Luke from the so-called Gospels of Saint Augustine (of Canterbury, the Apostle to the English) attests to the kind of Italian manuscript that could have come to late sixth-century England (Cambridge, Corpus Christi College, MS 286, fol. 129v; fig. 58). These manuscripts document the influence of Italian classical forms on the book painting of the early Middle Ages across the English Channel from the Frankish lands.[67] To a certain degree, the task fell to northern Hiberno-Saxon missionaries such as Saint Boniface (d. ca. 755) to remind the formerly Roman territories, including Gaul, of their classical artistic roots.

The vastly inferior efforts of a northern French artist named Gundohinus, who arguably worked in a Burgundian monastery known as Vosevio around 754, demonstrate an intriguing yet merely conceptual connection to the lost Roman heritage of Gaul (Autun, Bibliothèque municipale, MS 3 [S 2], fol. 12v). Around the central medallion depicting the enthroned Christ, a laurel border frames and circumscribes a scene of veneration. The spirit of late antique and classical painting had been lost among the Franks.[68] This point, however, should not be exaggerated. There were multicolored and innovative artistic productions executed in the Frankish lands during the second half of the eighth century, leading up to the great revival of lavish manuscript painting at the court of Charlemagne. Jean-Pierre Caillet has noted in this context another gospel book in Autun, which was made in Flavigny

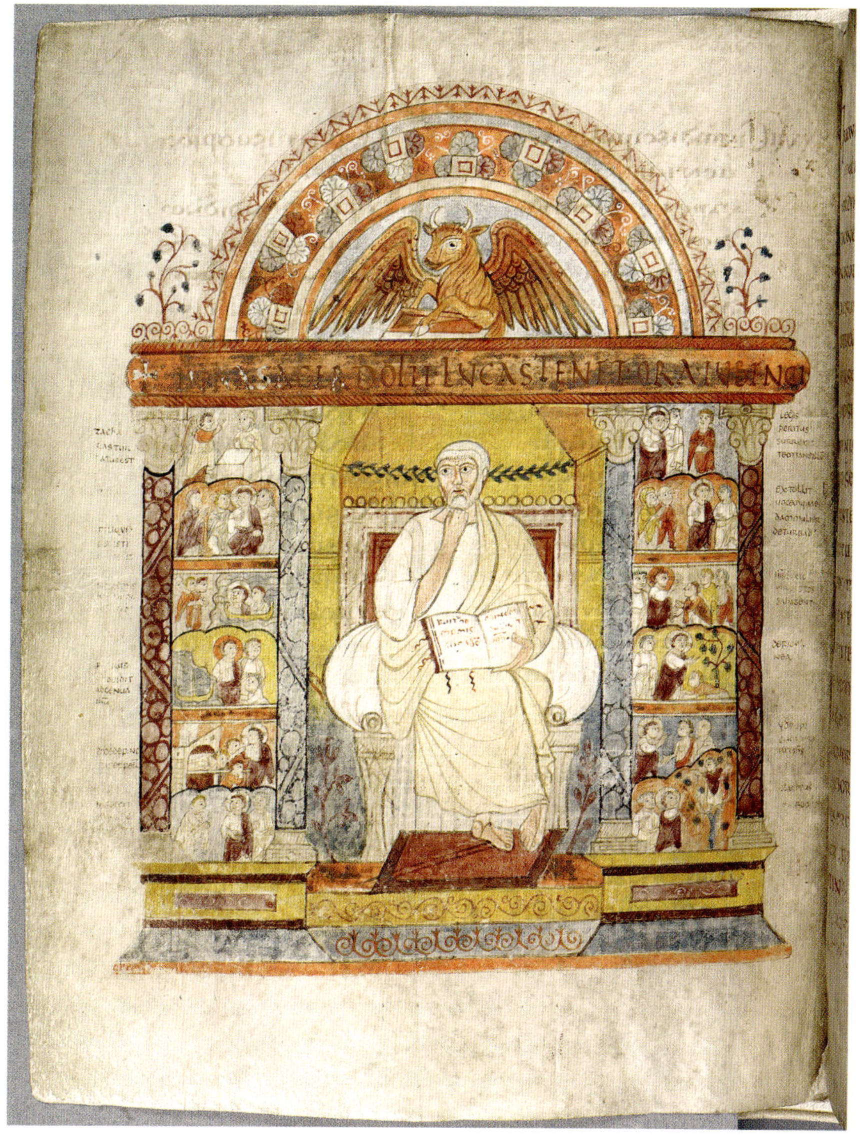

FIGURE 58

Gospel-author portrait of Luke. Gospels of Saint Augustine, Cambridge, Corpus Christi College, MS 286, fol. 129v. Photo courtesy of the Master and Fellows of Corpus Christi College, Cambridge.

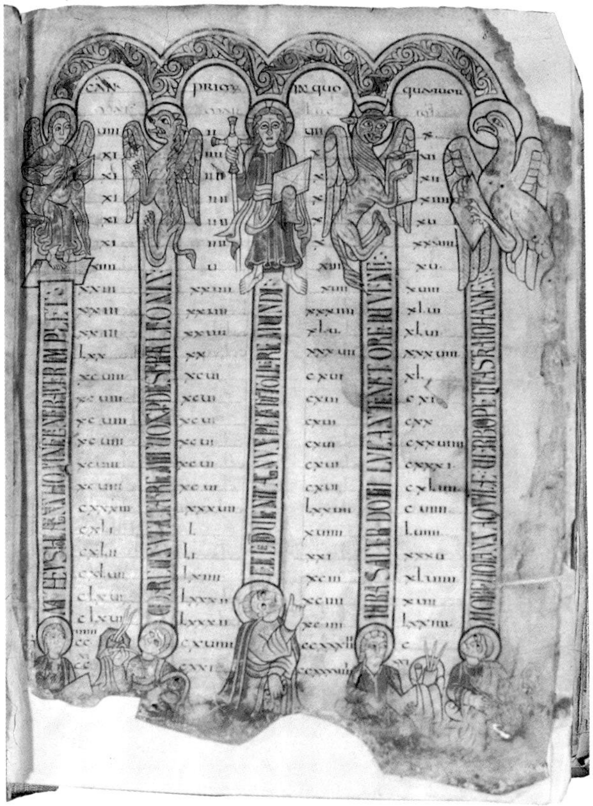

FIGURE 59

Canon tables. Gospels of Flavigny, Autun, Bibliothèque municipale, MS 4 (S3), fol. 8r. Photo courtesy of the Bibliothèque municipale, Autun.

under Insular influence (Autun, Bibliothèque municipale, MS 4 [S3], fol. 8r; fig. 59).[69]

An influx of Italian manuscripts has reasonably been considered a *sine qua non* for the development of Carolingian painting. This is, of course, why the artistic achievements of the Carolingian book arts have been labeled a "rapprochement," or prototypical Renaissance works.[70] Manuscripts ever or formerly on hand in important eighth-century Italian centers, such as Bobbio, Milan, Ravenna, Rome, and Verona, supplied the northern artists of the ninth century with examples of classical artistry and suitable models for a people aspiring to establish an *imperium*. It is important to make a methodological point here: artistic comparisons during the early medieval period should be considered informative about generic stylistic types of cultural exchange within a rich intellectual environment, unless artistic or textual documentation permits more refined and focused analyses. The overdetermined lines of cultural exchange during the late eighth century following the Lombard conquest are particularly nebulous, but as the cultural frontiers shifted, so did the viable possibilities for creation.

Jean Porcher has described the Italian precursors of the style of manuscript painting that flourished in the Frankish lands.[71] The style of wall paintings created under the influence of Byzantine art in the mid-eighth century may have influenced the development of the archetypal pictorial program of the handbook through the direct intervention of Adalhard of Corbie. For example, at Santa Maria in Valle at Cividale del Friuli, a fresco of a martyr from the so-called Tempietto of Desiderius was made roughly contemporary with the moment in 774 when Charlemagne subjected the Lombard king Desiderius to captivity, more likely than not at Corbie, entrusting the community with the deposed potentate's care until his death.[72] This inauspicious end to the Lombardic line nevertheless underscores the monks' exposure to northern Italian culture. The martyr also manifests a familiar facial configuration, recalling the gospel-author portraits in the Vienna Gospels. Semicircular browlines descend along an accentuated aquiline nasal ridge in both the wall painting of the martyr from the Tempietto and the miniature of Mark in the Vienna Gospels (fig. 29), dated by Bernhard Bischoff to a transitional period following the reign of Charlemagne and during the early reign of Louis the Pious, and hence after 814.[73] The graceful refinement of the painting of the martyr contrasts with the rugged masculinity of the evangelist, however, revealing the ways that Carolingian painters personalized their Italian influences during the reign of Charlemagne's successor.

Byzantium's legacy[74] to Italian artists of the eighth century can still be felt in such wall paintings, but also in manuscripts with a refined sense of ductus and draughtsmanship, as documented by extant pen-and-ink drawings. These linear forms betray their joint parentage in Italo-Byzantine styles. Such manuscripts probably influenced the stylistic interpretation of the visual forms of the constellations in the archetypal *Handbook of 809*. Circumstantial evidence for this (discussed in detail below) also comes from the prevalence of linear drawings and occasional washes in the princely luxury copies of the *Handbook of 809* at Monza and located in the Vatican (Reg. lat. 309, Monza f–9/176, and Vat. lat. 645; figs. 20, 36–38). Roughly contemporary with the production of the miniatures of the constellations in the archetypal copy of the *Handbook of 809*, an anthology containing Canon Law provides a rare record of the sketchy history of northern Italian drawing at the outset of the ninth century (Vercelli, Biblioteca Capitolare, Cod. CLXV/6, fol. 3r; fig. 60).[75]

Hans Belting has noted the relationship between these drawings and a group of manuscripts made in and around Nonantola from the late eighth to the first third of the ninth century. For Belting, this circle of manuscripts provides graphic records of a rich but lost early medieval Italian manuscript tradition that remained within the stylistic dictates of a late

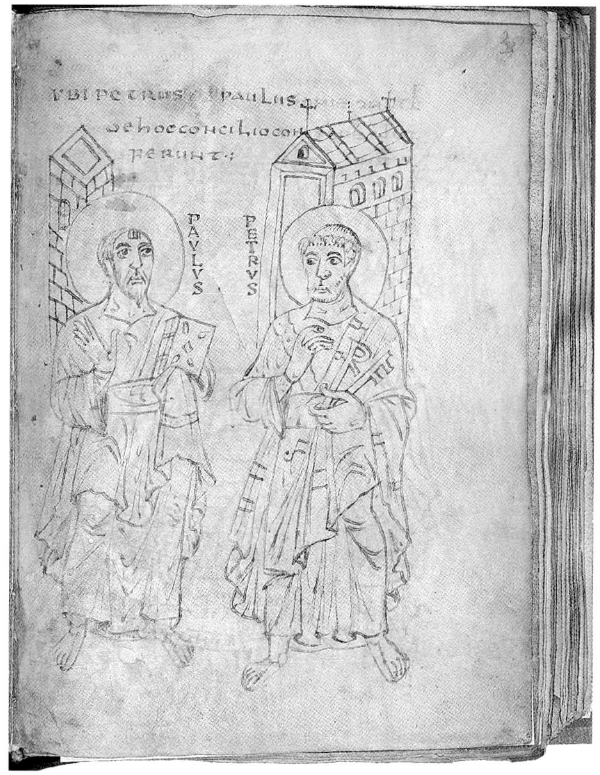

FIGURE 60

Saints Peter and Paul. Compendium of Canon Law, Vercelli, Biblioteca Capitolare, Cod. CLXV/6, fol. 3r. Photo: Fondazione Museo del Tesoro del Duomo e Archivio Capitolare.

antique repertoire. These specific manuscripts are, however, of ancillary importance and receive only passing treatment. Belting focuses instead on the refined miniatures in the Egino Codex, which was made in Verona in the closing years of the eighth century (Berlin, Preußischer Kulturbesitz, Staatsbibliothek zu Berlin, MS Phill. 1676, ca. 796–99). Belting considers the image of Gregory the Great (fig. 61) on folio 25v from the homiliary made for Bishop Egino of Verona (d. 802), a manuscript miniature that

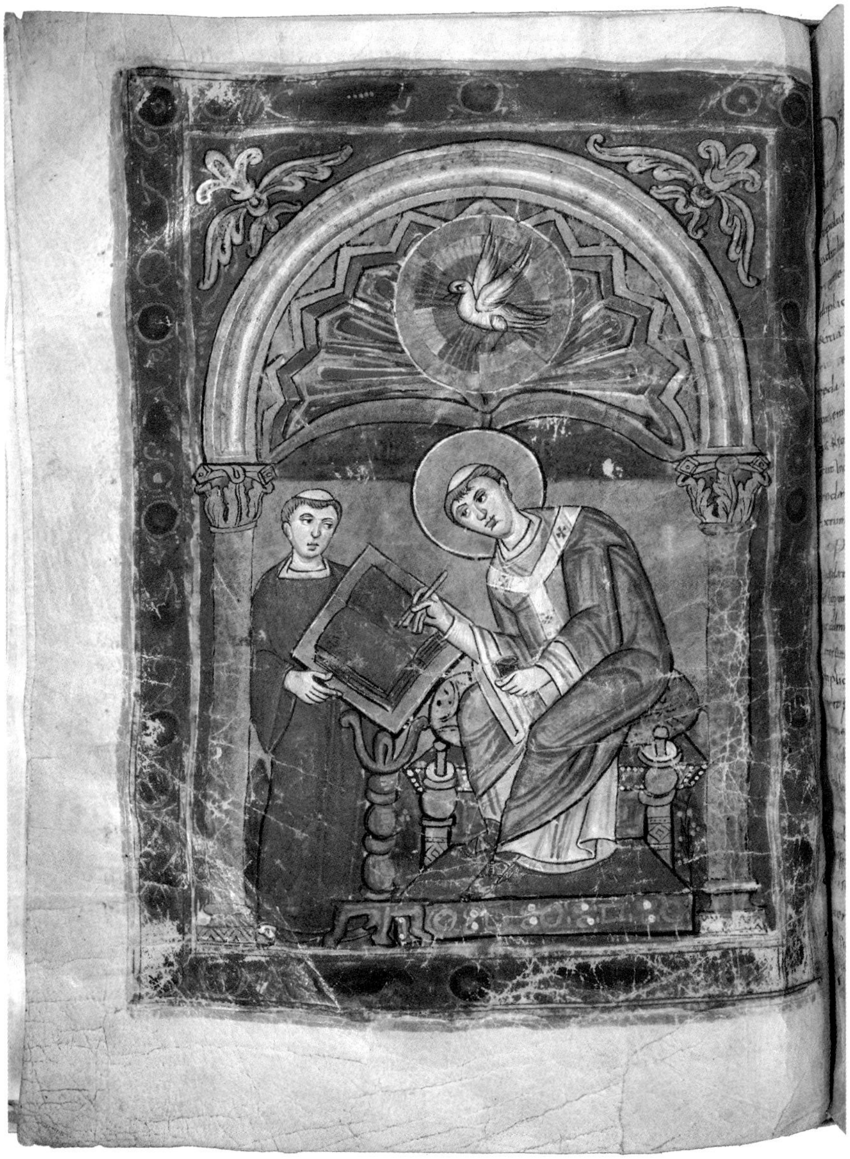

FIGURE 61

Gregory the Great. Egino Codex, the Homiliary of Bishop Egino of Verona, Staatsbibliothek zu Berlin, Stiftung Preußischer Kulturbesitz, Hs. Phillipps 1676, fol. 25v. Photo: bpk, Berlin / Staatsbibliothek zu Berlin / Art Resource, New York.

provides evidence for an earlier Italian manuscript tradition for which there is scant direct evidence. According to Belting, the early Carolingian gospel-author portrait of Saint Mark from the Godescalc Evangelistary (fig. 2), on folio 1v, was based on a related model but replaced the Italian spirit of its precursor with a Frankish vision of the form, and this opinion has been echoed more recently by John Mitchell (Paris, Bibliothèque nationale de France, Nouv. acq. lat. 1203, ca. 781–83).[76] The evangelistary, penned by the scribe Godescalc for Charlemagne and his queen Hildegard (d. 783), includes a cycle of illustrations derived from Byzantine precursors. For example, the image of the Fountain of Life is a novel introduction into Western manuscripts (Paris, Bibliothèque nationale de France, Nouv. acq. lat. 1203, fol. 3v).[77]

Belting proposes the highly plausible thesis that the image from the Egino Codex is a more representative and faithful copy of the sort of eighth-century model that would also have influenced the image in the Godescalc Evangelistary (fig. 2). The balanced and rhythmic compartmentalization of drapery folds around the evenly modeled form of Gregory the Great differs from the dynamic geometricization of Mark's form and robes in the evangelistary. Furthermore, the structural logic of the arcade over Gregory disintegrates into a linear series of stylized striations in the evangelistary. The upshot is that the portrait of Gregory provides more insight into the sort of lost Italian manuscripts that came to the Frankish lands in the 770s and 780s than does the Frankish interpretation of such an Italian model in the Godescalc Evangelistary, which was unequivocally linked to the most important early cultural activity of the court, perhaps inaugurating a long-standing tradition at Aachen, or alternatively at Worms, which has reasonably been suggested as a likely setting for momentary and fluctuating activity, given the testimony of Godescalc's confrère Adam.[78] As Lawrence Nees has argued, Godescalc's moniker "ultimus famulus Godescalc(us)," found in the dedicatory poem to the evangelistary, can effectively be translated as "most humble servant."[79] Godescalc's feigned humility in the colophon, however, belies the courtly role he probably played as member of an avant-garde class of early medieval "professional artists" devoid of clerical duties and traveling with the then-itinerant court on permanent "retainer."[80] It is worth keeping in mind that the creation of the Godescalc Evangelistary was roughly contemporary with the redaction of the *Revised Aratus* in Corbie executed in the latter portion of the eighth century.[81]

The incarceration of the Lombardic king in Corbie at precisely this turning point, circa 775, is hardly a coincidence. The historic confluence of post-Ravennate painterly traditions north and south of the Alps in northern France also helps explain another conduit for cultural exchange and painterly transformation that has led Nees to attribute both the Egino Codex and the Godescalc Evangelistary to the humble master's hand.[82] According to Nees, Egino and Godescalc imported northern influence, including a local Veronese variation of Caroline minuscule script, when they journeyed to the Apennine peninsula in 796. If so, then Nees surmises that Godescalc honed his nascent artistic skills in the milieu of Charlemagne's royal court chaplain, Fulrad (d. 784), after his appointment in 771, although Fulrad simultaneously served as Saint-Denis's abbot. If this story carries weight, then suitable evidence of late antique Ravennate or Roman manuscript illumination must have been at hand at Saint-Denis, and Nees relies on the probable presence of the Vergilius Romanus (Vatican City, Biblioteca Apostolica Vaticana, MS lat. 3967, before 525) in Saint-Denis during the early medieval period, supplying his carefully considered art-historical linchpin.[83]

Regardless of whether Godescalc had anything to do with the direct manufacture of the Egino Codex, Nees's remarks confirm what this discussion of the Aratean star pictures has repeatedly underscored: that in the third quarter of the eighth century in northern France, classical pictorial and textual resources were at hand, and that they hailed predominantly from originally Italian libraries. In other words, in Belting's exegesis, a lost record of Italian painting is manifest through the murky mirror of the Egino Codex; for Nees, the Egino Codex manifests late antique Italian influence on a Frankish painter. Either way, the Egino Codex and the Godescalc Evangelistary reveal the record of late antique Italianate painting as absorbed with wonder through late eighth-century eyes.

The Italian influence on Carolingian painting in the eighth century could be even more extensive and direct, then, than standard accounts of early Frankish painting suggest. It is necessary, however, to retrace the formal vestiges of these lost Italian drawings in the scant remaining evidence. The style used to render faces in the Vercelli drawings resembles the style used by the first master painter to create the facial features of the constellations in Drogo's copy of the handbook (Madrid 3307). For example, the draughtsman who drew the face of Saint Peter on the right in the compendium of Canon Law (fig. 60) designed the basic form of the face using the same set of conventions relied upon to draw the visage of Amphion, the left-hand musical brother of the Gemini group, in Drogo's copy of the *Handbook of 809* (plate 5). And in the pictures of both Peter and Amphion, a Z-shaped line joins the left brow to the nasal bridge, which descends to the relatively flat base of the nose before hooking back to conclude the definition of the proper left nostril. A pronounced shadow above the chin defines the lower lip in both images. Diagonal lines, identifying the transition from the mouth to the fleshy cheeks of the figures, supply Peter and Amphion with austere grimaces resembling dimpled frowns. In Drogo's copy of the *Handbook of 809*, the recessed region between Amphion's upper lip and nose, known as the philtrum, is delineated naturalistically with two quick ticks, whereas the image from the compendium includes one solid line as an index identifying the shadow between the two fleshy labial ridges of Saint Peter's upper lip. This similarity in actual style of execution is the sort of evidence that suggests that one or more Italian Aratean manuscripts arrived in Corbie and provided the linear models for the forms of the constellations that were brought by Adalhard to Aachen in 809.

In early medieval Italian manuscripts, as well as the books influenced by them, artists often rendered faces in the same way as those of Peter and Amphion. This conventional technique suggests a rudimentary form of foreshortening. The convention attempts to convey the spatial effects of a rotating neck upon the facial features, displayed in three-quarter poses. The face of Saint Mark in the Godescalc Evangelistary, although manifesting apparent incongruities, displays a related linear construction (fig. 2) that betrays the interlocking nexus of Italian and northern formal models for figural construction in the early years of the ninth century. The image of Gregory the Great from the Egino Codex alternatively models the form of the pope rendered in static magnificence, through what would later be known as a typically colorist painterly technique, rather than an asserted emphasis on privileged linear *disegno* (fig. 61).

The painters illuminating the books made within the Frankish lands, from Frisia to Farfa, during Charlemagne's reign adapted the linear conventions of Italo-Byzantine wall painting and northern Italian drawing to their own pictorial needs, apparently adopting stylistic conventions as demonstrative markers of antiquity and a suitable pictorial sensibility. This decorative component of formal presentation was itself redolent of the pervasive Carolingian spirit of reform, as concrete conventional formal recipes (such as the prescription for facial rendering just noted) supplied aesthetic and semiotic indices of well-rendered presentations of the human body, including their use to designate the constellations. As Hrabanus noted, formal properties could convey more than sensory satisfaction to a Carolingian audience. Even the juxtaposition of linear elements in compliance with antiquated formal codes elevated the art of the past, while the artwork as performance revealed the talent of the painter in the present together with his perspicacious view of the history of his own painterly traditions. The techniques of the painters who perhaps executed the archetypal *Handbook of 809* in the second decade of the ninth century—and certainly those of the four painters who made Drogo's copy in Metz—reveal a complex working praxis that presupposed a self-conscious compliance with historic artistic style, rendered palatable and current through creative engagement with evolving local traditions. Certain pictorial conventions, like the Carolingian approach to constructing facial features, remained an important component of the Frankish painter's skill set into the second half of the ninth century.[84] The painter of Amphion in Drogo's copy of the *Handbook of 809*, the first master painter, used this convention for his star pictures and taught his technique to his assistant.

Rather than painted models alone, it is far more likely that cycles of drawings also influenced the creation of images for the archetypal *Handbook of 809*. The linear interpretation of the figural form, intended to display the general contour outline of the constellations, facilitated the dissemination of star catalogues through cycles or samples of drawings. It is highly plausible that a cycle of Aratean drawings came to Corbie as a direct result of Desiderius's imprisonment at that center of manuscript production. Such a series (or series of series) of star pictures would have influenced Adalhard as the team leader, entrusted with managerial oversight in his decisions for the pictures introduced into his text, the *De ordine*, and therefore into the archetypal *Handbook of 809*. Perhaps he brought examples of pictures with him from Corbie to Aachen when he came to participate in the synod of 809. The fact that so many of the extant copies of the handbook and other Aratean texts include finished line drawings rather than painterly miniatures lends support to such a hypothesis. In other words, the star pictures found in Drogo's copy of the *Handbook of 809* were probably linked to northern Italian styles of drawing as well as painting. This only underscores the painterly achievement of the star pictures in Drogo's book.

Interpretative Artistic Decisions in the *Handbook of 809*

Otto Pächt has argued that the possible transfer of image cycles from illustrated *rotuli* (or scrolls) made of papyrus to illuminated codices occasionally resulted in changes to their organization and arrangement, ultimately resulting in novel text-image pairings. The wide column of unframed space remaining between text passages in a manuscript such as Drogo's copy of the *Handbook of 809* is, for Pächt, the result of a relatively unimaginative process of visual translation, relaying the star pictures that could have appeared in the archetypal Aratean Latin translation with imagery executed circa 300. Nevertheless, the alternative use of parchment folios in a book such as the archetypal *Handbook of 809* permitted artistic embellishments—namely, changes—to the imagery.[85]

In the creation of the archetypal *Handbook of 809*, however, certain images, such as Cancer, were actually expanded according to the desires of the Carolingian artists.[86] In the case of Cancer, the donkeys and feeding trough, or manger, were added to the image. As argued in chapter 2, these details were a common component of the archetypal pictorial cycle belonging to the *Handbook of 809*, but were apparently added to the image by the team working for Adalhard of Corbie. Adalhard's original drawing (possibly executed in tandem with an artist active ca. 808 in Corbie) more likely than not provided at the very least a working prototype for consideration by the artists who painted the archetypal *Handbook of 809*. The interest in applied meteorological concerns was a Carolingian innovation added to the pictorial program of the *Handbook of 809*, as revealed by the analysis of the iconography of the star pictures. The upshot is that iconographic peculiarities, such as the introduction of the donkeys at the manger into the image for the constellation of Cancer, should actually be considered the result of exegetical emendation manifest through the work of the painters. The fact that oddities such as the donkeys also appear in multiple princely copies of the *Handbook of 809* confirms this result (plate 5, figs. 36, 38). It is worth noting that the image of Cancer in Vat. lat. 645 prioritizes the donkeys flanking the trough at the expense of the displaced crab, who recoils into the gutter of the folio (fig. 38).

In contrast, the image of Cancer in Reg. lat. 309 discards the donkeys and returns to the older iconography, underscoring the way to identify individual artistic reinterpretation within a specific codex (fig. 62). The artists who painted the copy of the *Handbook of 809* made for Abbot Louis (d. 867) of Saint-Denis (Reg. lat. 309, ca. 859) also included within their pictures representations of the illuminated stars. In fact, in keeping with a practice employed by the illuminators of the book made for Louis, the miniatures of the constellations such as Cancer correlate better with the cycle of imagery contained within the revised third and final version (Borst's version type C) of the astronomical-computistical-pedagogical handbooks, such as Munich 210, ca. 818 (fig. 24).[87] Such oddities say little, however, about the iconographic models that Adalhard drew upon in Corbie when he supervised the creation of a new Carolingian set of star pictures for the lost archetypal *Handbook of 809*.

This is an important point. Kurt Weitzmann believes that the double-column layout and design of the third version (type C) of the handbook in Munich 210 provided an appropriate format for the diminutive renditions of the star pictures inserted within its text passages (figs. 23–24).[88] Presumably, such a format would also have been appropriate for the archetypal *Handbook of 809* and, by extension, for Drogo's copy. According to Weitzmann, the larger spaces in these books allowed a good deal of room for errors to work their way into the images. It would be in keeping with Weitzmann's general analytic program to see the addition of the donkeys to the miniature of Cancer, for example, as a Carolingian error that should be stripped away in search of the original Aratean cycle. Weitzmann presupposes a dynamic process of pictorial "emancipation" in which images were freed from their ancillary status to the texts that they illustrated and frequently clarified.[89]

Weitzmann's celebration of pictorial cycles fails to recognize, however, that certain picture books, such as the original *Handbook of 809* presumably compiled in Aachen, neither required framed miniatures nor offered their readers pedagogically suspect texts.[90] On the contrary, the addition of references to meteorological phenomena within the pictorial program of the archetypal *Handbook of 809*, which resulted in the addition of the donkeys to Cancer, was the result of a creative decision made by Adalhard and his team. Furthermore, the unframed scientific miniature continued to be considered an appropriate form for a depiction of the constellations in later copies of the *Handbook of 809*, such as Reg. lat. 309 (fig. 62) or Monza f–9/176 (figs. 36–37). The additions and omissions observed in the various princely copies of the *Handbook of 809* offer the best evidence about what was and was not important to the Carolingians interested in astronomy who assembled at Aachen in 809. The manuscript painters in Metz never considered the miniatures in Drogo's copy of the handbook to be ancillary to the passages that complemented and described them. Rather, the depictions and diagrams in the *Handbook of 809* supplied important supplemental information about the constellations and planetary astronomy.

Master Illuminators in Metz and Their Style of Painting

The paintings in Drogo's copy of the *Handbook of 809* benefited from the development of Carolingian painting since the creation of the archetypal *Handbook of 809* in the years 810–12. The four painters who painted Drogo's handbook in early medieval Metz probably worked near the bishop, and therefore (without any further information) it is reasonable to assume that they were probably affiliated with the cathedral of Saint-Étienne during the earlier years of his tenure. These artists had absorbed the lessons of Italian art discussed already and were likely aware to some degree of newly emerging Frankish styles, associated traditionally with various regional centers, including Reims. The paintings in Drogo's copy were emblematic of a pervasive interest in classical art that flourished during the reign of Louis the Pious (814–40).[91] Louis succeeded his father, Charlemagne, as ruler of the Frankish lands when Charlemagne succumbed to fever on 28 January 814.[92]

Manuscript illumination during the tenure of Archbishop Ebo of Reims (who served 816–35 and 840–41 and lived ca. 778–851) attained one of the highest levels of Carolingian artistic achievement. The biographies of Ebo and Louis are virtually parallel. Although Ebo had been suckled by the same wet nurse as Louis the Pious and had grown to manhood with uncommon advantages in Charlemagne's direct presence alongside the future emperor, he threw in his lot with Lothar I rather than Louis, rejecting the Carolingian David in favor of his Absalom. Unsurprisingly, once the political tensions had settled, Louis the Pious denied the pallium to his former court librarian and childhood companion as part of the aftermath of the 835 gathering at Thionville. In 840, Louis's death gave Ebo his chance to reassert control over the see of Reims, but Charles the Bald stripped Ebo of his metropolitan privileges at Reims in 841. Ebo could only obtain a final post at Hildesheim four years later, serving in this new see for six years.[93] It is

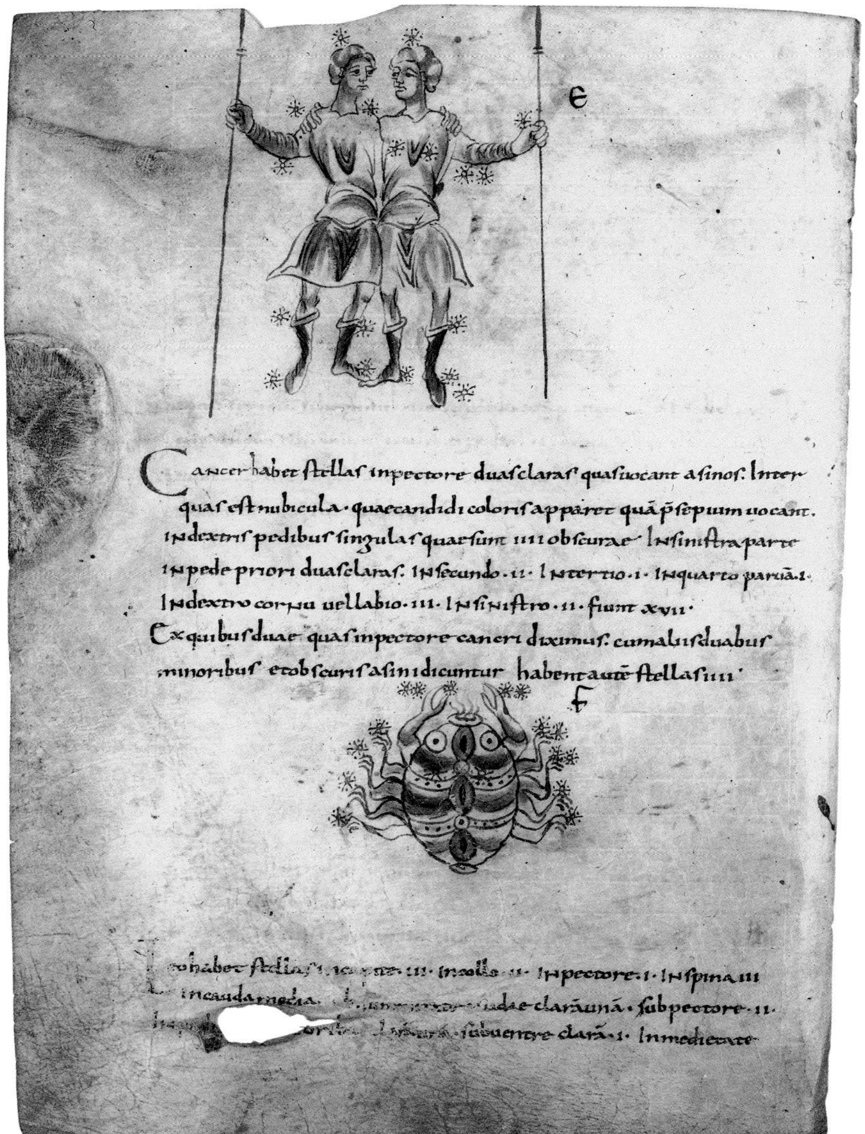

FIGURE 62

Gemini and Cancer. Vatican City, Biblioteca Apostolica Vaticana, Reg. lat. 309, fol. 93v. Photo © 2017 Biblioteca Apostolica Vaticana. Reproduced by permission of Biblioteca Apostolica Vaticana, with all rights reserved.

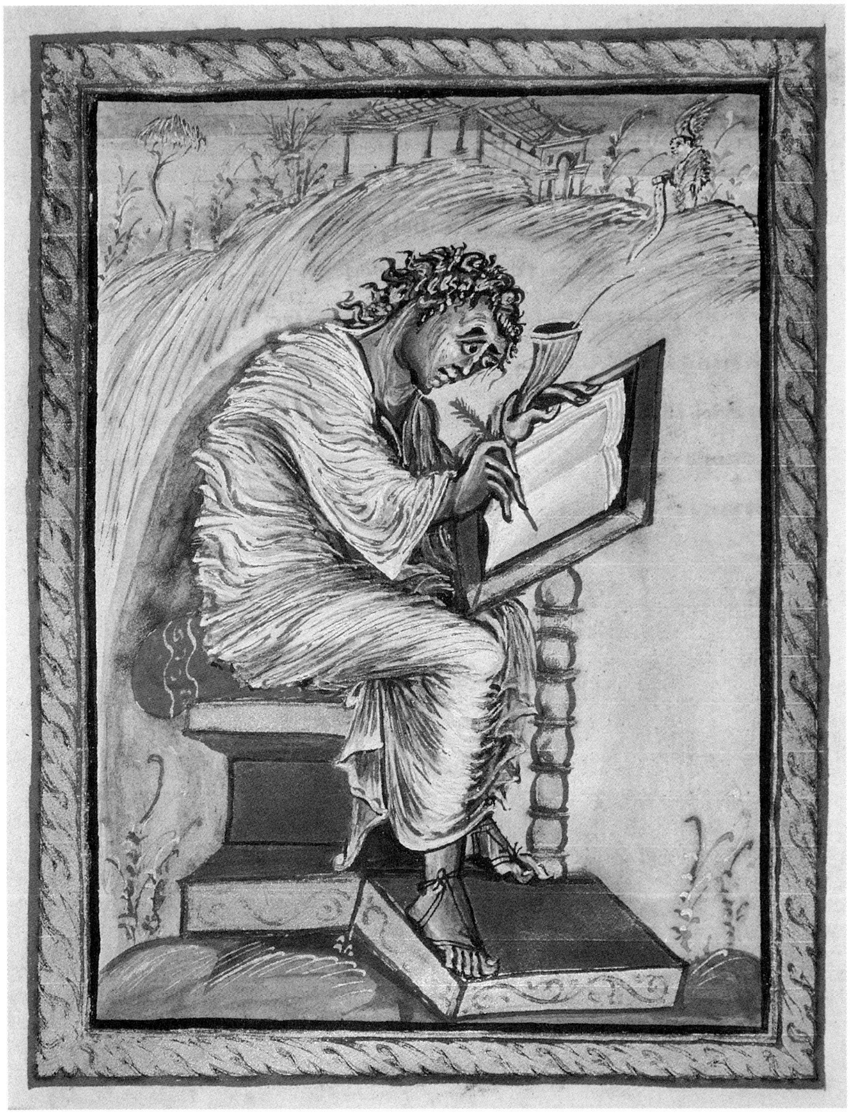

FIGURE 63

Gospel-author portrait of Matthew. Ebo Gospels, Epernay, Bibliothèque municipale, MS 1, fol. 18v. Photo: Bridgeman-Giraudon / Art Resource, New York.

important in the present context to emphasize that Drogo remained devoted to his older half-brother, Louis the Pious, throughout the crisis.[94]

Wilhelm Koehler and Florentine Mütherich argue that the formal conception of the gospel-author portraits from the Vienna Gospels influenced the nearly contemporary construction of the new expressionistic vision of painting that emerged in the Ebo Gospels (Epernay, Bibl. mun., MS 1). For example, one could say that on folio 76v of the Vienna Gospels (fig. 29), Mark wears a white classical robe and sits at work on his gospel book before an open landscape. Similarly, on folio 18v of the Ebo Gospels (fig. 63), Matthew wears a white robe and sits at work before an open landscape. It must be underscored, however, that apart from such general similarities, the artists painting manuscripts in Reims favored the extraordinarily eccentric and frenetic quality of line seen in the portrait of Matthew from the Ebo Gospels.

Alongside this powerful stylistic innovation in Carolingian manuscript illumination, there remained an attention to the intriguing classical forms also found in the Vienna manuscript (fig. 29). Rather than seek some idea of direct influence in this context, it is preferable to merely note the classicizing trends often associated with the manuscripts made under Louis. Since the expressionistic style of painting in Reims appeared seemingly *ex nihilo*, there is nothing to preclude the possibility that a team of original and extremely innovative itinerant painters (and not an isolated master) brought their unique artistic vision to Reims.[95] These possibilities have led Charlotte Denoël to propose that there was no established scriptorium at Reims, but multiple workshops in and around Reims where manuscripts could be made as the need arose, when funds were available and when there were artists on hand to be paid.[96]

Such sites are arguably less significant than the patrons with whom they are affiliated and the painters who passed through them, and this approximates the situation envisioned for Drogo and his four painters in Metz. This interpretative maneuver might seem disconcerting, but need not be so. Telling the stories of the production of these manuscripts—while prioritizing the books themselves as nodes in an interconnected framework of Carolingian writing centers and sites for manuscript illumination—permits the vagaries of creative innovation that emerge within any given book to be better appreciated. Falling back into a discussion of books from Metz, rather than seeking cross-currents in Drogo's handbook, actually precludes such inquiry (and therefore discovery).

The maverick style of the Ebo Gospels (fig. 63) also found expression in the masterpiece of Carolingian drawing made in Reims known as the Utrecht Psalter (Utrecht, Bibliotheek der Rijksuniversiteit, MS 32, fol. 36r; fig. 49). The manuscript has traditionally been dated to the period 820–35, during the halcyon days of Abbot Ebo.[97] Celia Chazelle has mounted the best case for the manufacture of the Utrecht Psalter during the tenure of Ebo's successor at Reims, Archbishop Hincmar, who served 845–82. Under this highly persuasive interpretation, the manuscript would have been created around 850, or in any case sometime shortly after the inauguration of Hincmar's service to Reims.[98] The idea of a later Utrecht Psalter has dramatically altered the way generations of scholars have thought about the development of the book arts in Reims, and it is a thorny issue in need of resolution. Stylistic similarities between certain illuminated or painted manuscripts in Reims and the copy of the *Handbook of 809* made for Drogo suggest that the artists who worked on his copy of the handbook in Metz learned from the style of book arts revealed by early Reimsian manuscripts. There is nothing more that can be said with certitude. It is equally telling, however, that the painterly innovations encountered in Drogo's manuscript reappear in books associated with Reimsian production, emphasizing the level of creative collaborative exchange and fluidity between Carolingian regional centers for manuscript production, rather than a predilection for so-called house styles. A lack of firm dates for Carolingian codices precludes any conclusions at this time, but the following observations may clarify certain unconsidered aspects of Carolingian painting in the second quarter of the ninth century.

One of the manuscripts from Reims, which shares stylistic affinities with the imagery found in the princely copy of the *Handbook of 809* made for Drogo, is the so-called *Physiologus*, an intriguing record of early Christian popular literature, homiletics, and to a certain extent zoology (Bern, Burgerbibliothek, Cod. 318; hereafter *Physiologus*).[99] Haecpertus, or Egbert(us), signed his manuscript in a colophon on folio 130r. Although Haecpertus's identity remains elusive, general consensus considers him to have painted the Bern copy of the *Physiologus* at an uncertain atelier somewhere in Reims (figs. 64, 66–68). There do not appear to be other works attributable to him. Stylistic analyses suggest that the book's creation is compatible with the years of Ebo's service as archbishop of Reims but leave open the possibility of a date during the era of Hincmar (ca. 845–82).[100] Florentine Mütherich attributes the image cycle in the *Physiologus* to a lost fourth- or fifth-century manuscript that was creatively but faithfully reinterpreted by Haecpertus, or possibly his team, working in a typical style for Reims before 850.[101]

Curiously, the Bern *Physiologus* belongs to the Latin translation (type C, which incidentally has nothing to do with the C version of the Carolingian handbooks) of the text, which, through complex paths of textual transmission, joined learned clerics in Frankish Europe to their Christian forebears in Ethiopia, who earlier had mulled over these moralizing texts. Bern is the oldest extant example of so-called redaction C, and it is believed that this Latin translation from Greek of the *Physiologus* was carried out in the eighth or ninth century, revealing widespread Carolingian interest in diverse modalities of early Christian or Constantinian cultural artifacts. This manuscript displays a development roughly parallel to that of the Latin translation of the *Revised Aratus* in Corbie during the eighth century, once again revealing that Greek literary works affiliated with Alexandrian and Hellenistic Egyptian precedent lay the foundation for a Carolingian Christian appropriation of science.[102] Nikolaus Henkel has explained that the text of the *Physiologus* treated animals, minerals, and insects along with their corresponding characteristics, supplying the metaphoric or metonymic topoi that allowed the introduction of moralizing sermon exempla as early as the second century. Early Christian exegetes appropriated this secular work of entertaining natural philosophy and turned it to their own populist, homiletic uses, transforming the text along the way.

It is in this vein that Ursula Treu has presented the most compelling explanation for the diachronic development of the *Physiologus* genre, namely, that it is better to think of this as a category of manuscripts rather than as an individual text. It is highly likely that the moralizing Christian allegories for which the *Physiologus* and later bestiaries are renowned were added at a later date than the original creation of the Greek literary tradition. For example, since the teachings of Origen (ca. 185–ca. 254) influenced some of the allegorical statements in the *Physiologus*, these statements could not have been added before the third century. Such memorable moralizing interpolations could be conjoined nearly at will to the animal tales, as spiritual advisers kept revealing their ordered cosmological "truth" about nature.[103] Latin translations were reportedly available as early as the fourth century.[104] The type C redaction of the *Physiologus*, including the copy from Reims now in Bern, provided an abridged twenty-four-chapter version of the text.[105] In summary, the Bern manuscript, like the *de luxe* copy of the *Handbook of 809* made for Drogo, was probably made during the late 820s or early 830s during the reign of Louis the Pious, who fostered an efflorescence of classically inspired artworks throughout the Frankish lands. In both the *Handbook of 809* and the *Physiologus*, an emphasis was placed on Christian uses for ancient erudition.

Many of the miniatures in the Bern *Physiologus* from Reims adopt the convention of a red frame surrounding a blue or black frame (fig. 64), reportedly harking back *inter alia* to the red-framed celestial imagery of the early Latin translation (with scholia) of Aratus.[106] In terms of precedents, it is useful to look at early Christian and late antique book illustration, such as the miniature of the battle of the lovesick bulls from the *Georgics* on folio 5v of the very late fourth- or early fifth-century Vatican Vergil (Rome, Biblioteca Apostolica Vaticana, Vat. lat. 3225; fig. 65). In this miniature, the frame defines the window through which the viewer looks onto a bucolic landscape where (as Kurt Weitzmann explains) two bulls duel to win the affections of a cow.[107]

In another framed bucolic scene on folio 7r from the Bern *Physiologus*, Jacob addresses, and perhaps

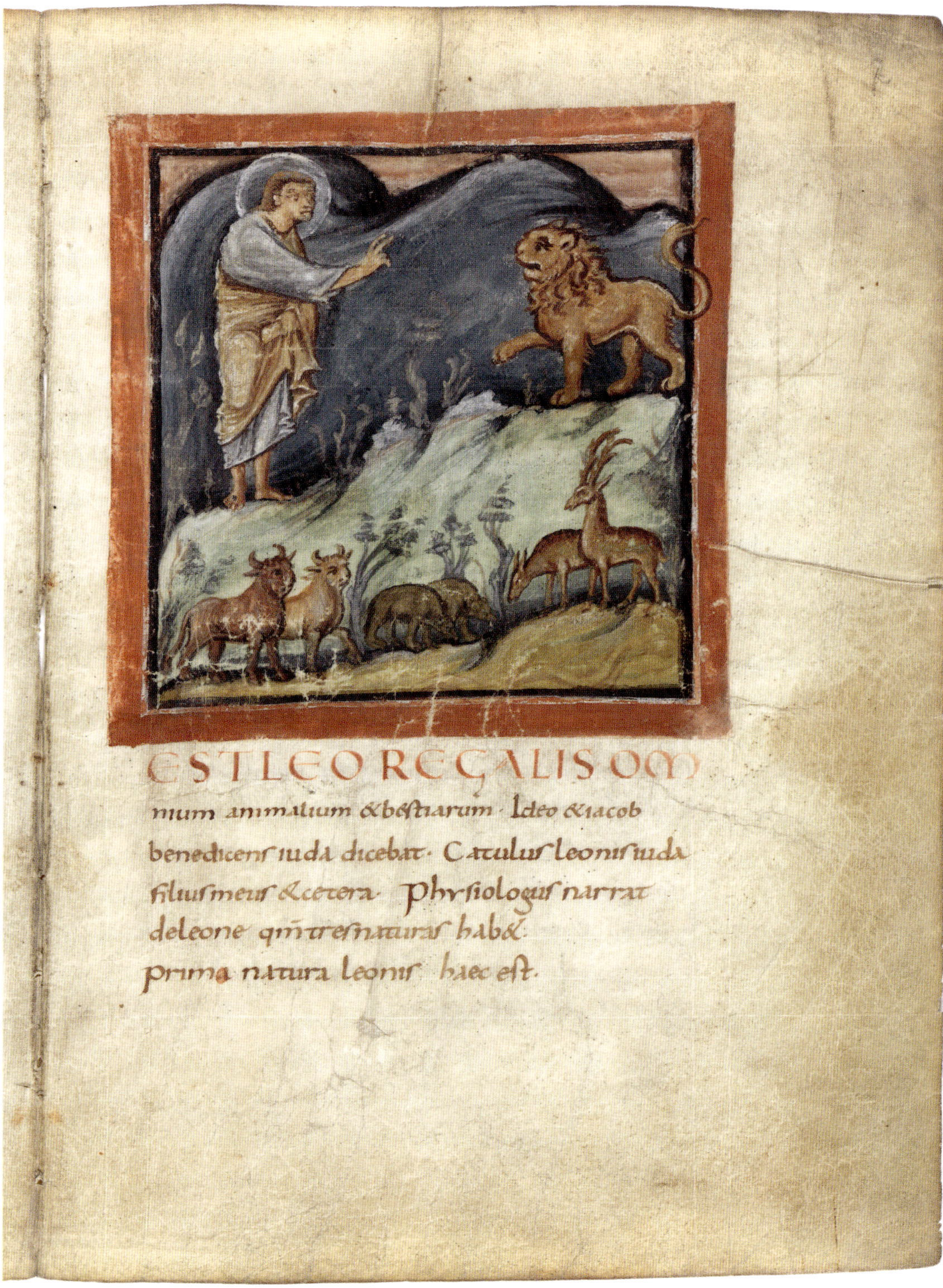

FIGURE 64

Jacob and the Lion of Judah. Bern *Physiologus,* Burgerbibliothek Bern, Cod. 318, fol. 7r. Photo: Codices Electronici AG, www.e-codices.ch.

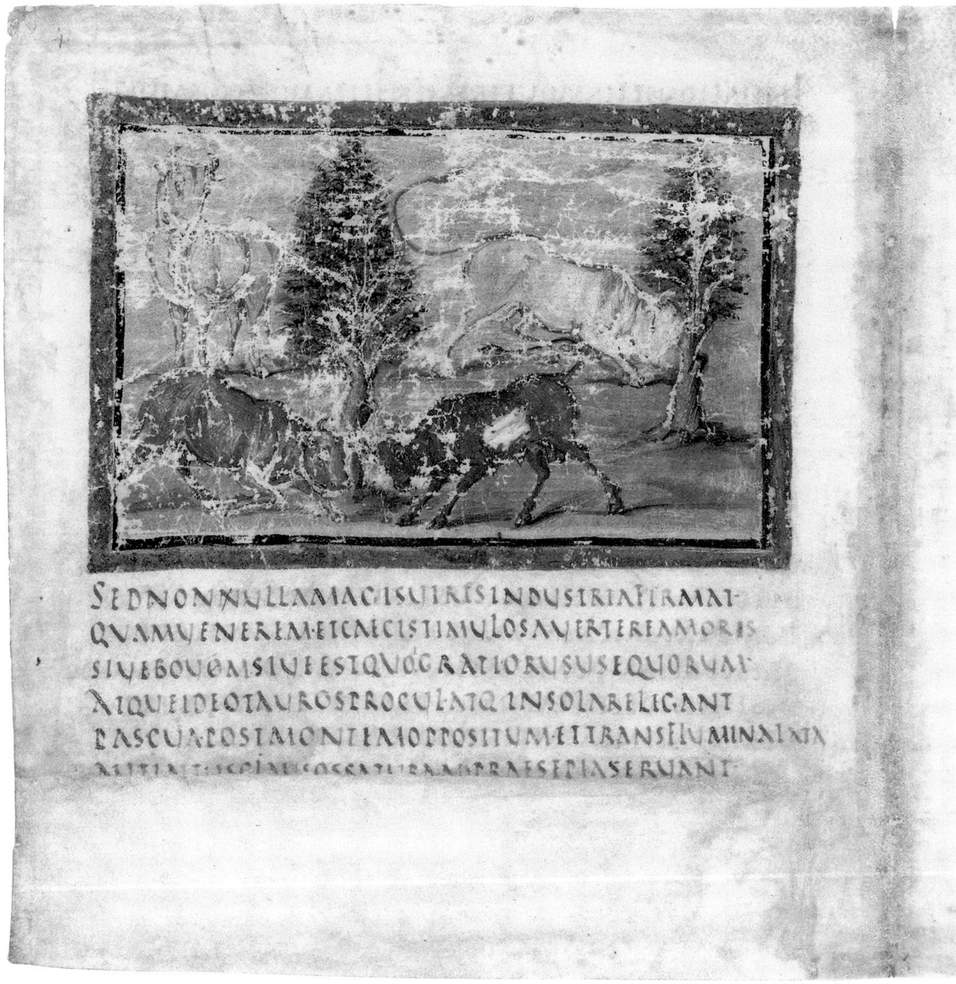

FIGURE 65
Jealous and lovesick bulls. The Vatican Vergil, *Georgics*, Rome, Biblioteca Apostolica Vaticana, Vat. lat. 3225, fol. 5v. Photo © 2017 Biblioteca Apostolica Vaticana. Reproduced by permission of Biblioteca Apostolica Vaticana, with all rights reserved.

even renders benediction upon, the "lion of Judah" depicted as the king of the beasts before a group of disinterested quadrupeds (fig. 64), as Helen Woodruff and Otto Homburger observe.[108] (The bears in this miniature recall the forms of Ursa Major and Minor in Drogo's copy of the *Handbook of 809*). The lion in the upper right corner of the Bern miniature has a decorative but refined mane that resembles Leo's feathery, wavy locks in Drogo's book (plate 6), while the lion in "First Property of the Lion," on folio 7v of the Bern manuscript (fig. 66), has the large, dilated pupil and heavy brow seen in the image of Leo from Drogo's book. According to the *Physiologus*, the lion dragged its tail along the ground in order to cover its tracks and survive. This was an allegory of Jesus's need to appear merely human while on the earth.[109] As can be seen here, the text is placed after the picture in the *Physiologus*, unlike the imagery in

Drogo's copy of the *Handbook of 809*, in which each description of a single constellation precedes its respective picture.

The first, blue horse pulling the chariot of Auriga on folio 57r of Drogo's copy (plate 6) resembles the general outline of the horse in the Bern *Physiologus* on folio 22r (fig. 67). On folio 22r, the depiction is of the "Caballus/Horse for Riding," representing one of the varieties of steeds described in Isidore of Seville's *Etymologies*, book XII.1.42–43, that was excerpted and added to the type C redaction of the *Physiologus* text after roughly the mid-seventh century.[110] The pose and arrangement of the legs for each of the galloping steeds are nearly identical, although compared with Drogo's copy of the *Handbook of 809*, the horse in the Bern *Physiologus* reverses the left-right orientation of the horse's hind legs.

There is also a connection between the image of Serpens in Drogo's copy (Madrid 3307) on folio 54v (plate 1) and the unframed miniature showing the fourth nature of the Serpent in the Bern *Physiologus* on folio 12v (fig. 68). In both of these unframed miniatures, the three-color tonal treatment of the scaly flesh adds a sense of texture and verisimilitude. The mottled stippling suggests the radiant effects of light splaying across a snake's scales. This was also the technique used later to create the serpent in the gospel-author portrait of John from a gospel book made for Drogo mentioned in chapter 1 (Paris, MS lat. 9388, fol. 150v; fig. 34). A related technique was adapted to the small scale of the historiated initial *O* from the Drogo Sacramentary on folio 43v (fig. 48). There, a coiling serpent writhes in anguish at the foot of the cross. The similar painterly technique used for each of these images casts light on workshop practices in Metz and Reims during the ninth century—or at least on the artists who worked in both cities.

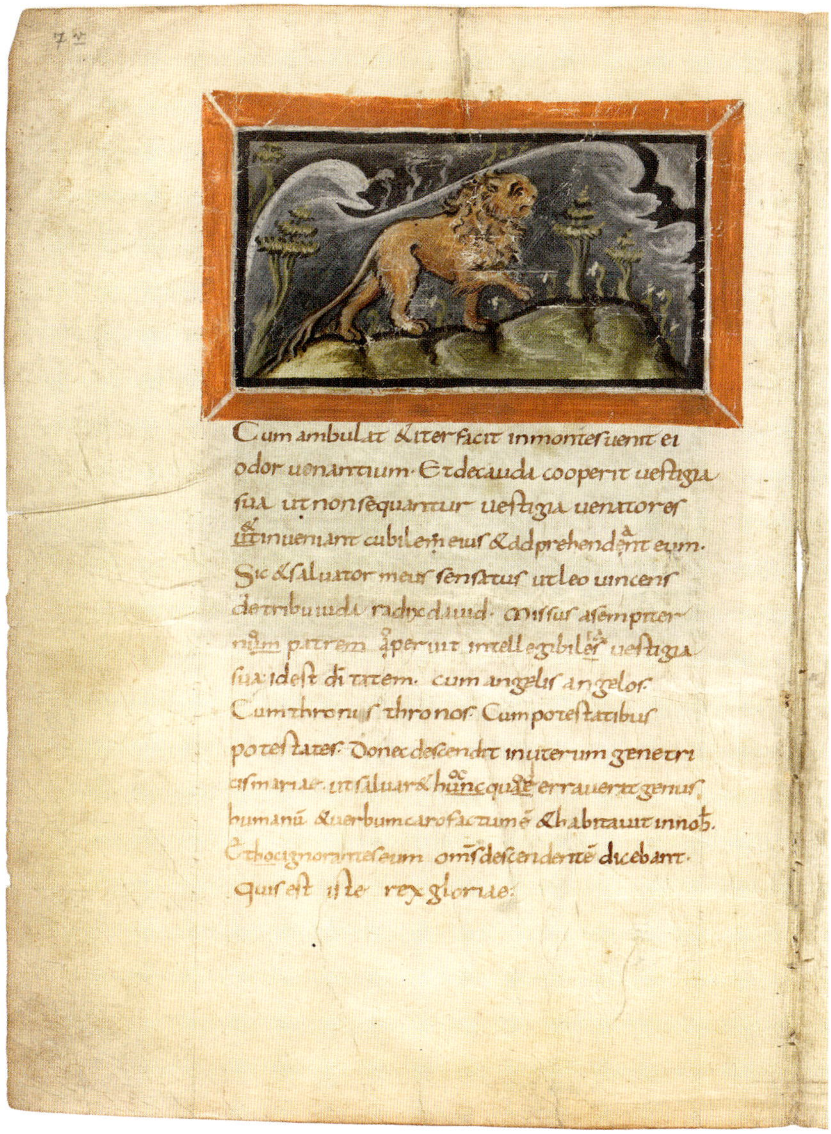

FIGURE 66

First property of the lion. Bern *Physiologus*, Burgerbibliothek Bern, Cod. 318, fol. 7v. Photo: Codices Electronici AG, www.e-codices.ch.

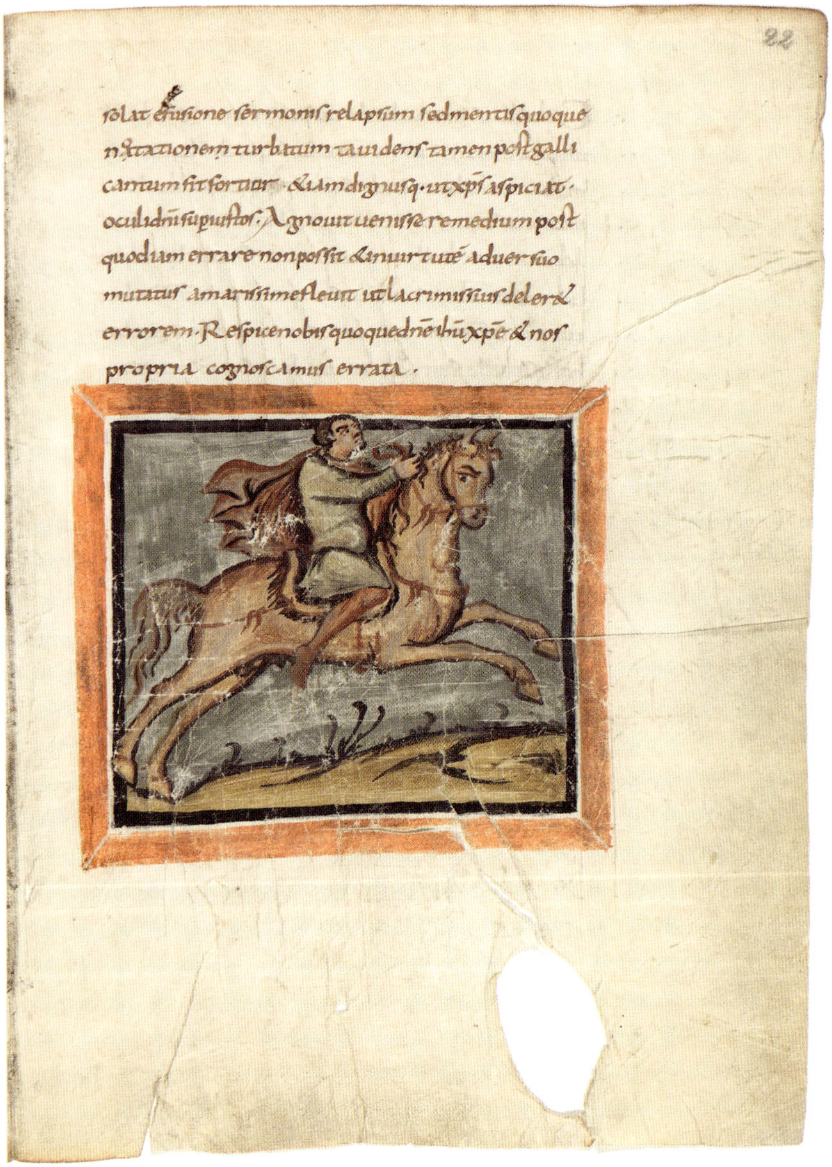

FIGURE 67

Caballus/horse for riding. Isidore of Seville, *Etymologiarum*, book XII.1, the Bern *Physiologus*, Burgerbibliothek Bern, Cod. 318, fol. 22r. Photo: Codices Electronici AG, www.e-codices.ch.

It is important to emphasize the profoundly innovative solution seen in the image on folio 12v of the Bern *Physiologus*, revealing the homiletic potential of serpents through a moral reminder of the sacrificial martyr's body that sustains Christ as the head of the Church Triumphant (fig. 68).[111] In order to offer a panegyric to spiritual devotion, this painter expanded the creative potential of the folio and ruptured the frame. It is interesting that Carolingian painters often dismantled their frames when achieving their most creative results, as in Drogo's copy of the *Handbook of 809*.

It is unlikely that an older copy of the *Physiologus* recombined both the framed and unframed miniatures in Bern. Rather, it becomes clear that in the ninth century the frameless text space in some cases permitted increased interpretative freedom and creative latitude for the Carolingian painters. The frame provided a window onto the classical world, but breaking the frame occasionally opened up new visual possibilities for Carolingian manuscript illuminators. The creativity seen in the image of Cancer from the archetypal *Handbook of 809*, as well as in its princely copies, such as the expanded miniature by the first master painter in Drogo's copy (plate 5), likewise benefited from this omission of the frame.

Another difference is the use of opaque colors in the Bern *Physiologus*, at odds with the splendid use of washes and the soft tones that unite the painterly style of Drogo's copy of the *Handbook of 809* and the style of books made later in Metz. For example, the Drogo Sacramentary displays lavish historiated initials (figs. 33, 47–48) embellished with gilded acanthus winding through jewel-like teal, violet, azure, cerulean, and viridian tones. These colors were also used in Metz by the second master painter of Drogo's copy in such markedly classical depictions of the constellations as Heridanus/River Po and Piscis Magnus on folio 61v (plate 15). The controlled painterly quality of the Bern *Physiologus* includes layered opaque colors appropriate to the style of the later artists working in Reims during the initial Ebo era, perhaps after 830 but before his first deposition and departure from Reims in 835. Again, this is offered exclusively as a proposal that fits the evidence.

It is possible that the first master painter of Drogo's handbook, or his assistant, later participated somehow in the manufacture of the *Physiologus*, either alongside or as part of the team supervised by Haecpertus. Exactly where Haecpertus supervised this team, was himself trained, or imparted information to his putative assistants is unknown. The fact that the best correspondences between the style of painting in the Bern *Physiologus* and in Drogo's copy are confined to folio 57r, displaying Leo and Auriga (plate 6), leaves such a speculative hypothesis an open possibility. In any case, the shared painterly techniques employed by artists in both early medieval Reims and Metz are already highly informative. In fact, the examples discussed here argue in favor of scientific and artistic stylistic diffusion and workshop integration or reciprocal influence. Considerations of Carolingian style cannot separate manuscripts meaningfully into schools so much as permit the researcher to track painterly transformation along complicated networks of mutual artistic influence and interaction. Hrabanus's description of symbolic connotation (mentioned above) can actually be applied to a macrocosmic assessment of motivations for and results of painterly change within Frankish ateliers, justifying a semiotic analysis of Carolingian painting itself.

As noted in chapter 1, the three great liturgical books in Metz were made in the following order,

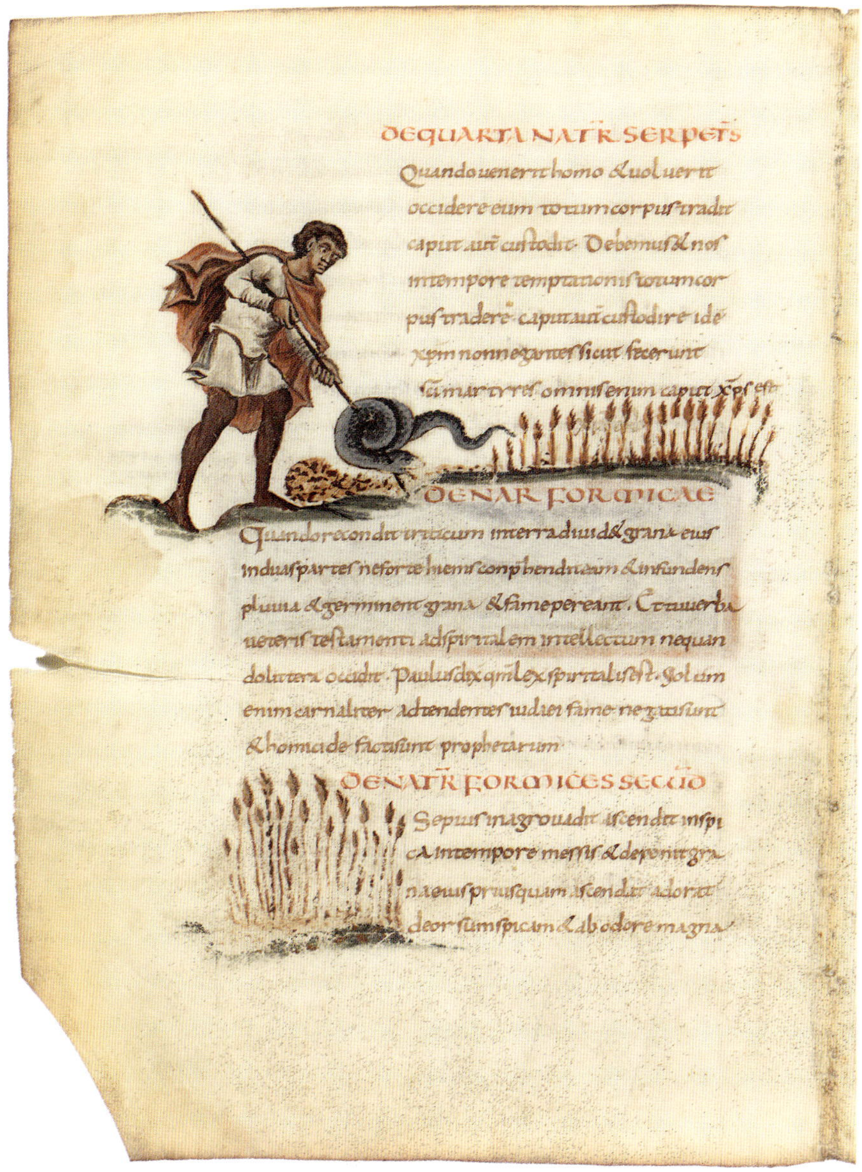

FIGURE 68

Properties of the ants and the snake hunt. Bern *Physiologus*, Burgerbibliothek Bern, Cod. 318, fol. 12v. Photo: Codices Electronici AG, www.e-codices.ch.

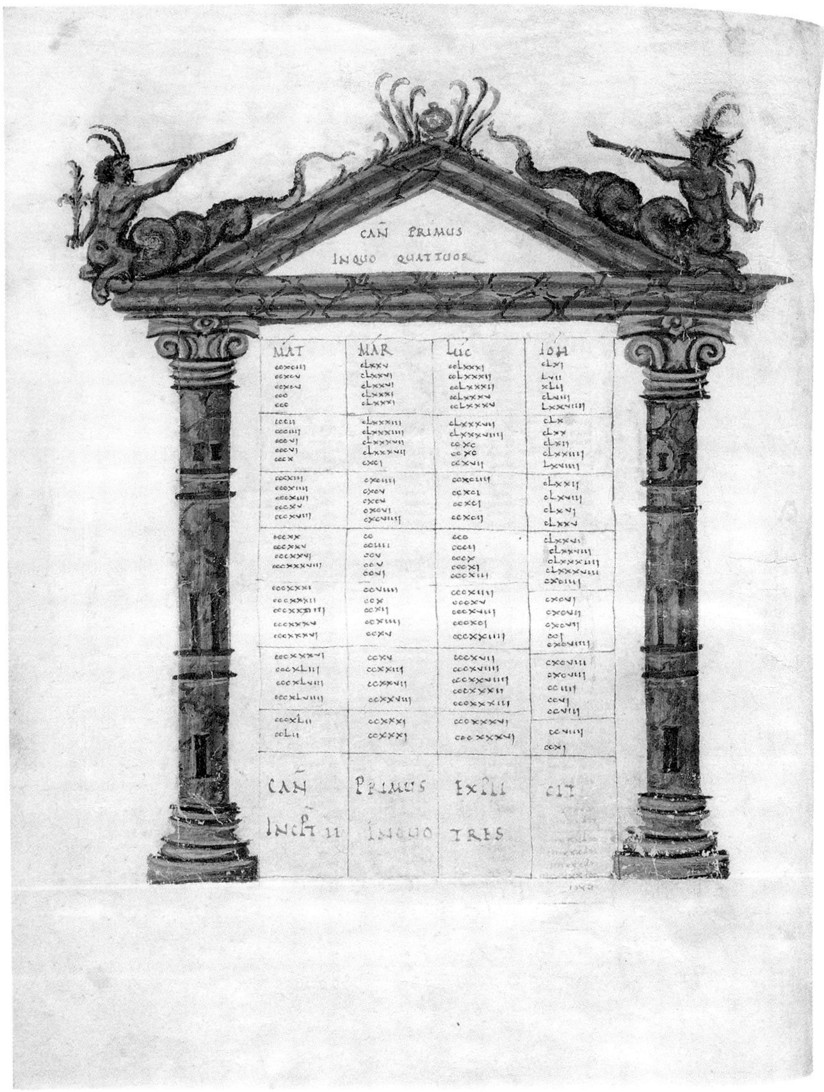

FIGURE 69

Canon table. Loisel Gospels, Paris, Bibliothèque nationale de France, MS lat. 17968, fol. 6v. Photo: BnF.

according to a fairly standard relative chronology that remains open to reinterpretation:

> Gospel book, Paris, Bibliothèque nationale de France, MS lat. 9383, ca. 835–45 (figs. 30–31)
> Gospel book, Paris, Bibliothèque nationale de France, MS lat. 9388, ca. 845–55 (fig. 32)
> Drogo Sacramentary, Paris, Bibliothèque nationale de France, MS lat. 9428, ca. 845–55 (fig. 33)[112]

Wilhelm Koehler has identified a number of similarities between the style of painting exemplified by Drogo's copy of the *Handbook of 809* and that of other manuscripts from early medieval Metz. The general tonal qualities of the water in an image such as Aquarius (plate 11) recalls the use of cerulean blue for a background hue in initials like the *O* of folio 43v in the Drogo Sacramentary (fig. 48). Lastly, Koehler links the Drogo Sacramentary Ascension initial on folio 71v (fig. 47) and the image of Heridanus from Drogo's copy of the *Handbook of 809* on folio 61v (plate 15). The painters in Metz carefully laid down their chosen hues within dark, heavy contours that define the crest from which Christ departs the earth as well as the rippling musculature of the river god Heridanus. The face of Heridanus even resembles the face of Christ.[113] In this case, therefore, the work of the second master painter is demonstrably related to the later manuscripts made in Metz, whereas the work of the first master painter is closer to the work executed in Reims immediately before and after the production of Drogo's copy of the *Handbook of 809*, particularly the illustrations in the *Physiologus*. The work of the second master painter, in contrast, laid the foundation for the style of illumination that Drogo would continue to cultivate for the remainder of his life. Recognizing the recurring expressions of

Drogo's desire for great books matters more than focusing primarily on the precise date of manufacture for each of these volumes. Drogo's choices vis-à-vis his preferred painterly styles recur in the later books associated with him and with medieval Metz, regardless of how the exact commissions for those books were fulfilled.

The work of the second master painter in Drogo's copy of the *Handbook of 809* is echoed in the motifs from another gospel book that was made in Reims in the first half of the ninth century but had made its way to Beauvais by the twelfth. That manuscript entered into the collection of a prominent local member of the legal profession, Antoine Loisel (1536–1617), from whom it received its common name, the Loisel Gospels (Paris, Bibliothèque nationale de France, MS lat. 17968).[114] The Loisel Gospels shares cultural proclivities likewise visible in the Hellenistic painterly style of the Ebo Gospels. For example, on folios 6 through 11 of the Loisel Gospels, fanciful figural motifs perch atop the architraves on either side of their sloping pediments from the series of canon tables.[115] The finial at the pinnacle of the pediment on folio 6v (fig. 69) resembles the rounded modeling of the constellation Crater on folio 62v (plate 17) of Drogo's handbook, painted by the assistant to the second master. The mottled use of paint, which modulates the textures of the visible flesh in the torsos of the hybrid creatures above the architrave, resembles the work of the second master painter in Drogo's handbook. There is a particularly strong resonance between the tritons in the Loisel Gospels and the depictions of the Centaur on folio 62r (plate 16) as well as the satyr form of Sagittarius seen on folio 60r (plate 12).[116]

In all these images, the exaggerated, heavy contours of the torsos define the pronounced musculature, while the deep eye sockets reveal, from within their shadowy reserve, a defiant indifference to their pagan past. The protruding horns of the tritons in the Loisel Gospels also echo the horns in the image of Capricorn on folio 59v of Drogo's copy of the *Handbook of 809* (plate 11). The textural modeling of the curvilinear horns is representative of the work by the second master in Drogo's handbook; in both cases the horns are rendered with flair, as a flurry of repeated, seemingly freehand painted dabs in the Loisel Gospels or as confident cones with articulated illusionism in Drogo's book. The coiling tails of the hybrid tritons (fig. 69) likewise resemble the tails of the hybrid creature Capricorn (plate 11). As was common practice for the second master painter in Drogo's copy, a strong contour line distinguishes the edge of the figure's form. It is also significant that the textural effects of the second master's style evoke the expressionistic facture seen in the figures of the evangelists in the Ebo Gospels (fig. 63).[117]

Another interesting connection to the styles of painting popularized in Metz during the 840s is the decorative insertion of vegetal forms into the columns sustaining the architrave of the canon table on folio 11r of the Loisel Gospels (fig. 70).[118] This is a delicate device reminiscent of the coiling acanthus that circles the columns of the canons on folios 2v and 3r (figs. 30–31) of a gospel book made in Metz circa 840 (Paris, Bibliothèque nationale de France, MS lat. 9383).[119] Perhaps the presence of foliate forms in the canon tables documents yet another artistic exchange between these painters associated with Metz and Reims. The highly individualized form of canon tables associated with Metz manuscripts from the 840s relies upon this pictorial device, which is shared with the artists of the canon tables of the Loisel Gospels.

The links between the style of the second master painter in Drogo's copy of the *Handbook of 809* and the work in the Loisel Gospels suggest that the same artist may even have traveled from Metz to Reims to work on commissions. Comparison with the work of the second master painter in Metz in the 840s could also help explain the watery, light pastel colors that recur in the Drogo Sacramentary. It is highly likely that the old Austrasian ties that had once united Metz and Reims fostered easy travel and the successful transfer of small bands of artists during the Carolingian period from one city to the other, and perhaps even back again. Ebo's departure from Reims under duress in 835 may have prompted artists in his employ to depart Reims as well and seek work elsewhere, fostering rather than discouraging stylistic exchange. This is not to suggest that the Reims team left as a group, but rather to point to specific moments when artistic trends in the two cities intersected. The relocation of some individuals is the easiest way to account for this phenomenon.

Lawrence Nees has argued that independent artists worked in various cultural centers and ateliers throughout the ninth century.[120] The present stylistic analysis of the manuscripts made in Reims and Metz during the years 830–50 suggests that the two centers were bound together artistically much more tightly than previously thought, even though the political situation was at times quite turbulent. Drogo remained true to Louis the Pious, whereas Ebo found himself among the cadre of rebellious ecclesiasts.[121] Rather than linking specific schools of manuscript illumination with specific monastic or cathedral sites and assigning univocal house styles to those ateliers, it can be more informative to trace the development of painterly techniques and to track in which manuscripts they appear. This approach offers clues to the working styles of individual early medieval artists. Following these traces also permits a richer and more accurate analysis of how manuscripts can document influence and participation from individuals who formerly were considered separated by workshop conditions and professional local demands.

This approach permits the recognition of dramatic differences such as the expressionism of Reims versus the more tongue-in-cheek, playful classicism of Metz, while emphasizing stylistic continuities as well. It would be problematic to derive from these considerations a heavy-handed formalism grounded in working techniques. Instead, the semantic, or hermeneutic, approach to the interpretation of working styles proposed here gives us the opportunity to follow the production of various anonymous medieval artists by grounding historical claims about their lives in their brushstrokes. Style is, ultimately, a choice in a world of influences. The precise nature of stylistic exchange during the first half of the ninth century has vexed researchers to date; frankly, clarification of the connections linking Metz to Reims has awaited close examination of Drogo's copy of the *Handbook of 809*.

Celestial Globes and Manuscript Models: Creative Polycyclic Compilation

Another image at the intersection of a shared Reimsian-Messine Austrasian artistic impulse is the historiated initial *O* on folio 43v from the Drogo Sacramentary, in which a figure holds an object resembling a blue globe beneath a scene of the Crucifixion (fig. 48), as discussed by Chazelle. In her iconographic analysis of the liturgically significant initial *O* for Palm Sunday, Chazelle recounts earlier explanations of the imagery and identifies the white-bearded figure to the lower right of the cross as Nicodemus. Other iconographic exegeses have argued unpersuasively that the figure is a conflated image of Melchizedek/Drogo, depicts Moses or Hosea, or symbolizes the holy city of Jerusalem, Jewish tradition synecdochically embodied by the synagogue, or even the Mosaic Law and Covenant of the Old Testament.[122] This minuscule rendition of Nicodemus does indeed offer a view of a humble figure who lived during a period of historic and theological change shared with other members of a hegemonically Jewish population inside the Roman Empire. The Crucifixion provided the pivotal turning point in salvation history, and the cross was the instrument of torture that inaugurated the new Christian spiritual age.

Nicodemus rests as an attendant witness in the company of Ecclesia, gathering Jesus's blood within her sacred cup, as well as Mary and beloved John, lamenting the martyrdom of the sacrificial Lamb of God.[123] For this reason, above Jesus's head, the *corona* placed between normative lunar and solar personifications awaits him as his laurel Crown of Righteousness in the heavens (2 Timothy 4:6–8).[124] This possibly infused the novel imagery of the historiated initial O with a form of symbolic *interpretatio christiana,* derived from the image of the constellation painted in Metz and found within Drogo's copy of the *Handbook of 809* (plate 2). According to Chazelle's exegesis of the historiated initial, Nicodemus holds a globe signifying the world sphere, otherwise referred to as the *orbis terrarum.* The general iconography, she recognizes, harks back to classical ruler portraiture. In this context, the globe draws attention to Nicodemus's political clout and elevated social standing within his first-century Jewish community,

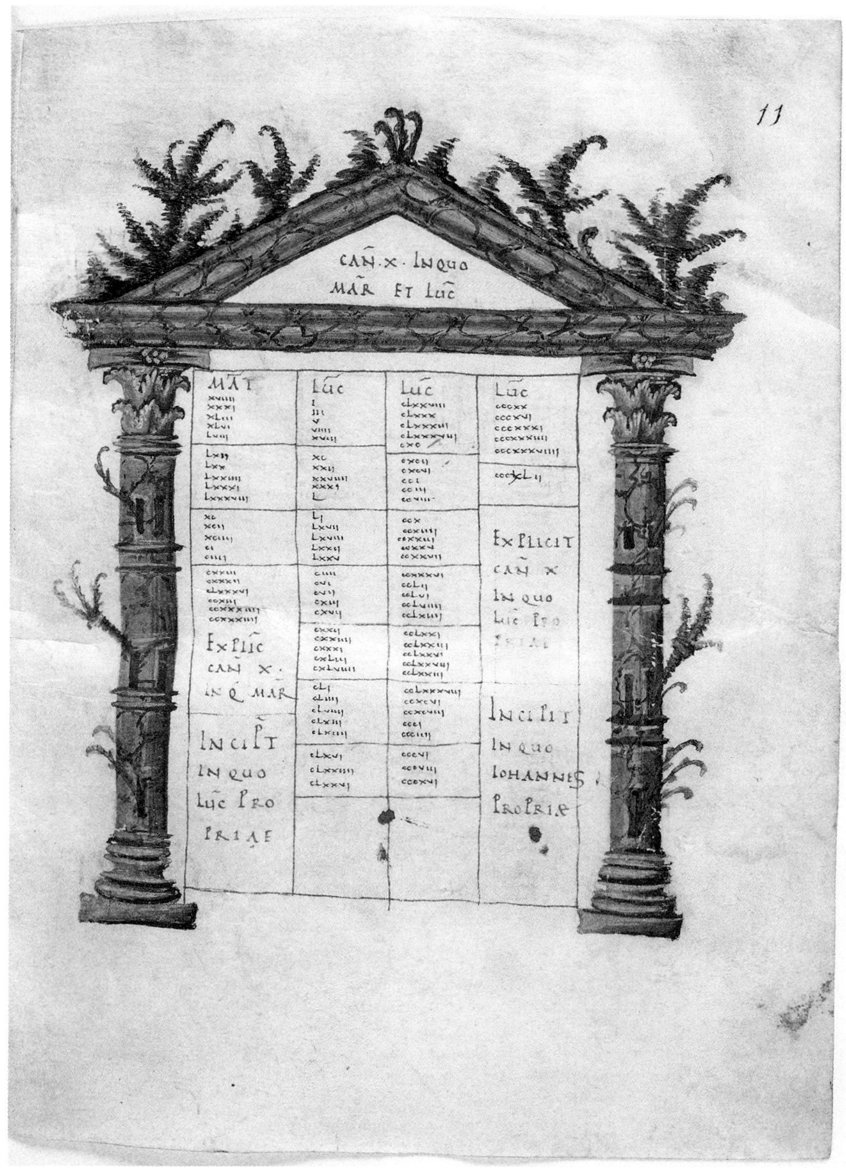

FIGURE 70

Canon table. Loisel Gospels, Paris, Bibliothèque nationale de France, MS lat. 17968, fol. 11r. Photo: BnF.

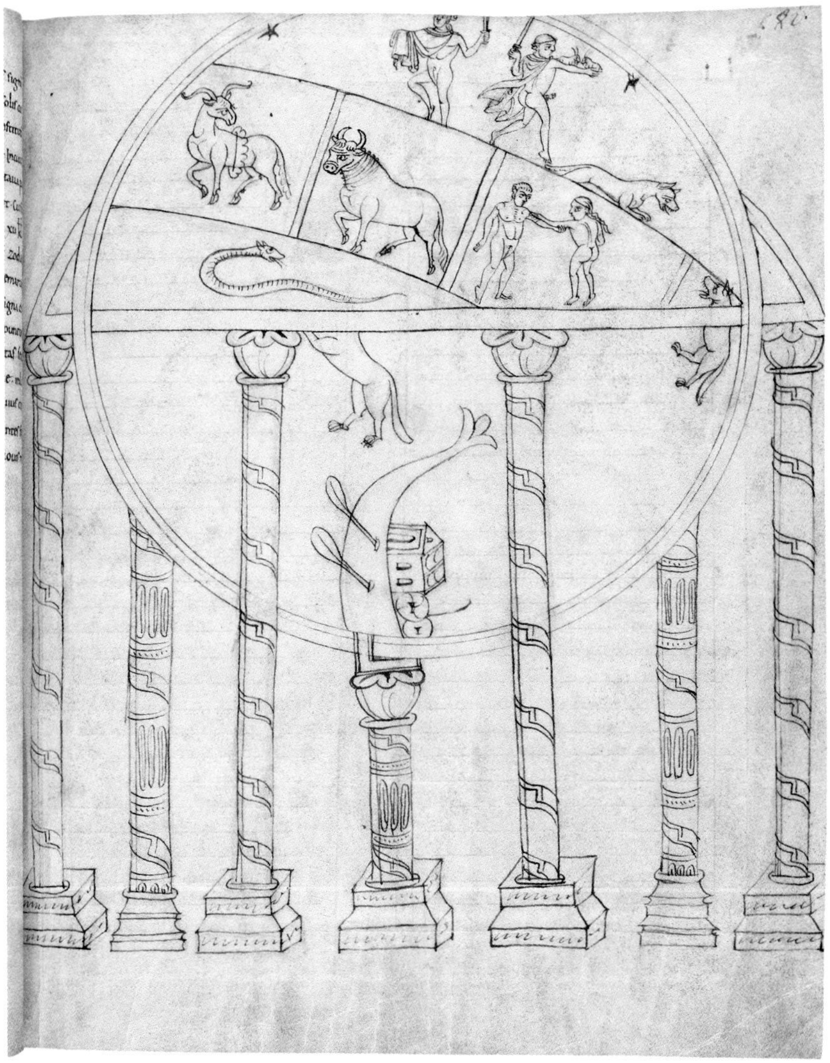

FIGURE 71

Celestial globe. Stiftsbibliothek St. Gallen, Cod. Sang. 902, page 81. Photo: Stiftsbibliothek St. Gallen.

as juxtaposed with the everlasting authority and religious wisdom of the dying Christ.[125]

That Nicodemus sits at the feet of his dying master holding a globe emphasizes the rabbinical role Jesus had played for his disciples. The master had attempted to explain the mysteries of earth and the heavens to Nicodemus and his putatively blind cohort.[126] The Carolingian artist who painted the initial *O* may also have been alluding to the role that such globes played in the pedagogical advancement of educational reforms in the ninth century.[127] In this respect, the image of Nicodemus also recalls classically inspired images of Urania, the muse of astronomy, with or without Aratus and with a celestial globe.[128]

Celestial globes provided important early medieval models of the heavens, though few examples of them remain, for they were made of precious metals (which have not withstood the lust for raw materials) or of perishable substances that fell prey to the ravages of time. On page 81 of Saint-Gall 902 (fig. 71), belonging to the *Recensio interpolata* of the *Revised Aratus* tradition (for a review of the nomenclature, see chapter 2), there is a diagram of a celestial globe still in its stand. Another reliable copy of the *Revised Aratus* text, which is also a ninth-century Corbie manuscript, shows such a globe on folio 63v (Paris, Bibliothèque nationale de France, MS lat. 12957; fig. 35).[129] Elly Dekker has catalogued such medieval representations of celestial globes, which often accompany the *Involutio sphaerae* (Turning of the sphere). Under Dekker's perspicacious analysis, the image from Corbie supplies yet another layer of circumstantial evidence for the sophistication of astronomical study at Adalhard's home monastery in the first half of the ninth century. The Corbie presentation of a celestial globe accurately includes not only a visualization of the pin used to stabilize the globe and designate true north, called a clamping screw, but uncommonly sets it within the vertical meridian ring at right angles to the horizon ring that surrounds it. The meridian ring rests squarely upon the simulacrum of a central reinforcing leg, complementing the other six laid out in hexagonal format around the circumference of the horizon ring. The celestial globe from Saint-Gall, by contrast, fails to preserve a sensible orientation of the horizon ring relative to the meridian, offering a model that rests upon a foundation that is too shaky for any practical use.[130] Dekker explains that roughly two hundred years after the creation of the archetypal *Handbook of 809*, the star pictures (which were already familiar to painters at work on Aratean manuscripts) could be affixed to a spherical wooden core in the preparation of a medieval celestial globe, according to extant letters from Gerbert of Aurillac (946–1003).[131]

The blue hue of the globe from the Drogo Sacramentary (see fig. 48) may approximate either the blue of heaven or the blue of the sea. Either type of globe could be resting in Nicodemus's lap in the historiated initial O. Claudius Ptolemy prescribed the manufacture of a star globe in his masterful mathematical description of the heavens from Hellenistic Egypt, ca. 150, in the *Almagest* VIII.3:

> We make the colour of the globe in question somewhat deep, so as to resemble, not the daytime, but rather the nighttime sky, in which the stars actually appear. . . . At that point we mark the position of the star; then we apply to it a spot of yellow colouring. . . . As for the configurations of the shapes of the individual constellations, we make them as simple as possible, connecting the stars within the same figure only by lines, which moreover should not be very different in colour from the general background of the globe. The purpose of this is, [on the one hand], not to lose the advantages of this kind of pictorial description, and [on the other], not to destroy the resemblance of the image to the original by applying a variety of colours, but rather to make it easy for us to remember and compare when we actually come to examine [the starry heaven], since we will be accustomed to the unadorned appearance of the stars in their representation on the globe too.[132]

In this context, it is fascinating to consider the extent to which the original drawings and linear forms used to create the ancient prototypical cycles of stellar illustrations actually are derived from the celestial globes, since the Aratean tradition of texts, including the *Excerptum*, describes the constellations on just such a sphere.

A poignant example is found in the archaic astronomical drawings found in the Eudoxus Papyrus, made circa 190 B.C.E. (Paris, Louvre 1, inv. 2325), so called because a clever acrostic in the scroll yields through successive initials a nod to "Eudoksou techne," or the "art/skill of Eudoxus." Despite this promising manifestation of wit, the effort at an homage to the fourth-century B.C.E. astronomer failed, since the pictures were probably added from an alternate source and, rather than *illustrating* the text, they merely complement it. Another possibility is that the pictures remained in a text that had been extensively abridged. The images, which are hopelessly out of synchronization with the text, thus impede rather than aid exegesis of the particular text.[133]

Neither straightforward simplicity nor realistic reportage was the primary artistic goal of classical or early medieval scientific illustration. From the beginning, the pictures were opportunities for creative expression. As a result, it is highly unlikely that globes ever provided a complete model that was copied exactly during the creation of a pictorial program for either a papyrus or a codex. All that would ever have influenced the creation of pictorial programs of the constellations would have been features, characteristics, or iconographic peculiarities that in their qualities informed the transformations of certain star pictures because they were of interest to a team of painters such as Drogo's four illuminators in Metz.

In this respect, it is fascinating to note the blue linear drawings of the copy of the *Handbook of 809* now in Monza (Monza f–9/176; figs. 36–37). On the one hand, these artists in Lobbes adhered to the Ptolemaic prescription for lines that were close in hue to a dark blue background color, which would have been appropriate for a celestial globe and demonstrated individual artistic reinterpretation within a specific codex. On the other hand, Ptolemy had called for linear forms that lacked embellishments so that the student's ability to identify the constellations in the night sky would not be jeopardized. Yet every cycle of star pictures, whether they accompanied a text by Aratus or Hyginus or were painted into one of the princely copies of the *Handbook of 809*, included some sort of artistic embellishment throughout the Middle Ages. When the draughtsmen in Lobbes, working on Monza f–9/176, created their pictures of the heavens, they followed a semantically charged ancient practice attached to the manufacture of celestial globes by creating a cycle of drawings in blue. It is hard to say, however, whether there is a distant connection between the blue ink of the Monza copy and an ancient star globe, given the sophistication of the detailed drawings in the handbook. The choice of blue ink gave the drawings a certain classicism that the Carolingian artists rendered in a ninth-century, contemporary style, in keeping with the connotative possibilities for hues that we have seen advocated by Hrabanus at the outset of this chapter. Blue signaled not only the cerulean sky over the Frankish heartland but the empyrean haven of the Heavenly City, attainable for the living through the intellectual journey or elevated spiritual rapture that contemplation of the stars engendered.

On the evidence supplied by the Paris copy of the *Revised Aratus* (MS lat. 12957; fig. 35), which was made in Corbie, we know that Adalhard and medieval Franks were familiar with celestial globes. The odd form of the satyr used for Sagittarius (plate 12) as well as the use of the form of Engonasin for Hercules (plate 2) in the archetypal *Handbook of 809*, which subsequently found their way into the princely copies of the text, could have been derived from an ancient Aratean source for the star pictures.[134] There were, of course, the iconographic connections between passages of the Aratus poem, the *Phaenomena*, and the images of these constellations.

Another possibility is that certain images from the *Handbook of 809*, such as that of the satyr form of Sagittarius, came to be included in the handbook as a direct result of the Carolingian observation of actual celestial globes or representations of them in drawings, mediated by late antique precedent and perhaps even facilitated by pilgrim traffic.[135] The image of Sagittarius in Drogo's copy of the *Handbook of 809* resembles the disfigured image on a copy of the oldest recorded star globe, housed in the National Archaeological Museum of Naples (inv. 6374). The monumental Farnese Globe is a first- or second-century

Roman celestial globe with a diameter of 65 cm, which with moderate reliability records a lost Hellenistic celestial globe that could have been created as early as the second century B.C.E. (after 128 B.C.E., according to Dekker), provided that the less than systematic distribution of equinoctial colures and depicted constellations are considered in conjunction with the concept of precession (figs. 54–55).[136] The current appearance of Atlas resulted from successive restorations ending in 1562 and 1796, causing losses to the surface of the globe *inter alia* around the poles, as noted by Dekker and others. Martin Folkes (1690–1754) executed the stereographic projection of the Farnese Globe, supplying a problematic illustration of the Atlante Farnese for Richard Bentley's Manilius of 1739 containing the *Astronomica* (fig. 53).[137] As the engraving more easily reveals, however, the visible image of Sagittarius could have been interpreted, perhaps even "incorrectly," as that of a bipedal satyr (figs. 53–55), and hypothetically this detail could also have found its way into a tradition of decorative celestial globes from this or another source.[138]

In other words, to label the Farnese Globe as the source of this image in the *Handbook of 809* ignores the rich variety of iconographic types available to ninth-century artists, as discussed in chapter 2.[139] Instead, it is vital to reflect upon the connotative symbolic contents alluded to by certain iconographic forms and the metonymic associations they would have fostered within ninth-century Frankish minds. Egyptian precedents carried less cultural freight, the authority of Ptolemy notwithstanding, than discernibly Roman or early Christian, especially Constantinian, models. The panoply of variant forms resulting in multifarious copies of copies of zodiacs and star pictures throughout the early medieval world does not entail an entirely indifferent selection process: certain forms became hegemonic by association, and the archetypal *Handbook of 809* was one important Frankish standard-bearer in this respect.

Throughout this chapter, the semiotic possibilities for a scientific and visual *translatio imperii* have been underscored, and yet another instance of this Carolingian phenomenon is the decision to adopt the satyr form of Sagittarius for the archetypal pictorial program of the *Handbook of 809* and most princely copies of it. This classically inspired alternative form of Sagittarius as a satyr was also, however, considered to be an (even if perhaps not *the*) accurate form for the constellation. Consideration of additional Frankish examples of the satyr as Sagittarius helps clarify the iconography's origin, metonymic associations, and additional connotations. It is important to recall, however, that in the copy made for Abbot Louis of Saint-Denis (Reg. lat. 309, ca. 859), the artists had already reverted to the traditional Centaur form in their rendering of the constellation on folio 97r (fig. 72). The Carolingian artists who painted the archetypal *Handbook of 809* in Aachen, and the Messine masters who followed their lead, drew upon countervailing polycyclic options funneled through northern French and Frankish monasteries such as Corbie and based on Italian early medieval precedents.

The satyr was an astronomically or scientifically apposite form in Carolingian images of the autumnal and winter celestial hemisphere (bottom, fig. 73) or the zodiac wheel (fig. 52) from the Saint-Gall *Recensio interpolata* manuscript (Saint-Gall 902) discussed above.[140] The images for constellations of the zodiac such as Aquarius, Aries, Capricorn, Pisces, and Virgo in Drogo's copy of the *Handbook of 809* also bore iconographic features that linked them closely to the zodiac wheel from the *Recensio interpolata*. The *Recensio interpolata* of the *Revised Aratus* greatly

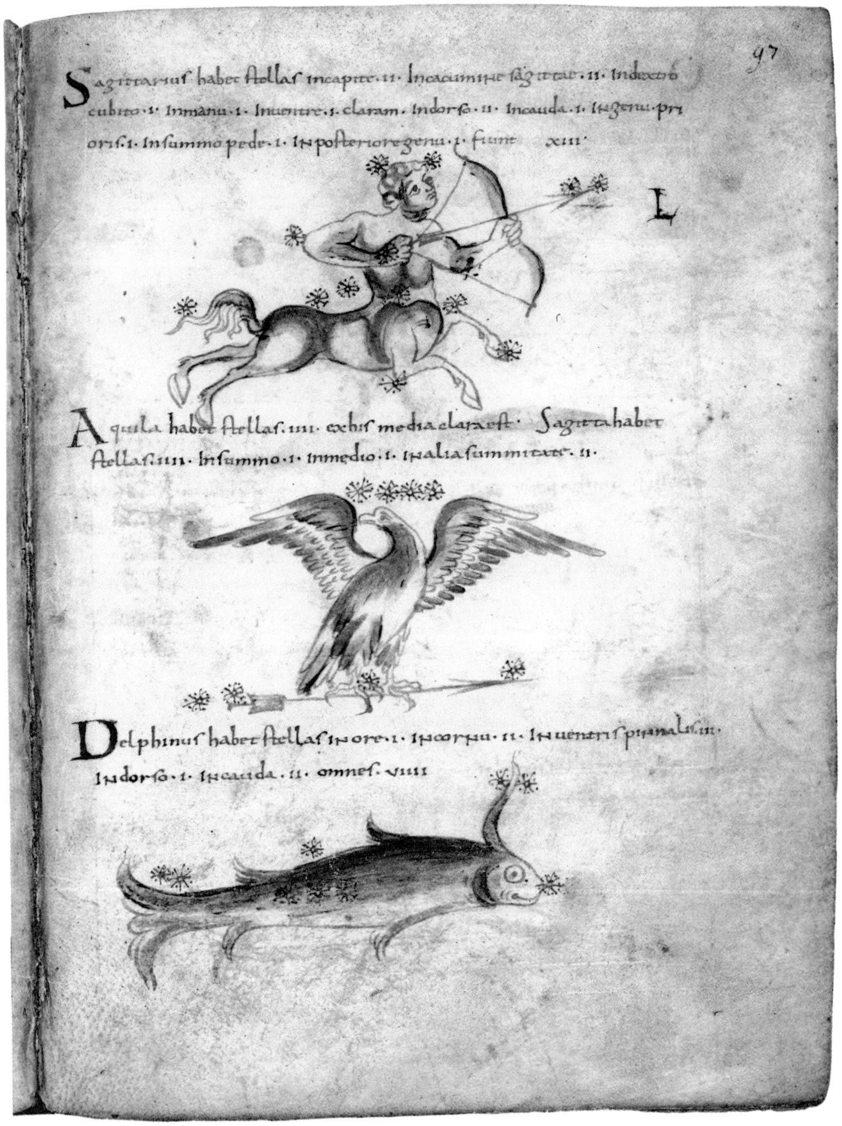

FIGURE 72

Sagittarius, Aquila, and Delphinus. Vatican City, Biblioteca Apostolica Vaticana, Reg. lat. 309, fol. 97r. Photo © 2017 Biblioteca Apostolica Vaticana. Reproduced by permission of Biblioteca Apostolica Vaticana, with all rights reserved.

influenced Carolingian star pictures and the further development of star pictures in medieval art. The depictions of the zodiac on the celestial hemispheres in the manuscripts of the *Recensio interpolata* share an evident heritage with the star pictures on lost, ancient celestial globes. Such charts provide important records and planar projections of the lost globes, which are in turn linked metonymically to a lost and authoritative astronomical tradition. Resurrecting these forms in the archetypal *Handbook of 809* encouraged its reception as an authoritative book and reflected the endemic form of exegetical emendation throughout Frankish scriptoria fostering the Christian acculturation and appropriation of *Latinitas*.

Archetypal Considerations

One requirement of the star pictures displayed on an ideal celestial globe was that they be depicted from the perspective of a divine observer in outer space, not someone standing on the earth. Elly Dekker has explained that the celestial globe maker's job is thus the opposite of that of the professional stargazer, who must adhere to a principle she labels "Hipparchus' Rule," after the classical Greek astronomer Hipparchus (fl. 130 B.C.E.), who was instrumental in the early propagation of mathematical observation and presented star placements with calculated and verifiable precision in his *Commentary to Aratus*. The mathematical astronomer seeks to adduce by reproducible means where stars are situated in the night sky as seen from the vantage point of the terrestrial horizon. A celestial globe capturing the external view of the sphere of the fixed stars, in contrast, displays the dorsal sides of the figures of the constellations.[141] The upshot is that a viewer who wishes to assess the appearance of constellations on earth using a celestial globe must mentally unpack the globe's visual information. First, as explained by Dekker, the imagery on an ideal globe needs to be transposed in its left-right orientation like a reflection in a mirror. Second, the images can require back-to-front reversal for reinterpretation.[142] For example, the depiction of Leo in Drogo's copy of the *Handbook of 809* (plate 6) leaps to the left, as the image of the constellation would appear to do beyond the dome of the heavens; from the earth, however, the lion would appear with its head to the right and tail toward the left. In other words, the image of Leo in Drogo's copy displays the left-right orientation appropriate to an image on a star globe, as seen on the Farnese Globe (figs. 53–55).

In the case of Sagittarius, the picture in the handbook relies on the satyr (rather than centaur) form arguably associated with star globes, yet the second master painter of Drogo's copy of the *Handbook of 809* oriented the satyr figure facing right (plate 12). In other words, the iconographic form of the constellation Sagittarius in Drogo's copy of the *Handbook of 809* and its archetype, according to this interpretation, came from a tradition associated with celestial globes, but the artists *reversed* the typical left-right orientation appropriate to that iconographic tradition. This suggests that these early medieval artists did not always consistently follow their own editorial decisions—or perhaps that such consistency was not an artistic or iconographic priority for ninth-century artists, who adapted their medieval programs *ad hoc*, drawing polycyclically on competing precedents with an eye to the classical authority of the pictorial source, interpreted creatively.

The depictions of the constellations in Drogo's copy of the *Handbook of 809* that do conform in their left-right orientation to the depictions on an ideal

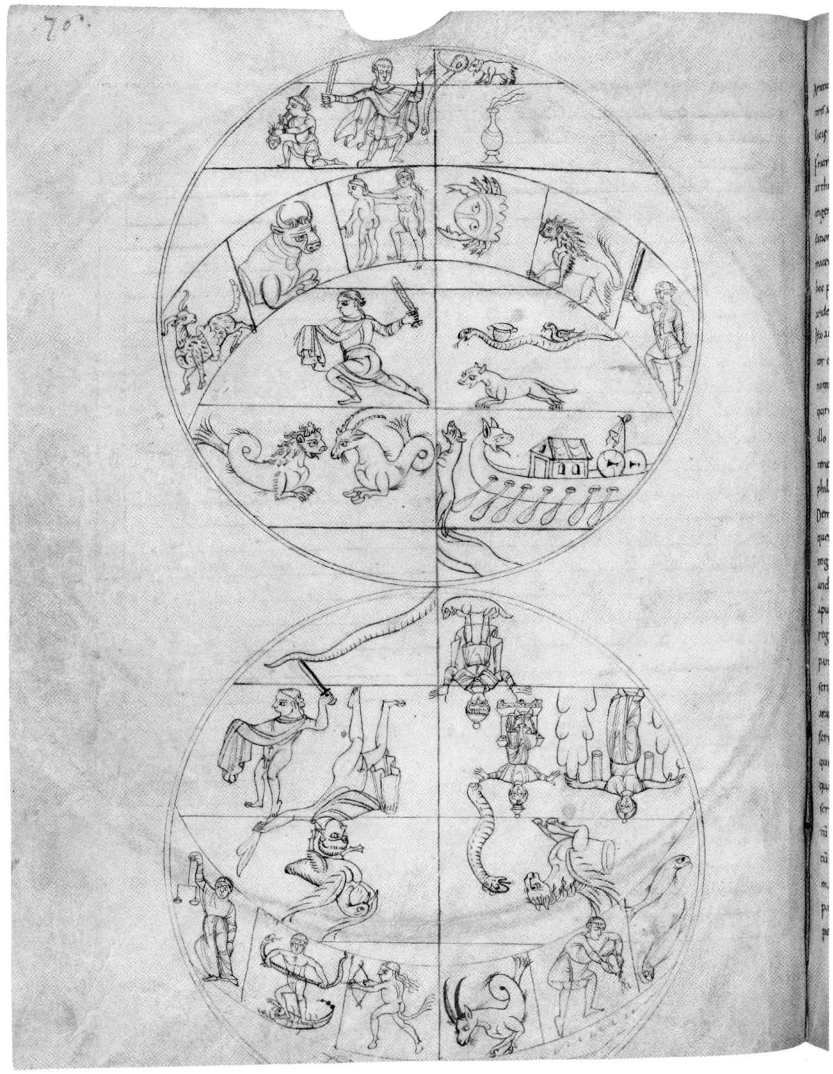

FIGURE 73

Summer and winter celestial hemispheres. Stiftsbibliothek St. Gallen, Cod. Sang. 902, page 76. Photo: Stiftsbibliothek St. Gallen.

star globe are Ursa Minor, Serpens, Serpentarius, Cancer, Leo, Auriga, Pegasus, Capricorn, Navis, Heridanus, Centaurus, and Corvus. Cepheus, Cassiopeia, and Andromeda tend to be portrayed in somewhat frontal poses, whether treated as constellations viewed from the earth or from the heavens (fig. 53). In addition, some constellations have fairly generic shapes that look alike whether seen from heaven or earth: Corona, Deltoton, Pisces, Lyra, Ara, and Crater. At best, then, roughly twenty-one out of forty-two depictions, or half, of the constellations in Drogo's *Handbook of 809* are related directly to the sorts of images found on ideal celestial globes. It is preferable to investigate when and why certain pictures, such as the satyr form of Sagittarius, are related to and derived from the imagery on celestial globes than to postulate an original set of Aratean pictures derived from globes alone. Each cycle of star pictures displays a significant individuality, which permeates its imagery and suffuses the depictions of the constellations with both artistic and scientific meaning.

Rather than seeking a lost pristine cycle of star pictures that could be somewhat artificially foisted upon the lost archetypal *Handbook of 809*, we would do better to interrogate the individual differences that arose and manifested themselves in discrete pictorial programs. In the case of Drogo's copy of the *Handbook of 809*, Carolingian prelates had a predilection for meteorological matters. Adalhard of Corbie probably adopted the atypical form of the satyr for his historic image of Sagittarius that was included in the original *Handbook of 809* made in Aachen. The form added classical cachet and asserted a sophisticated, scientific formal position about Sagittarius. The way the second master painter in Metz, whose work is related to painting in Reims on the Loisel Gospels, executed his depiction of the satyr resulted in a visually arresting rendition to which additional symbolism could be attached in keeping with larger Carolingian concerns and themes in a form of exegetical emendation; more will be said about this aspect of the Frankish star pictures in chapter 4. At each stage of the creation of these astronomical picture books, there was an opportunity for creativity. The forty-two paintings of the constellations and the diagrams in Drogo's copy of the *Handbook of 809* are thus best remembered as the fruits of four men working in Metz.

Although the text and diagrams belonging to book V of the copy of the *Handbook of 809* made for Drogo and the one made for Abbot Hubert of Lobbes (Monza f–9/176) resemble each other closely, there is less general consistency displayed by the cycles of diagrams found in other princely copies of the book. This was probably the result of the rapidly developing interest in planetary theory during the Carolingian period. As texts continued to circulate, diagrams considered to be old-fashioned were either embellished artistically or replaced, continuously treated as works-in-progress subject to emendation in a period of intellectual ferment. The diagram of planetary order within the solar system in Abbot Hugo of Saint-Quentin's copy (Vat. lat. 645) on folio 67v (fig. 74), for example, includes a personification of Terra in the first radial circle of the diagram (treated as a roundel), unlike the corresponding diagrams in the copy made for Drogo or in Monza f–9/176. This should be considered an example of individual artistic reinterpretation within a specific codex.

The diagram of Pythagorean intervals on folio 66v of Vat. lat. 645 precedes, rather than follows, the diagram of the planetary order in the solar system. That diagram of intervals also incorrectly identifies the interval between Venus and the sun as three tones rather than three semitones, even though the distance between Saturn and the fixed stars is correctly listed as three semitones. Before the appearance of these diagrams, Vat. lat. 645 also includes an intriguing but atypical diagram of a wind rose on folio 66r (fig. 9), as explained in chapter 1. There was not necessarily a wind rose in the lost archetypal *Handbook of 809*, but such a diagrammatic illustration might have invited a complementary meditative function, and therefore its inclusion would be appropriate in the book, perhaps at the very beginning as an overarching frontispiece. The texts for the diagrams in Vat. lat. 645 follow an initial set of diagrams clustered together in a package. The scribe for each of the texts left no space on the respective parchment leaves for diagrams to be inserted. Someone probably excised the lost diagrams of the apsides and latitudes at a later time. As noted already, the text for the diagram of the apsides referred its reader to an accompanying diagram *below/subiecta*.[143] As a result, it can be inferred that the archetypal arrangement of the diagrams proceeded according to the layout and design scheme seen in Drogo's copy of the *Handbook of 809* and not the one seen in Vat. lat. 645. The text preceded its illustration, and both were intended to create complementary syntactic units, advancing the trend for a paratactic presentation of pictorial imagery in the handbook. These textual and visual syntactic units were therefore intended to come in pairs in the original *Handbook of 809*. Lastly, the two unfinished diagrams in Drogo's copy of the *Handbook of 809* on folios 67v and 68r are insertions that bear no relationship to the original *Handbook of 809*.

In terms of design and iconography—setting aside the fact that Drogo's copy of the *Handbook of 809* was painted—the pictorial cycles in the princely copies

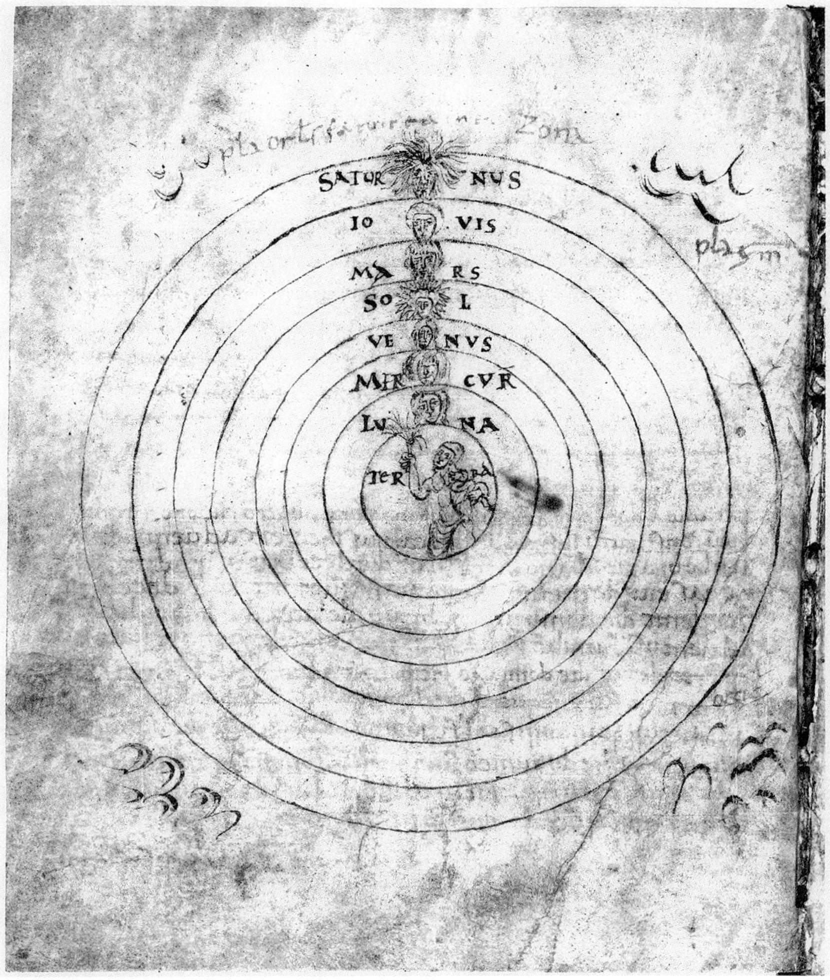

FIGURE 74

"The Order of the Planets" diagram. Vatican City, Biblioteca Apostolica Vaticana, Vat. lat. 645, fol. 67v. Photo © 2017 Biblioteca Apostolica Vaticana. Reproduced by permission of Biblioteca Apostolica Vaticana, with all rights reserved.

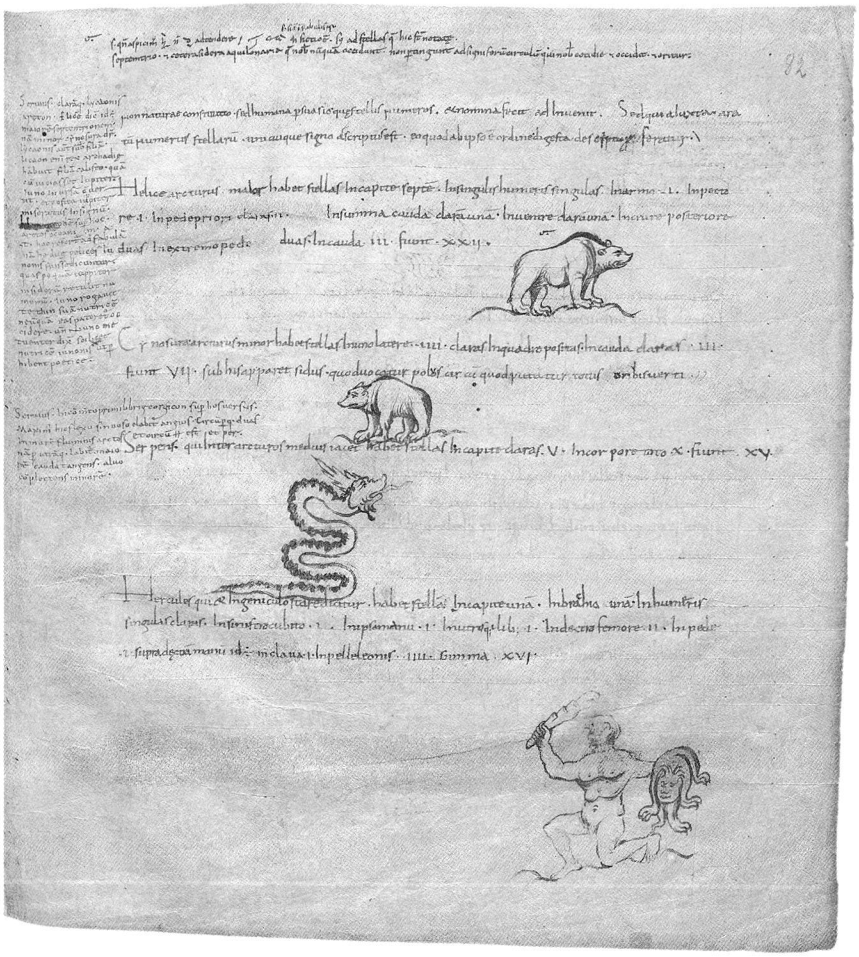

FIGURE 75

Haelice/Arcturus Maior, Cinosura/Arcturus Minor, Serpens, and Hercules. Staatsbibliothek zu Berlin, Stiftung Preußischer Kulturbesitz, Hs. Phillipps 1832, fol. 82r. Photo: bpk, Berlin / Staatsbibliothek zu Berlin / Art Resource, New York.

made for Drogo, Hubert, and Hugo (Madrid 3307, Monza f–9/176, and Vat. lat. 645, respectively) formally resemble one another with an overall degree of consistency. The iconographic peculiarities displayed by the imagery in the copy of the *Handbook of 809* made for Abbot Louis of Saint-Denis (Reg. lat. 309, ca. 859) are often explained by appeal to the third of Borst's redactions (type C) of astronomical-computistical-pedagogical handbooks, such as Munich 210. It is preferable, however, to consider the format, layout, and arrangement of Drogo's copy of the *Handbook of 809*, reflected in the preponderance of the other princely copies, as closest to the painted archetypal cycle of imagery contained within the lost original *Handbook of 809*, which was probably compiled under Adalhard's supervision in Aachen by 812.

Although creative interventions can appear with the manufacture of any codex, it is possible to be more precise about certain aspects of the lost archetypal program of illustration in the *Handbook of 809*. The left-right orientation of the figures of the constellations can be determined according to the following criteria. Princely copies of the *Handbook of 809* are treated as the chief witnesses of this astronomical tradition. Abbot Louis's copy (Reg. lat. 309) includes a cycle of illustration that recombines various examples of Aratean star pictures into a new image cycle. This rendered that copy an innovative work of Carolingian astronomical art and therefore an unreliable record of the archetype. Whenever a verifiable agreement between form and orientation presents itself for the figures of the constellations found throughout the other three princely copies of the *Handbook of 809*, those images probably reflect the forms of the star pictures from the archetypal manuscript. It is impossible to determine details in the lost archetype, such as whether the image of the

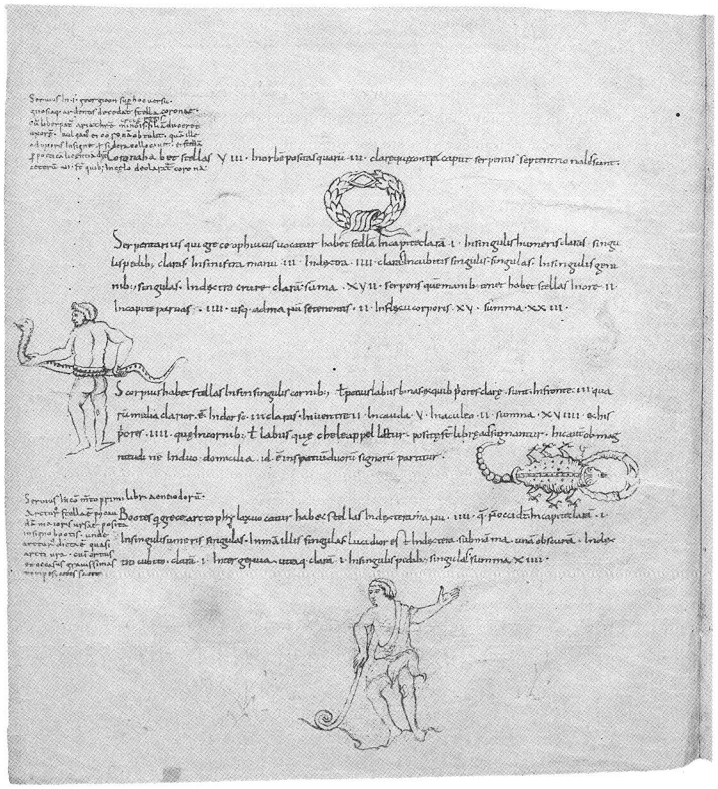

FIGURE 76

Corona, Serpentarius/Ophiuchus, Scorpio, and Boötes/Arctophylax. Staatsbibliothek zu Berlin, Stiftung Preußischer Kulturbesitz, Hs. Phillipps 1832, fol. 82v. Photo: bpk, Berlin / Staatsbibliothek zu Berlin / Art Resource, New York.

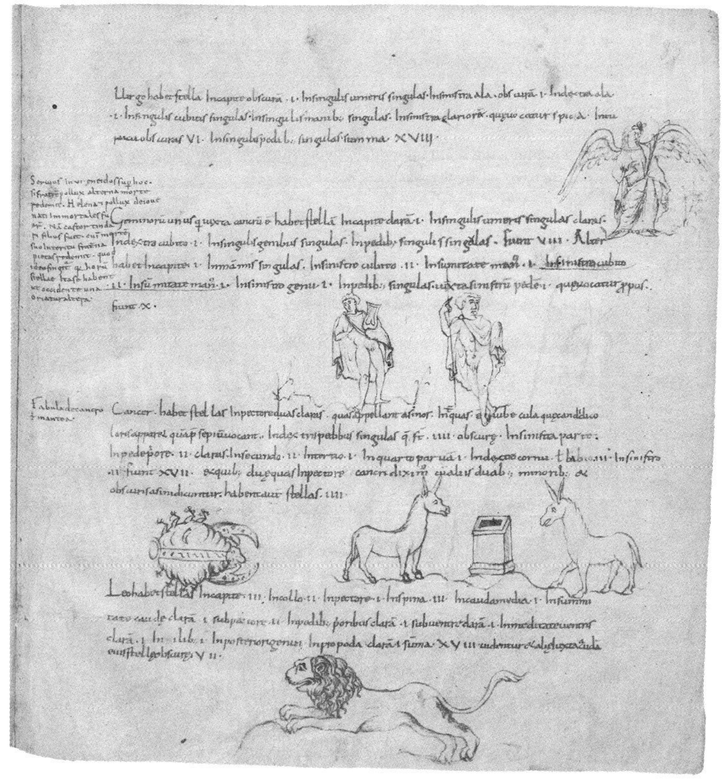

FIGURE 77

Virgo, Gemini, Cancer, and Leo. Staatsbibliothek zu Berlin, Stiftung Preußischer Kulturbesitz, Hs. Phillipps 1832, fol. 83r. Photo: bpk, Berlin / Staatsbibliothek zu Berlin / Art Resource, New York.

lyre in the archetype had five, eight, or ten strings. But, given these criteria, the following depictions of the constellations in Drogo's copy of the *Handbook of 809* more likely than not closely resemble those of the archetype prepared in Aachen: Ursa Major, Ursa Minor, Serpens, Hercules, Serpentarius, Scorpio, Boötes, Cancer, Leo, Auriga, Taurus, Cepheus, Cassiopeia, Andromeda, Aries, Deltoton, Pisces, Perseus, Lyra, Aquarius, Capricorn, Sagittarius, Delphinus, Orion, Canis Major, Lepus, Navis, Coetus, Heridanus, Piscis Magnus, Ara, Centaur, Hydra, Corvus, Crater, and Canis Minor. The differences among the remaining depictions are limited to only six out of forty-two constellation pictures: Corona, Virgo, Gemini, Pegasus, Cygnus, and Aquila.

A manuscript in Berlin, containing a copy of the *De ordine*, probably resulted from a ninth-century literary exchange between Laon and Corbie. This

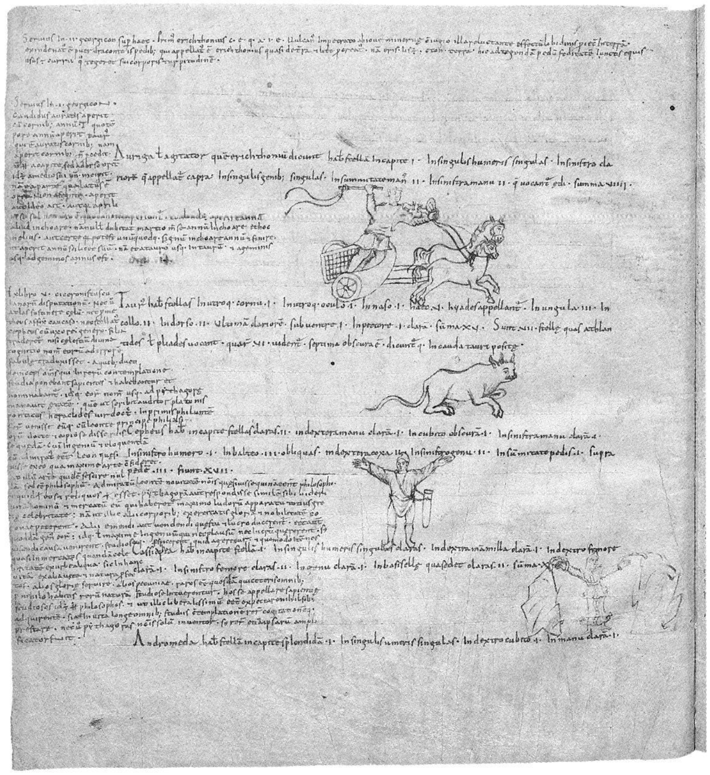

FIGURE 78

Auriga, Taurus, Cepheus, and Andromeda. Staatsbibliothek zu Berlin, Stiftung Preußischer Kulturbesitz, Hs. Phillipps 1832, fol. 83v. Photo: bpk, Berlin / Staatsbibliothek zu Berlin / Art Resource, New York.

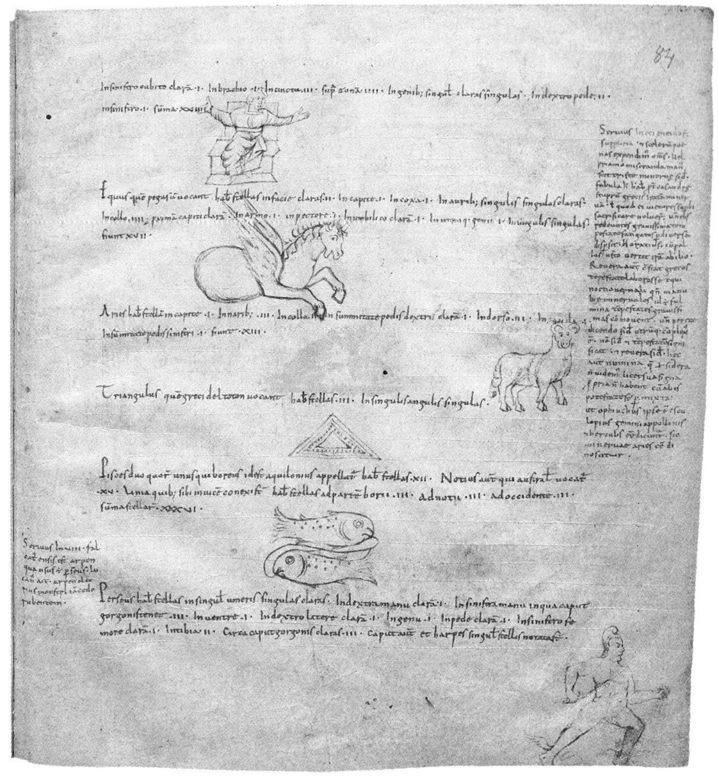

FIGURE 79

Cassiopeia, Pegasus, Aries, Deltoton, Pisces (Aquilonalis and Australis), and Perseus. Staatsbibliothek zu Berlin, Stiftung Preußischer Kulturbesitz, Hs. Phillipps 1832, fol. 84r. Photo: bpk, Berlin / Staatsbibliothek zu Berlin / Art Resource, New York.

manuscript can help resolve lingering questions about the archetypal program of star pictures from the manuscript in Aachen. As I have already suggested, the archetypal program of illustrations for the *Handbook of 809* was probably created in rough form by a team of astronomically minded artists alongside local monks, working at Corbie under the supervision of Adalhard. Additional support for this hypothesis comes from the Berlin *Ordine* and another with which it was formerly conjoined (Preußischer Kulturbesitz, Staatsbibliothek zu Berlin, MS Phill. 1830).[144] Both manuscripts, which have been studied by Contreni, include the personal notes of Martin Hiberniensis (819–875) of Laon and were made in the third quarter of the ninth century. Martin's personal library included a commentary on Matthew by Paschasius Radbertus (786–860, a deacon and abbot of Corbie) that attested to the exchange of books

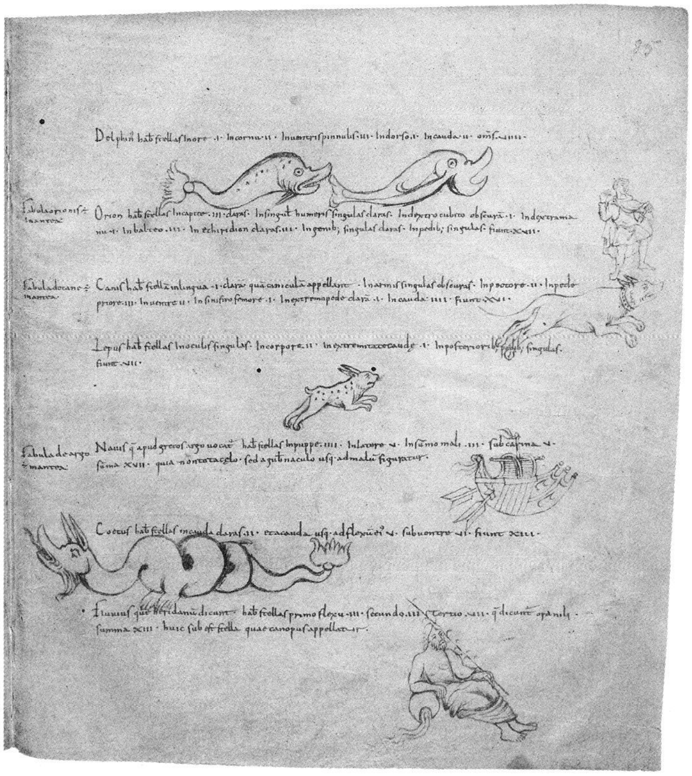

FIGURE 80

Delphinus, Orion, Canis Major, Hare, Argo, Coetus, and Heridanus/River Po. Staatsbibliothek zu Berlin, Stiftung Preußischer Kulturbesitz, Hs. Phillipps 1832, fol. 85r. Photo: bpk, Berlin / Staatsbibliothek zu Berlin / Art Resource, New York.

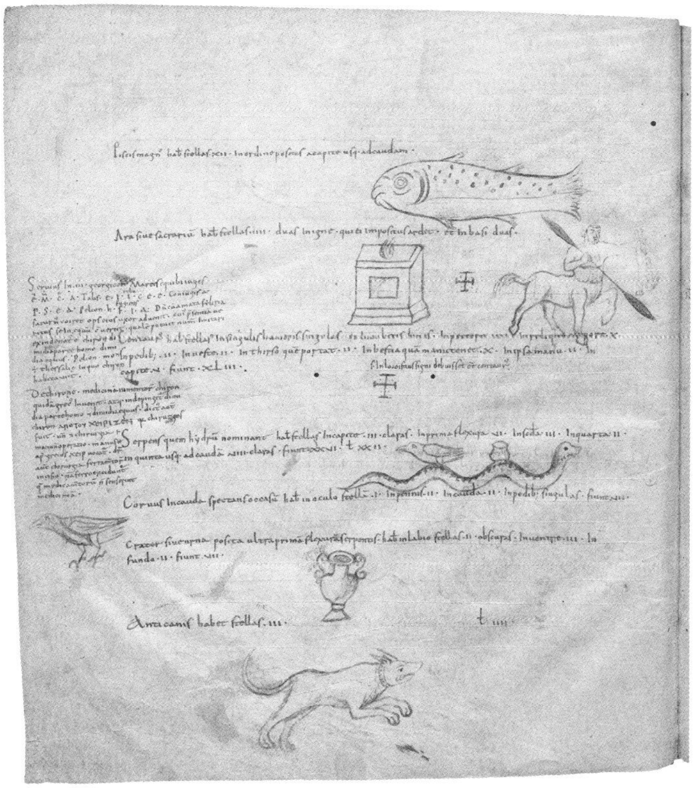

FIGURE 81

Piscis Magnus, Ara/The Shrine, Centaur, Serpens/Hydra, Corvus/The Raven, Crater/The Urn, and Anticanis/Canis Minor. Staatsbibliothek zu Berlin, Stiftung Preußischer Kulturbesitz, Hs. Phillipps 1832, fol. 85v. Photo: bpk, Berlin / Staatsbibliothek zu Berlin / Art Resource, New York.

between Laon and Corbie in the ninth century.[145] During the 870s, it is likely that a copy of the pictorial program that Adalhard had included within the copy of the *De ordine* could still be found in Corbie. This cycle of illustrations also presumably inspired the pictorial program of the Berlin *Ordine* (figs. 41, 75–82), which so closely resembles the pictures in Drogo's copy of the *Handbook of 809*.

As a result, the collection of star pictures in the Berlin manuscript provides another window onto the original pictorial program Adalhard recommended for inclusion in the archetypal *Handbook of 809*. The pictorial program in the Berlin manuscript closely resembles the imagery in Drogo's book and in the copy now in Monza made for Abbot Hubert. Rare differences include the backward-turned head of Capricorn on folio 84v of the Berlin manuscript (plate 11

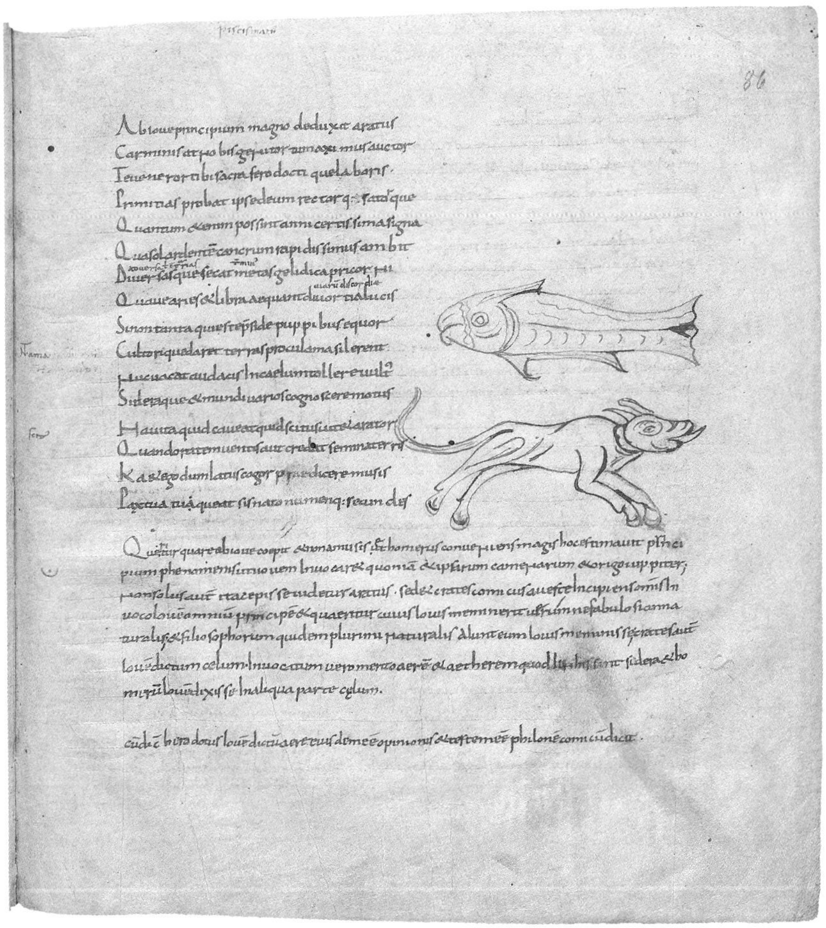

FIGURE 82

Piscis Magnus and Canis Major or Anticanis/Canis Minor. Staatsbibliothek zu Berlin, Stiftung Preußischer Kulturbesitz, Hs. Phillipps 1832, fol. 86r. Photo: bpk, Berlin / Staatsbibliothek zu Berlin / Art Resource, New York.

and fig. 41) and the lack of a lyre in Amphion's left hand in the depiction for Abbot Hubert on folio 63v (figs. 36 and 77). These are, however, exceptions that demonstrate the utility of the Berlin *Ordine* for the reconstruction of the archetypal pictorial program of the *Handbook of 809*. Since the pictorial cycles in the two codices are not identical for the *De ordine*, neither is a straightforward copy of the other. Overall, the pictures in the Berlin manuscript resemble the forms of the six remaining problematic star pictures as they appear in the copies made for Drogo and Hubert. In the case of the Gemini (fig. 36), the appearance of the lyre in the Berlin manuscript (fig. 77), the copy of the *Handbook of 809* made for Drogo (plate 5), and the copy for Abbot Hugo (Vat. lat. 645; fig. 38) suggests that the reference to Amphion was originally intended. This provides additional evidence that those forms were directly related to the archetypal program prepared by Adalhard in Corbie and then introduced into the archetypal *Handbook of 809* in Aachen. Consequently, the miniatures and diagrams of Drogo's copy of the *Handbook of 809*, minus the unfinished sketches for additional diagrams on folios 67v and 68r, provide a faithful record rather than an exact copy of the lost archetype of the *Handbook of 809*. The images of the constellations in Madrid 3307 implied their authority through the introduction of erudite classical references. The iconography on display permitted their ideological exploitation and exegetical emendation, revealed in Metz for Drogo by two master painters and their assistants.

CHAPTER FOUR

RESTORING WHAT WAS LOST

Astronomy and Natural Astrology in the Carolingian Era

When Einhard (ca. 770–840), Charlemagne's biographer, reported that the emperor "learned the art of calculation and with deep purpose and great curiosity investigated the movement of the stars,"[1] he was not merely emphasizing Charlemagne's respect for the liberal arts. Einhard embedded his encomium between remarks about the ruler's moderation at table—he rarely drank "more than three times during a meal"[2]—and a reminder of Charlemagne's pious patronage of the "church of stunning beauty at Aachen."[3] According to Einhard, Charlemagne was an appropriate type of Frankish ruler, refining himself through the study of liberal arts such as astronomy, which he undertook with the assistance of his learned Anglo-Saxon teacher Alcuin (see the introduction and chapter 1), first at court and later through written correspondence after Alcuin's elevation as abbot of Tours in 796.[4]

In his personal pursuit of learning, Charlemagne sought to exemplify the traits of a wise spiritual patriarch, ready to lead his people by example. In this he was following the example of Abraham, who allegedly taught the "Egyptians" the ways of "astrology" discovered by the Chaldeans—a point made much earlier by Isidore of Seville in book III of the *Etymologies*, where he summarizes the origins of astronomy and astrology.[5] Isidore had in fact begun the *Etymologies* for his Visigothic patron and sovereign, Sisebut (d. 621), to elevate the king's intellect and satisfy his early medieval courtly interest in arcane mysteries "compiled from [Isidore's] recollection of readings from antiquity."[6] Charlemagne's intellectual refinement benefited the Frankish lands under reform, and a similar reward awaited students of the liberal arts in the monastic and cathedral schools.

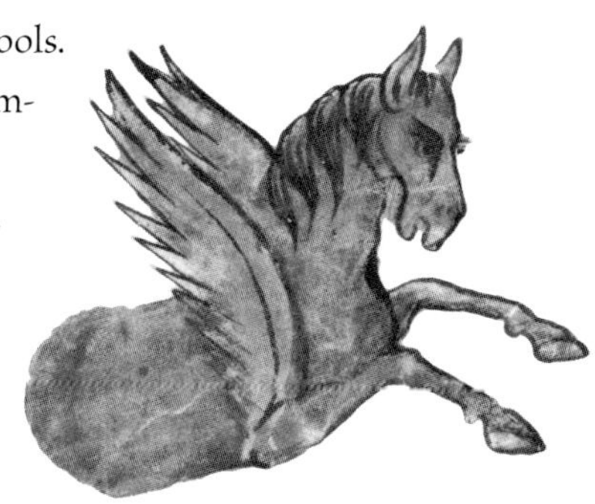

This was not just an epiphenomenon of the need for administrative laborers, scribes, the *consiliarii* who drafted capitularies, and the *missi* who relayed them to the people.[7] As Janet Nelson and David Ganz have both argued,

Charlemagne assumed the responsibility of the Old Testament patriarch. He embodied through his royal office the role of sacred vicar of Christ to his people and upheld theological dogma as a matter of personal responsibility, supervising conciliar decisions and calling for public fora where doctrine or the limits of *scientia* could be established, such as at the Aachen synod of 809.[8] This complete intersection of political power and spiritual responsibility is also reflected from the standpoint of the liberal arts in Drogo's copy of the *Handbook of 809*. In addition, a thorough review in the schools of the opinions of the patristic fathers concerning astronomy and astrology, as well as the original texts of Carolingian exegetes, would have aimed at nothing less, in Walter Ullman's classic formulation, than "the attempt at a regeneration, at a *renovatio* or renewal, at a rebirth of the Frankish people."[9]

The situation was more complicated during the reign of Louis the Pious, whose public penance in 833 resulted from a short-lived usurpation of the throne, until Ebo of Reims and the other perfidious clerics saw the king reinstated in 835 at Drogo's cathedral of Saint-Étienne in Metz. This period of royal tribulation coincided with the classicizing era in the visual arts during which Drogo's handbook was created in Metz. Still, discipline in what Mayke de Jong has labeled "the penitential state" played a vital role in undergirding the spiritual sanction of the king's political authority recognized by the episcopal hierarchy.[10] And, at the intersection of power and knowledge, prelates oversaw the modes of inculcation in the ecclesiastically backed cathedral schools. The discipline of the liberal arts was also an integral component of the individual sanctification of the Frankish people, as John Contreni has argued. According to Contreni, John Scottus Eriugena (ca. 800–ca. 877), whose writings contributed to the intellectual climate of the Franks during the reign of Charles the Bald (who was emperor between 875 and 877 but was king of western Francia since 840), made the fundamental point explicitly clear when he argued that "learning was not only an aid toward the achievement of Christian wisdom, but a means of salvation itself."[11]

Such study, however, required the complex negotiation of pagan spiritual traditions.[12] The constellations in the heavens, Latin translations of the poetic *Phaenomena* of Aratus, and the *Naturalis historia* of Pliny all relied on pernicious mythological references to the pagan past. As part of the larger Carolingian reform projects, it became necessary to assess what Frankish boys (and girls) should be studying about *astronomia* or *astrologia*.[13] This resulted in the illustrated compendium, the *Handbook of 809*, that provided a normative collection of ninth-century ideas about the heavens. As demonstrated in chapters 2 and 3, a critical examination of the text and star pictures from Drogo's copy of the *Handbook of 809* (Madrid 3307) reveals the complex early medieval attempts to understand the meaning and significance of *scientia*, *astronomia*, and *astrologia* that culminated in the handbook, a concerted effort at scientific sacralization within the Carolingian schools.

The knowledge of antiquity, inscribed upon the hearts and minds of Carolingian pupils, would not only transform the cultural wealth of classical *paideia* into an attractive trove offered by theology's ninth-century scientific handmaiden.[14] A close reading of Hrabanus Maurus's exegesis of the Isidorean distinction between *natural* and *superstitious* astrology in his *De institutione clericorum* (On the education of clerics, 819; hereafter the *Education*) clarifies these terms as they were used by ninth-century Carolingians. Rather than adopt a polemical or disinterested stance before star pictures such as those in the *Handbook of 809*, Carolingian prelates justified their study of astronomical constellations and zodiac signs on practical as well as theoretical grounds. Both *theoria* and *praxis* played an important role in the development of Carolingian principles, regulating that which should be studied about astronomy in Frankish schools. As will be argued below, the codification of certain normative, permissible pathways to knowledge about the heavens also resulted in technologies of social regulation that would shape the ways in which Carolingians monitored the night sky, divided spatially into the twelve signs of the zodiac.

In this final chapter, the main theme of part 2 (and the book as a whole) is carried to completion: Drogo's copy of the *Handbook of 809* was a Carolingian tool facilitating the sacred study of astronomy through the creative, exegetically rich pictorial program painted by four illuminators in medieval Metz. The significance of Drogo's copy of the *Handbook of 809* within the intellectual climate of the ninth century can be evaluated through a critical appraisal of three images: those of Hercules, Cancer, and Chiron the Centaur. In order to assess what did and did not cause Franks anxiety before the Carolingian star pictures and zodiacal cycles, I examine the *De cursu stellarum* of Gregory of Tours. A critical reading of the passage on the liberal art of astronomy from Hrabanus Maurus's *Education* offers a firsthand avowal of the benefits of astronomical study. This requires a brief history of the liberal arts before we evaluate Hrabanus's sources in Cassiodorus, Augustine, and Isidore. The critical reading ends with an exegesis of Hrabanus's conclusion to his text, arguing for the practical benefits of astronomical study. In what follows, the equally important spiritual benefits of study are underscored, with an emphasis upon Alcuin, Charlemagne's tutor, and—as a temporal bookend to the Carolingian period—Remigius of Auxerre. These considerations complement the ensuing semiotic discussion, which assesses the aesthetic and semantic status of the images of the constellations in Drogo's copy of the *Handbook of 809*. The union of period Frankish concerns and a Carolingian semiotic theory for the star pictures reveals that the images of the constellations in Drogo's handbook participated in widespread technologies of scientific soteriology through exegetical emendation.

Shadowboxing Serpents: Teaching the Liberal Arts in the Frankish World

Three illustrations from the *De ordine*, book V.2 of the *Handbook of 809*—Hercules, Cancer, and Chiron the Centaur—reveal the ways in which *astronomia* and *astrologia* were variously interpreted in the ninth century. Miniatures of these constellations from Drogo's copy of the handbook display, through artistry that encompasses both the creative omission and the insertion of visual details, a concerted Frankish effort to separate the acceptable from the deleterious aspects of astronomical study. In book V of the *Handbook of 809*, the two new texts, the *Excerptum* (V.1) and the *De ordine* (V.2), were derived from recent translations of the third-century B.C.E. *Phaenomena* of Aratus, executed at Corbie during the eighth century. The *Excerptum* harks back to the second of these translation projects in Corbie, known as the *Revised Aratus*. The *De ordine* derives in part from a set of Greek scholia, translated into Latin in the late third or early fourth century, that accompanied a Latin translation of the Aratean *Phaenomena* (which itself relied heavily on the first-century translation, of 15 B.C.E.–19 C.E., attributed to Germanicus Caesar).[15]

In the *De ordine* from the *Handbook of 809*, Adalhard's team pays homage to the Aratean literary tradition, from which the star pictures came: "[I]n accordance with Aratus the number of stars belonging to each sign has been appointed; the description should follow according to the order, which is his own."[16] The anonymous author(s) had emphasized that the constellations were "admitted into the heavens from some models or fables" on account of "human conviction."[17]

In the first of three case studies that illustrate how Frankish exegetes and artists grappled with the pagan legacy of their constellations, the miniature of Hercules from Drogo's copy (plate 2) displays slightly modeled musculature, the pelt of the Nemean lion (bearing an anthropomorphic visage), and the hero's club. The shape also gathers together the missing stars, which were arbitrarily chosen by ancient stargazers and grouped together into the cluster identified as the constellation. As the introduction to the *De ordine* claimed, the ingenuity of earlier sailors, farmers, and astrologers had created the forms of the constellations. These star groupings participated in a socially codified nexus of symbolic and mnemonic associations as inventory fables, which have been described by Mary Carruthers. Carruthers claims that star pictures like those included in the *De ordine* offered students and teachers consulting astronomical compilations such as the *Handbook of 809* a definitive set of constellations operating as "mnemotechnical tools."[18] The star patterns identified with specific constellations were selected for practical purposes. They were determined neither by necessity nor capriciously; they gave form to certain protagonists from myths and fables—forms that could then be "discovered" in the firmament. In this respect, ninth-century *astronomia* had as much to do with ancient literary traditions as it did with natural history or science.

The Carolingian antiquarian, translation, and copying projects all participated in the dissemination of acceptable texts. This unspoken but endemic mode of regulation, resulting in projects such as the Frankish Latin translations of the Aratus and the introductory passage to the *De ordine*, exemplifies the concern throughout the Frankish lands for textual accuracy. Without overstating the sweeping nature of Charlemagne's *Admonitio generalis* of 789, it is vital to underscore that the entrenched biblical belief in an ascertainable truth revealed through the correction and revision of exegetical and classical texts (*emendatio*) enabled clerical educators to monitor and make suggestions about what their students should study in the Frankish schools.[19] Ongoing refinements in Frankish authorial comfort and competency with Latin took place alongside the efforts to arrive at true copies of complete texts.[20] The Carolingian *renovatio* was, at its core, simultaneously (1) an effort to resuscitate what eighth- and ninth-century Franks considered useful early Christian and classical contributions to their nascent Christian society and (2) a programmatic justification for the suppression of countervailing discourses. Lawrence Nees has emphasized the Christian import of the correction and emendation projects carried out in the Frankish lands. He argues that in lieu of salvaging classical *auctores* simpliciter, a preferable way to understand the Carolingian renewal is with respect to the reform of the individual through the careful recovery of the patristic texts and exegetical literature restored and preserved for posterity by Carolingian scribes.[21]

New texts such as the *De ordine* established the approved limits of *scientia*.[22] As texts were corrected and treatises such as the *De ordine* promulgated official statements about specific subject matter such as astronomy in the Frankish schools, monasteries emerged as cultivated centers of erudition, supplanting the reliance on imported scholars that had been crucial to the development of Carolingian Latin, as Mary Garrison has remarked.[23] In this respect, the contribution of Adalhard's Corbie to the Frankish development of astronomy and the star pictures cannot be overstated. Under the new Carolingian interpretation, astronomy's utility for young prelates in training within the monastic schools did more than "transmit the remnants of ancient science through the early middle ages until the twelfth century" via inept "epigones," in the words of Richard Lemay.[24] Quite the contrary. The *De ordine* advanced a novel Christian monastic outlook that drew upon a classical Aratean view of the heavens.[25]

Within the Carolingian empire, the constellations could be relied upon to resolve practical problems such as the appropriate times for waking the monastic brethren to sing their nightly prayers of the Divine Office, following the principles established by Gregory of Tours (d. 593–94) in his treatise *De cursu stellarum* (On the course of the stars; composed after 573).[26] In his treatise Gregory had explained how monastics could readily discern the time of night for prayers by using the relative placement of the constellations in the sky. Gregory's text offered general guidelines to follow so that there would be ample time for the singing of the Psalms. As the visibility of the constellations changed throughout the course of the year, monks needed to know which constellations arose on the eastern horizon throughout the night.[27] The night office of nocturns (now labeled matins) posed the greatest problem. Gregory adapted the constellations and rearranged them into Christian forms, however, and in this his text differs from Isidore's later reclamation project. Rather than indicate the placement of Cygnus, the swan, for example, Gregory encouraged the observation of the Crux Maior (Greater Cross), with its arms (rather than wings) outstretched. Gregory's text relied on actual observation, and his reported stellar positions agree with precessional models for the latitude of Tours.[28] Isidore reclaimed instead the historic "names" of the constellations as terms, which had etymologies linked to astronomy.

The complex early medieval understanding of the constellation Hercules revealed by the Carolingian text *De ordine* and the miniature in Drogo's handbook relied upon suspect pagan literary traditions such as the *Phaenomena* and its aforementioned translated scholia, the Basel *Scholia*.[29] These scholia also included references to the mythological origins of the constellations recounted in the *Catasterisms* of Eratosthenes (who died at the beginning of the second century B.C.E.), as Hubert Le Bourdellès has argued.[30]

Two points merit attention here. First, the constellations and their fables were considered valuable pieces of astronomical information. The *De ordine* (text V.2) immediately followed the *Excerptum* (text V.1) in the *Handbook of 809*, suggesting— erroneously, perhaps, to modern readers—that the mythopoetic descriptions and forms of the constellations such as Hercules belonged to superstitious strategies of prognostication. In fact, as Wolfgang Hübner has argued persuasively, apart from nuanced nomenclatures employed by specific authors, during the late antique and early medieval periods the terms *astronomia* and *astrologia* were synonymous.[31] Indeed, André Le

Bœuffle determined that roughly 80 percent of all uses of the term *astrologia* in Latin literature denote the English word "astronomy."[32]

Second, reliance upon the fables which gave rise to the constellations, including Hercules, is noted in the introduction to the *De ordine*, even though the painter in Drogo's copy of the handbook saw fit to omit Hera's serpent and the Tree of the Hesperides (plate 2). The serpent and tree do appear in the scene of combat depicted in other Carolingian presentations of Hercules, such as the Cologne *Aratus* (fig. 8) and the Basel *Scholia* (fig. 7). The Basel *Scholia* and Eratosthenes's *Catasterisms* thus clarify the reason for Hercules's defensive posture, which in the image from the *Handbook of 809* seems incomplete, like that of a shadowboxing fighter. After Hercules's victory, Zeus conferred the constellation, his catasterism, upon the hero.[33] Yet even though the critical introduction to the Carolingian catalogue of the stars, the *De ordine*, refers to the pagan sources of the fables that gave rise to the images of the constellations, the Frankish painter removed significant references to the Herculean myth from the hero's image.[34] This reveals the tension at the core of Carolingian discussions of *astronomia/astrologia*: the practical and pedagogical utility of the liberal art of astronomy for young Christian prelates was paramount, while aesthetic considerations were to remain ancillary.

A similar evasive maneuver, coupled with an apology for the classical names of the planets and zodiacal constellations, can be found in a telling poem by Theodulf of Orléans (d. 821), who is discussed in greater detail below. In *De septem liberalibus artibus in quadam pictura depictis* (On the seven liberal arts shown in a certain picture), astronomy occupies the apex of a mnemotechnic image in verse recounting the interlocking utility and effect of serious study.[35] The liberal arts rise like a tree, with study of the stars in the canopy above its boughs. The disciplines of the *trivium* and *quadrivium* of the medieval liberal arts curriculum offer the virtues inherent within a moral Christian life. The upright Carolingian scholar knows that study yields character, hence the disciplined person will invariably study harder. Theodulf warns such students not to fear the "pagan names" of the constellations associated with the zodiac or the planets: "This custom is assigned by the fathers of ancient times" (lines 83–96).[36] In addition, "These things the tree held, leaves and fruits hanging, / Thus it shed beauty and provided much mystical sense. / In the leaves, understand words, in the fruits, meaning; / The former grow often, the latter, used well, nurture us" (lines 99–102).[37] As Carruthers has argued, poems with extended metaphors such as these operate as "cognitive schematics, pictures whether in words or in paint that are made for the thinking mind."[38] Theodulf's words encourage further meditation or pensive reflection. In so doing, his art participates in the pedagogical and spiritual enterprise that constituted clerical reform of the liberal arts during the Carolingian period.[39]

Astronomy and Astrology: Hrabanus Maurus, Augustine, Cassiodorus, and Isidore

Hrabanus Maurus's (d. 856) treatise *De institutione clericorum* (the *Education*) provided the Carolingian realm with its best statement about the role of astronomical study in the intellectual and spiritual development of young prelates.[40] And just as Hrabanus outlined what was appropriate for young prelates to learn about the liberal arts, so well-trained Frankish clerics emerging from monastic and cathedral schools were entrusted with delineating acceptable pathways to knowledge (*scientia*) for their future pupils and parishioners. It is therefore no surprise to discover, at the core of Hrabanus's discussion of the liberal art of astronomy, an exhaustive effort to parse the distinction between *astronomia* and *astrologia* that draws on multiple *auctores*.

Hrabanus drafted the *Education* during the years that he served as schoolmaster in Fulda under the tenure of Abbot Eigil (818–22). He dedicated the treatise to Archbishop Haistulf at Mainz, thus ensuring that the collected wisdom of the excerpted patristic fathers contained in the *Education*, and Hrabanus's notes on their texts, remained in the repository of one of the most important towns of the Frankish lands.[41] Haistulf was in a position to disseminate Hrabanus's treatise beyond the walls of the schoolmaster's cloister in Fulda, transforming the *Education* from a useful compilation of guidelines for clerical service into a moral prescription for normative procedures and theological positions to be fostered in Carolingian monastic and cathedral schools. Book I of the *Education* addresses clerical responsibilities and liturgical vestments appropriate for the Roman Ordo of the Mass preferred by the Carolingian ecclesiastical reforms. Book II discusses heresy and orthodoxy, the hours of the Divine Office, confession, fasting, Latin chant, and the feasts of the liturgical calendar. Book III covers the liberal arts and their utility for a churchman's formation, complementing the mastery of sacred scripture and exegetical interpretation. The liberal arts, divided into the courses of study appropriate to the *trivium* and *quadrivium* (discussed below), provided preparatory training that would eventually assist the cleric in the performance of his spiritual duty.[42]

Within book III, Hrabanus provides excerpts culled predominantly from texts by Augustine and Gregory the Great. There are also citations from Isidore of Seville, although the majority of samples from Isidore's writing appear in books I and II, as attributed by Detlev Zimpel.[43] Book III includes seven chapters devoted to exegeses of the utility of the liberal arts for young clerics in training: grammar (III.18), rhetoric (III.19), dialectic (III.20), arithmetic (III.22), geometry (III.23), music (III.24), and astronomy (III.25). The purpose of the *Education* was to supply a standard compilation, like the *Handbook of 809* on astronomical study, offering young priests and monks a definitive resource for consultation.[44]

In the early Middle Ages, Martianus Capella's description of the liberal arts, the *De nuptiis Philologiae et Mercurii*, disseminated the structure and scope of the seven disciplines.[45] It is unclear who "invented" the seven-part system of the liberal arts. As recently as 2001, Bernard Ribémont maintained that the liberal arts were a product of classical learning and scholarship. Following Henri Marrou, Ribémont asserted the presence of a seven-part system of the liberal arts that had been formalized by the time of Marcus Terentius Varro (d. 27 B.C.E.). According to this view, astronomy formed a mathematical group of the liberal arts along with music, geometry, and arithmetic, with the roots of the fourfold division having been firmly planted no later than the time of Archytas of Tarentum (d. 347 B.C.E.).[46]

John Marenbon has argued instead that Augustine was the first to delimit the seven-part system of the liberal arts in his *De ordine* (not to be confused with the *De ordine ac positione stellarum in signis* in the archetypal *Handbook of 809*), setting the origins of this educational system much later, in the fourth century (ca. 387). The standard seven-part system of the liberal arts included the *trivium*, comprising grammar, logic, and rhetoric, and the *quadrivium*, comprising arithmetic, geometry, music, and astronomy. Marenbon emphasizes that Augustine's metaphor of a stairwell for reason to climb on its way to spiritual discernment transformed the liberal arts into tools facilitating meditation. In his *De ordine* (II.14.39), Augustine explained, "Lest [reason] should fall from on high, it sought for steps [*gradus*] and worked out for itself a way and an order through its possessions."[47]

Danuta Shanzer likewise has argued in support of the later origin of the seven-part system of the liberal arts, but disputes the attribution to Augustine. According to Shanzer, Augustine wavered about placing astronomy in the list of the liberal arts, even going so far as to replace astronomy with philosophy in his list from the *De retractiones* I.6 (ca. 428) at the end of his life.[48] But as Shanzer also notes, Augustine did include astronomy among the liberal arts described in his *De ordine* II.14.39–II.15.43, so there is little doubt that he had included astronomy in the canonical list of arts by the fifth century. The question is whether Augustine had (at times) adopted—as had many classical, late-classical, and other early Christian thinkers—an alternate and perhaps expanded list of the liberal arts. According to Vitruvius (*De architectura*, VII.Intro.14),[49] Varro had composed the *Novem disciplinae* (Nine disciplines). In this book the liberal arts reportedly also included architecture and possibly medicine. If so, then Varro would have followed a nine-part system of the liberal arts, contra Ribémont and other proponents of the seven-part system's classical origin.

In addition, according to Shanzer, Martianus Capella lampooned this nine-part system in his *De nuptiis*. Each of the female liberal arts at the marriage of Mercury and Philology offered an oration on her respective liberal art. Architecture and Medicine, however, were asked to remain silent because they pertained to "mundane matters."[50] For Shanzer, this is compelling evidence that in the *De nuptiis* the seven-part system had become canonical. Augustine apparently still employed an alternative system somehow involving philosophy in the hierarchy of disciplines, given the evidence in his *De retractiones* I.6. The *De nuptiis* provided the definitive statement about the seven-part system of the liberal arts, while it offered early medieval classrooms a thinly veiled celebration of paganism.[51]

This is one reason for Hrabanus to begin his chapter on astronomy in the *Education* (III.25) with an apology for its study by Christian clerics. He then turns to an analysis of the equivocation, ubiquitous throughout exegetical literature, that confounds the differences between astronomy and astrology. First, Hrabanus cites a passage from Cassiodorus's *Institutiones* II.6.4, but embeds a textual reference to Augustine's *De ordine* II.15.42 within the excerpt. Hrabanus followed Augustine when he protested that astronomy was "a matter worthy of pious people and a great torment to inquisitive folk."[52] Jacques Fontaine has argued that two troublesome aspects of Augustine's misgivings about astronomy in this brief passage had also vexed Isidore. On the one hand, the Psalmist declared the revelatory potential of stars in the firmament (Psalms 19:2). In the heavens, the signs of God's creation were filtered through "human conviction" by the establishment of normative sets of stars—and were thereby made memorable, as described in the introduction to the Carolingian *De ordine* and depicted in the miniatures of the *Handbook of 809*. On the other hand, the study of astronomy was a distraction that could divert "pious" minds from meditation upon sacred Christian writ, turning the attention of ecclesiastic trainees toward the literary "fables" likewise alluded to in the *Handbook of 809*.[53] The following excerpt from Cassiodorus (*Institutiones*, book II.6.4) cited in Hrabanus's *Education* supports the first of Fontaine's two readings. It would appear that Hrabanus (d. 856), Augustine (d. 430), and Cassiodorus (d. 585), taken together in the *Education*, suggest that celestial study could confer upon students a form of spiritual benefit: "There remains astronomy. If we seek after the knowledge of astronomy with a pure and moderate mind, it enlightens our understanding, as the ancients say, with great clarity. It is such a wonderful thing to approach the heavens mentally and to examine that entire celestial structure using rational investigation, and by theoretical speculation explore great hidden mysteries."[54]

One of the reasons Augustine advocated astronomical study, cleverly revealed by Hrabanus's juxtaposition of excerpts, was its evidentiary support of biblical miracles involving the heavens. Cassiodorus followed Augustine in this respect and began his discussion of astronomy in book II.7.1 of the *Institutiones* with a list of "mysteries" that was slightly emended by Hrabanus in the *Education*. Such cases of insertion or deletion are often most expressive of ninth-century beliefs about astronomy and evidence of the pervasive Carolingian creativity seen in the texts and images assessing the proper place of astronomy within Carolingian society.[55] Just as the artist who painted Hercules in Drogo's copy of the *Handbook of 809* preferred to omit the serpent that the hero was fighting, so Hrabanus changed Cassiodorus's sequence of miracles pertaining to the heavens.

After declaring *astronomia* the "law of the stars," the passage from Cassiodorus's *Institutiones* (II.7.1) excerpted by Hrabanus for book III.25 of his *Education* adduces a set of miracles that does not adhere to the ordinary movements of the celestial bodies. By splicing together the inadequate definitions offered by Augustine, Isidore, and Cassiodorus, Hrabanus proposed a new conception of the distinction between *astronomia* and *astrologia* that nevertheless drew heavily upon his predecessors. Cassiodorus's list of heavenly signs included Joshua's apparently supernatural control over the sun that remained steadfast in the sky over the Battle of Gibeon (Joshua 10), the Christmas star (Matthew 2), and the three-hour eclipse at the Crucifixion (Luke 23).[56] These visual manifestations of God's power were the exceptions supposedly confirming the rule: "these events are called miracles because wondrous things happen against the usual rules of circumstance."[57] Hrabanus also included Cassiodorus's qualification that "[stars] can neither remain at rest nor move except in the way in which their Creator arranged them."[58]

Stephen McCluskey has argued that such miracle lists demonstrated to their early medieval readers God's divine ability to sanction alternative courses of movement for certain celestial bodies at specific moments in time. The Christian God was reportedly able to cause atypical celestial phenomena, even if such aberrant displays contradicted ordinary human experience. The irregularity of such movements was precisely what made them miraculous and drew attention to a special case of divine foreknowledge—God's ability to comprehend in advance the nature of a heavenly body, including any nonstandard behaviors such stars or planets might manifest.[59]

Hrabanus opted to replace Cassiodorus's reference to the Christmas star with the biblical story of King Hezekiah, who witnessed the sun retrace its path along ten steps (in a form of solar retrograde motion) as an omen portending the extension of his life by fifteen years (2 Kings 20).[60] Rather than dismiss this as an indication that Hrabanus relied upon a corrupted text of Cassiodorus's *Institutiones*, the careful integration of disparate texts and ideas in the *Education* would argue instead that Hrabanus sought to nuance his account of the *astronomia/astrologia* distinction.[61] Having established the regularity of celestial movements, which fell within the purview of *astronomia*, Hrabanus proceeded to further delimit the acceptable bounds of human observation, perhaps drawing upon Augustine's critique of human experience, which held that it was an unreliable conduit to knowledge about the essence of a thing, according to book XXI.8 of *City of God*. Augustine had also adduced the miracle of Hezekiah's steps along with Joshua's victory at Gibeon as examples of celestial irregularity. He even included a pagan example of alleged irregular celestial movement, referring to a passage from Varro's (d. 27 B.C.E.) *De vita populi Romani* (On the race of the Roman people) describing the morning star, Venus (also known as Lucifer), changing in size, hue, and orbital path. As Augustine explains, "A portent, therefore, does not occur contrary to nature, but contrary to what is known of nature."[62]

Augustine was, in fact, setting up his argument for the plausibility of eternal hellfire from book XXI of *City of God*, for which he needed to argue that the fact that medieval folk had never seen incorruptible flesh (which could not be reduced to ash by devilish flames) did not mean that the Creator was incapable of modifying the usual properties of such substances.[63] Like Augustine, Hrabanus intended to convey in his *Education* that "laws" of motion governing the heavenly bodies were subject to change, and that the appropriate epistemic stance before the movements of the planets and stars was one of reserved appreciation for the regularity of their orbits. Even so, the natural proclivities of the stars and planets to move in certain ways were accidental rather than essential properties. God could alter them under exceptional circumstances. At the core of Augustine's and Hrabanus's appreciation of the astronomical "law of the stars" was a theological rather than natural-historical hesitation. Both considered the epistemological abilities of young clerics learning about astronomy to be curtailed by the interventions of a Creator who could change the ordinary courses of the planets and the stars. Yet such interventions were also miracles that could be marshaled as evidence of the power of the Christian God, according to the early medieval conceit advanced by Cassiodorus. Whether emphasizing the regular orbits of the stars and constellations or their miraculous deviations, Hrabanus (relying upon Augustine and Cassiodorus) affirmed the meditative benefit of the study of the liberal art of astronomy for Frankish schools.

Hrabanus justified the theoretical and theological study of astronomical orbits, such as the planetary movements depicted in the diagram from the *Handbook of 809* (fig. 3), in the first half of his chapter on astronomy in the *Education* (book III.25). He then turned to an excerpt from Isidore's *Etymologies* that, in the context of Hrabanus's compilation, should be considered an effort to resolve the equivocation apparent in early medieval uses of the terms *astronomia* and *astrologia*. At this point in book III.25, Hrabanus attempts to clarify for clerical students the theoretical and practical uses of astronomy in service of the Christian Carolingian reform programs. Isidore of Seville had argued in book III of the *Etymologies* that "[t]here is some difference between astronomy and astrology. Astronomy concerns itself with the turning of the heavens, the rising, setting, and motion of the stars, and where the constellations get their names" (XXVII.1).[64] In other words, *astronomia* here refers to the same functional descriptions and origins of the names of the constellations contained in the introduction to the Carolingian text of the *De ordine* cited above, as well as under the heading *astrologia* in the *Excerptum* from the *Handbook of 809*. Hrabanus recognized the need for clarification and inserted a caveat, resembling a parenthetical aside, into his text: "it is permissible for both [*astronomia* and *astrologia*] to pertain to one discipline."[65]

This reveals an important point about Carolingian compilations. Often, the rhetorical mode of the compilation is dismissed as a derivative, and therefore unimaginative, form of composition. Yet with his parenthetical aside Hrabanus altered the definition Isidore had espoused in the *Etymologies*. Hrabanus sought a carefully selected synthesis of mutually informative interpretations of the difference between astronomy and astrology in the works of Cassiodorus, Augustine, and Isidore, while grounding his discussion in the need to clarify which astronomical skills were and were not appropriate for students of theology in the Carolingian schools. Hrabanus nevertheless also included Isidore's novel seventh-century resolution to the *astronomia/astrologia* equivocation: "But astrology is partly natural, partly superstitious. . . . It is natural as long as it investigates the courses of the sun and the moon, or the specific positions of the stars according to the seasons; but it is a superstitious belief that the astrologers (*mathematicus*) follow when they practice augury by the stars, or when they associate the twelve signs of the zodiac with specific parts of the soul or body, or when they attempt to predict the nativities and characters of people by the motion of the stars."[66]

As Hübner has argued, there was a semantic overlap between the practices involving stargazing denoted by *astronomia* and *astrologia* in the seventh to ninth centuries. In fact, Isidore himself had adapted his definition of *astronomia* from a Ciceronian statement about astrology: "in astrology, the turning of the heavens, the rising, setting, and motion of the stars" (*De orat.* I.187).[67] Hübner nevertheless considers Isidore's attempt at clarification "unclear." He also rejects the possibility that Isidore was endeavoring to distinguish between theoretical discussions of the liberal art of astronomy and practical application of stargazing techniques to navigation, agriculture, and, by extension, meteorology.[68] Lynn Thorndike, the great historian of medieval magic and science, echoes from the past Hübner's discontent.[69] There is a consensus in the scholarly literature, however, that Isidore and Hrabanus opposed the creation of "superstitious" astrological birth horoscopes (known as genethlialogy) and the assignment of zodiac signs to the body for medical treatments (otherwise referred to as melothesia) by early medieval heretical groups. Such dissenters continued to follow the beliefs espoused by the fourth-century Priscillianists, who had been declared heterodox at the Council of Braga in the mid-sixth century.[70]

There is also reason to believe, however, that when Hrabanus Maurus read and then cited the passage by Isidore, he believed that a wedge had indeed been inserted between the theoretical liberal art of astronomy and astrological praxis. Max Lejbowicz has argued for just such an understanding of the Isidorean distinction between *astronomia* and *astrologia*.[71] Lejbowicz treats the opening paragraphs of Isidore's account of the origins of astronomy in book III of the *Etymologies* as a compositional whole that can be interrogated on its own terms. Isidore attributed the development of the concept of *astrologia* to ingenious efforts at system-building that had resulted from anonymous stargazers whose "observations of certain dimensions and numbers" tracked the stellar movements.[72] In book III.xxiv, Isidore had identified astronomy with the "law of the stars," as did Cassiodorus in his *Institutiones* (II.7.1), also included by Hrabanus in his *Education*. The regularity of celestial motion (*astronomia*) made possible its practical study and application (*astrologia*). According to Lejbowicz, it is superstitious astrology that was problematic, primarily because it overstepped the bounds of orthodox Christian practice.[73]

Throughout his *Education*, Hrabanus customarily appended his own elucidations to compiled texts. His personal exegesis at the end of book III.25 further examines the practical application of astrological observational techniques:

> At any rate this part of astrology which follows by natural investigation the orbits of the sun and moon and of the stars, and explores certain distinctions of time carefully, is necessary to be learned by the clergyman of the Lord through skillful study, in order that he may truly investigate through certain inferences of rules, and through established and genuine determinations of arguments, not only the previous passages of years, but also be able to calculate accurately future events, with the result that they know how to relate the beginnings of the feast of Easter and certain events of all solemnity and festivals which they ought to observe, and that he may be able to make known to the people of God what ought to be celebrated by them all, according to religious practice.[74]

Valerie Flint has noted Isidore's preoccupation with observational practices and their utility for the medieval yearly liturgical and agricultural cycle.[75] For Hrabanus, *natural* astrology, unlike astronomy, focused on the practical resolution of computistical problems, made possible by the *Handbook of 809*. The most vexing problem was the annual need to fix the date for the floating feast of Easter, which, as described above, required the coordination of lunar and solar calendar cycles. This fundamental feast day had to be observed consistently throughout the Frankish lands, in keeping with the general desire for a uniform liturgy. For Hrabanus, natural astrology, unlike superstitious astrology, provided acceptable techniques for predicting future events such as the first full moon after the vernal equinox, requisite for determining the date of Easter. When fixing the feast, Hrabanus and his astronomically minded students could rely upon the predictable and natural motions of the sun and moon for the spiritual benefit of the ninth-century Franks, thereby remaining comfortably within the pale of acceptable stargazing techniques (falling under the rubric of *astrologia naturalis*) and complementing the theoretical study of *astronomia*.

Superstitious astrologers similarly attempted to interpret the future, but tried to extrapolate from the stars information that was unwarranted. As Lejbowicz has argued, Isidore opened the possibility for an early medieval salvaging effort, allowing the reclamation from classical treatises of all that was not clearly heterodox.[76] Whereas *astronomia* for Hrabanus denoted the study of the liberal art of astronomy in the monastic or cathedral schools, *astrologia* treated applied interpretative practices, which relied on celestial information to predict various kinds of future events. In this respect, Hrabanus's reading, or misreading, of Isidore's passage from the *Etymologies* anticipated a modern understanding of the difference between Stephen Hawking's or Carl Sagan's astronomy, on the one hand, and Jeanne Dixon's or Walter Mercado's astrology, on the other.

Other acceptable applications of natural astrology for both Hrabanus and Isidore would have included meteorological prognostications, likewise reflected in the imagery in the *Handbook of 809*. In a second case study, Cancer was apparently displayed with donkeys between its claws for the first time in the *Handbook of 809* (plate 5), so as to emphasize the nonpernicious aspects of prediction, including the forecasting of inclement weather. In the description and star catalogue for Cancer, the text from the *De ordine* in the *Handbook of 809* clearly explains the presence but not the use of the donkeys at the manger before Cancer the Crab.[77] According to the rudimentary principles in play, a later text in the *Handbook of 809* (book V.12), derived from Pliny's *Naturalis historia*, book XVIII.80, par. 353, explains what Aratus had long before reported in his *Phaenomena*:[78] "In the zodiac sign of Cancer there are two little stars called the donkeys. Within a little space between them, there stands a haze, which is called the manger. When these donkeys cease to appear in a peaceful sky, a violent tempest follows. If the fog removes the one in the north, then the southern wind rages; if the fog removes the southern one, then the northern wind rages."[79] The Carolingian artists who painted the image of Cancer in princely copies of the *Handbook of 809* (such as Drogo's) creatively included the meteorological Aratean references alluded to by the northern and southern donkeys.[80]

The first master painter created Hercules in Drogo's copy of the *Handbook of 809* and in doing so omitted the Tree of the Hesperides from his image of the constellation. In the image for Cancer, the same artist adopted the acceptable addition of the donkeys, feeding at their manger, because of a preoccupation with discovering practical applications for natural astrology. This sophisticated emendation project probably also reveals a concerted effort by Adalhard, as planner of the hypothetical original illustrated *Handbook of 809*, working in tandem with his master painter to cross-reference the various components of book V in a comprehensive, encyclopedic visual display of information pertaining to Cancer.[81] By regulating the appropriate visual presentation of the zodiac sign of Cancer from the *Handbook of 809*, Adalhard and his team sanctioned the use of the heavens for meteorology, which relied upon the basic stargazing techniques associated with *astrologia naturalis*. Further uses of the zodiac sign of Cancer for birth horoscopes—belonging to *astrologia superstitiosa*—were forbidden.

Predicting rain and assessing wind direction based on the celestial donkeys' disappearance from the constellation of Cancer was one form of meteorological prediction in the *Handbook of 809*. Cancer would also have been linked by Hrabanus's students with the solstice and the beginning of summer, as it is today. This offers one practical agricultural use for following "the specific positions of the stars according to the seasons."[82] On the summer solstice, the sun attains its highest point (23.5 degrees north latitude) and calls farmers to their fruitful fields.[83]

Hrabanus probably also followed Isidore in his belief that the etiology of certain maladies and the diagnosis of appropriate medical treatments were bound up with seasonal changes and their impact upon the body. In this respect, medicine remained closely linked to meteorology and annual cycles in medieval humoral medicine. Roughly twenty-five years after the *Education*, Hrabanus relied extensively on the medical portions of Isidore's *Etymologies* for his statements about medical theory in his own encyclopedia, *De universo*, although the Christian moralizing tone of Hrabanus's work separates it from Isidore's.[84] As Isidore writes in the *Etymologies* (book IV.xiii.4), a physician should "be acquainted with astronomy, through which he may observe the logic of the stars and the change of seasons. For . . . according to their mutations our bodies are also changed."[85]

The Carolingian efforts at reforming clerical instruction in astronomy, and Hrabanus's reading of Isidore's distinction between natural and superstitious astrology, required principles for regulating which applications of *astrologia* were acceptable. Monastic medical practice and the health of young clerics were integral components of the communal care entrusted to high-ranking prelates. Such concerns would have been at the forefront of Hrabanus's mind during the years he served as schoolmaster at Fulda, and all the more pressing upon his elevation to abbot of the monastery (822–42). First and foremost, Hrabanus sought to cultivate a Christian community united by its uniform liturgical practice and liberal arts curriculum. Second, there was a practical advantage for medieval clerics who planned for seasonal discomforts, such as the colds and influenza that accompanied the onset of the fall and winter months.

Isidore, by contrast, staunchly supported the classical idea that the human body is a microcosm that reflects the structure of the macrocosm, or universe (fig. 1).[86] Fontaine has argued that Isidore provided his review of the significance of the liberal arts in books I–III of his *Etymologies* as a prelude to all further study, including study of the Creator who had inserted the order within the cosmos that was likewise to be found in the microcosm of the human being.[87] For Isidore, a cosmological application of *astrologia naturalis* to Galenic humoral theory legitimized his encyclopedic inquiries into the rapports between the body and its divine Creator, culminating ultimately in a meditative union between the human being and its maker. For Hrabanus, the same metaphysics buttressed his Carolingian desire to reclaim certain astronomical practices in the service of the Christian church. Although different in their aims and levels of sophistication, the two men were kindred spirits, prioritizing sanctification through science.

Evaluating a third case study reveals that a related effort to reclaim the mythical founder of medicine, Chiron the Centaur, can be found in the *Handbook of 809* (plate 16). In this instance, the inclusion of the pagan Bacchic *thyrsus* (a shaft or spear, embellished with pine cones, carried by the followers of Dionysus) in both text and image establishes the link between the constellation of the Centaur and Chiron, Achilles's mythic tutor. Folio 38v of the Basel *Scholia* explicitly states that the Centaur in question is Chiron, as discussed in greater detail in chapter 2.[88] Although the *De ordine*, the new star catalogue created for the *Handbook of 809*, does not name Chiron as the intended referent, the inclusion of the *thyrsus* and the mention of Chiron in the Basel *Scholia* make the attribution highly likely. The redactors of the *Handbook of 809* sought to underscore any and all connections between the constellations and acceptable scientific applications of *astrologia naturalis*, including medicine.

For Hrabanus, the practical application of natural astrology, with its computistical and stargazing techniques, made possible the proper ecclesiastical administration of the annual liturgical year, agricultural planning, and medical treatment. Mastery of both the practical side of celestial study and its conceptual underpinnings was a necessary part of the training of future clerics, and in this respect the distinctions offered by Hrabanus and Isidore between *astronomia* and *astrologia* clarified specific pathways to *scientia*, or knowledge, that were appropriate for young clerics to pursue. Practitioners of superstitious astrology, such as those who created birth horoscopes, exceeded the bounds of acceptability in their prognostications of certain future events. By instead basing their predictions concerning seasons for sowing, harvests, and medical treatments—and for the Paschal full moon of Easter—on results discerned from extrapolations of lunar and solar movements, Hrabanus's Frankish students of natural astrology justified their claims by observing the Creation.

The princely copies of the *Handbook of 809* (such as Madrid 3307, the manuscript made for Drogo in medieval Metz) reveal painterly efforts to address similar concerns. Adalhard of Corbie and whichever artists assisted him with the preparation of the handbook's archetype at Aachen worked together to regulate the ways in which young Frankish students could visualize the constellations they were studying in the schools. This required interpretative decisions and exegetical clarity, as the pictures in the *Handbook of 809* coalesced into coeval documents of scientific and cosmological import. In his *Education*, Hrabanus had recommended a way to assess what young clerics should know about *astronomia* or *astrologia*. Carolingian painters did the same in the images from the *Handbook of 809* made for Drogo. These complementary visual and textual projects were marshaled for their educational value in the service of ninth-century Frankish reforms, fulfilling Isidore's hope, expressed at the end of book III of his *Etymologies*, that teachers "might draw minds tangled in secular wisdom away from earthly matters and set them in contemplation of what is above," as mentioned in the introduction to this book.[89]

In addition, Hrabanus's emphasis on computistical skills drew attention to the need for the ecclesiast to schedule future events and observe the appropriate feasts accurately. This is itself a form of prognostication. Hrabanus distinguished between the deleterious form of forecasting, which superstitiously anticipated human events, and the kind that was made possible precisely because the stars and planets followed their natural paths. In the end, Hrabanus in the *Education* deemed natural astrology acceptable, because it submitted to the divinely created order. Superstitious astrology fell outside the purview of acceptable ways to acquire knowledge. Its predictions (a) treated as unchangeable the natures of the stars, which God could alter by performing miracles; (b) extrapolated knowledge from the stars that was not contained within their natures; and (c) were contrary to orthodox practice. In other words, superstitious astrology demanded more from the stars than was theologically or *epistemologically* warranted.[90]

The fundamental Isidorean distinction between natural and superstitious astrology established the acceptability of computistics, which made possible accurate time reckoning as it was presented in the *Handbook of 809*. The emphasis on celestial observation in both Hrabanus's and Isidore's version of the *astronomia/astrologia* distinction grounded the study of the constellations in the *Handbook of 809* as an important skill for young clerics in training and as a normative technology of truth production: it led to legitimate clerical practice and thus resulted in the dissemination of orthodox belief.

Hrabanus saw the liberal arts as useful insofar as they enabled someone to learn about the Christian faith. A careful analysis of the *Education* and other works has frustrated scholars, such as Ribémont, who have sought a coherent philosophical view of the liberal arts in Hrabanus's writings.[91] That was never his goal. What mattered to Hrabanus, especially in his manual for young clerics, was to provide a primer of rigorously justified, standard Carolingian religious opinions through refined arguments and comprehensive appeals to the appropriate patristic fathers. Those fathers, in turn, in a sort of circular reasoning, undergirded the legitimacy of the liberal arts through their concordance with and reliance upon the sound reasoning taught to students of the *trivium*. In the end, however, what mattered was the advancement of the ninth-century church's goals. Even though Hrabanus became a learned prelate, throughout his life the liberal arts remained ancillary to the orthodox teachings they helped elucidate.

This is made explicit in Hrabanus's discussion of grammar in book III.18 of the *Education*. Carruthers has discussed Hrabanus's justification for the study and adoption of pagan stylistics in the service of Christian teaching out of respect for the *florem eloquentiae*, or "flower of eloquence," contained within pagan books. The beautiful turns of phrase used by Germanicus in his translation into Latin of the Greek *Phaenomena* of Aratus are excellent examples of compositional stylistics worthy of emulation. Hrabanus understood that "when we read the gentile poets . . . whatever in them we find useful, we convert to our teaching."[92] At the end of this passage, Hrabanus concludes with what appears to be an original personal appeal: "[D]o not let a brother perish who is weak in our science, on account of whom Christ died, as he would if he were to see us backsliding into idolatry."[93] According to Hrabanus, the intellectual value of astronomical information remained secondary to the clerical utility of such studies. Just because clerics such as Drogo could use the computistical principles in the *Handbook of 809* to reckon the date for Easter and other feast days did not mean that certain aspects of the pagan stories attached to the constellations would not hurt the faith of some parishioners or of younger brethren reading in the cloister.

Hrabanus shifted the focus of Augustine's well-known apology for adopting what was deemed useful in ancient Egyptian erudition and Platonic philosophy in chapter 40 of *De doctrina christiana* (*Christian Instruction*). Augustine had argued in favor of "despoiling the Egyptians":

> [A]ll the teachings of the pagans have counterfeit and superstitious notions and oppressive burdens of useless labor, which anyone of us, leaving the association of pagans with Christ as our leader, ought to abominate and shun. However, they also contain liberal instruction more adapted to the service of truth and also very useful principles about morals; even some truths about the service of the one God Himself are discovered among them. These are, in a sense, their gold and silver. They themselves did not create them, but excavated them, as it were from some mines of divine Providence which is everywhere present, but they wickedly and unjustly misuse this treasure for the service of demons.[94]

Augustine thus advocated a repatriation project that sanctified anew all of the ideas that pagan thinkers had rightly surmised when contemplating the Creator's universe.

Hrabanus and Augustine sought to turn the value of these pagan ideas to some Christian, pedagogical good. Augustine's model emphasized the intellectual perspicacity of the ancient scholars who followed the ordered, teleologically perfect design of nature. In so doing they gleaned truths about the cosmos from their flawed philosophies. For Hrabanus, the textual artifice and the information about the constellations in a poem such as the Germanicus translation were valuable insofar as they fostered Christian educational reform within the Frankish lands. Scientifically verifiable or philosophically valid ideas were only useful provided that students were not tempted to revert to the pagan practices and mysteries that had passed the texts on to the medieval cultures in the first place. For similar reasons, the redactors of the archetypal *Handbook of 809* under Adalhard of Corbie censored the mythological narratives in the textual descriptions of the constellations. Only ideas useful for stargazing, agriculture, and meteorology passed critical review. The sanctification of the liberal arts, according to Hrabanus Maurus in the *Education*, required vigilant clerics and teachers who had mastered the computus and understood the regularity of celestial movements.

The same guardians of Frankish ecclesiastical reform monitored their parishioners or students in order to guard them from intellectual excess and the deleterious temptations that alternate, and even magical, forms of knowledge could inspire. The study of ninth-century Carolingian astronomy was not a scientific, experimental discipline, but rather one component of an evangelistic enterprise. Science underwent a process of sacralization in early medieval culture, as prelates such as Hrabanus sought to make their religion more scientific.[95]

Dwelling on Higher Things: Renewal of the Mind Through Sacred Study of Astronomy and Natural Astrology

As discussed in chapter 1, the precise nature of Alcuin's participation in the early development of the court school at Aachen has been challenged.[96] There is, however, no question as to his expertise in astronomical matters or his active efforts at resolving Carolingian astronomical questions.[97] As already mentioned, under Alcuin's tutelage, Charlemagne "with deep purpose and great curiosity investigated the movement of the stars."[98] On the one hand, this statement by Einhard, Charlemagne's biographer, probably exaggerates the king's sophistication; Einhard describes him as being able to perform arithmetical operations but not to write letters very well.[99] On the other hand, this somewhat formulaic piece of praise (possibly derived from Cassiodorus) does emphasize the extent to which astronomical stargazing was a visual, observational praxis in the Latin West during the ninth century. Even without any facility at writing letters, the king could follow the movements of the planets and discern which constellations were visible during any season of the year.

Charlemagne's efforts at following the movements of the planets (known as wandering stars) and monitoring the paths of the constellations (comprising fixed stars turning with the celestial sphere) contributed to his intellectual well-being.[100] This was an important component of Alcuin's theory of the liberal arts, as discussed in his *De vera philosophia* (On true philosophy; before 804), which supplied an introduction to the *De grammatica* and texts on the *trivium* in dialogue format.[101] Marenbon has noted that Alcuin drew upon Augustine's idea that the liberal arts were steps to wisdom. Alcuin also, however, promoted the spiritual utility of the liberal arts. The arts were not only necessary for the cultivation of personal discernment, which enabled Christian scholars to separate what was useful in the pagan learning of the past from that which was harmful. Rather, under Alcuin's interpretation, they were integral (even essential) components of the sort of vatic refinement that became accessible through pedagogically beneficent aspects of divine revelation—and components that should thus be cultivated within the minds of young clerics. In other words, sacred study of the liberal arts was a catalyst for the unveiling of spiritual truths revealed through the complementary examination of exegetical literature and sacred writ: "Wisdom built herself a house; she put up seven columns. Although this statement refers to divine wisdom, which built itself a house—that is, a body—in the Virgin's womb, and strengthened it with the seven gifts of the Holy Spirit; or the church, which is the house of God and which he illumined with these gifts; wisdom is, however, strengthened, by the seven columns of liberal letters, nor has it led anyone to perfect knowledge unless he has been raised up by these seven columns or steps."[102] Sacred study and visionary practice reveal themselves as twin components of medieval mystical meditation. The pursuit of knowledge within the early medieval school was its own form of ascetic disciplined practice, with the related rewards of increased intellect and the deepening of spiritual sensibilities. More than just conferring knowledge of philosophy and natural history, books of astronomy closed the gap between the heavenly realms and the learned walls of the monastic or cathedral schools.

Mary Alberi has argued that this decisive turn in Alcuin's justification for the necessity of the liberal arts in early medieval society resulted from prevailing criticism of pagan learning throughout the Frankish lands. According to Alberi, proper instruction in the liberal arts also prepared a cleric to engage in polemics with heretics and to appropriately defend Christian orthodoxy.[103] Alcuin was deeply invested in the education of young, pious Franks; moreover, he sought to marshal the sorts of Christian soldiers called for in the *Epistola de litteris colendis* (Letter on the cultivation of learning, ca. 789–800).[104]

Alberi has noted the similar thematic interests that united Alcuin's various letters with his *De vera philosophia*. She draws particular attention to one letter on astronomy that is particularly useful for this study:

> My teacher [Ælberht, d. 780] was frequently in the habit of saying to me, "The wisest of men were those who had discovered these arts in the natures of things. The disgrace is great when we abandon [those arts] nowadays to go to waste. But now, the cowardice of many men does not take the trouble to know the explanations of things [*rationes rerum*] which the Creator inserted into their natures. You know best how pleasant arithmetic is in its principles [*rationes*]; how necessary it is for learning divine scriptures; how pleasant is the examination of the heavenly constellations and their movements. And nevertheless, he is rare who is of the sort that takes trouble to know such things."[105]

The liberal arts yielded a particular pleasure, but their study also played an important role in making people aware of the principles of science at work in the Creation and of the Creator who had set them there in the first place. In this letter and also in the *De vera philosophia,* Alcuin demonstrated the path of the early medieval intellectual: seeking to understand the inner workings of natural things was nothing less than a spiritual calling.[106] As Marenbon has also argued, this awareness of the inner workings of nature resulted in "perfect knowledge," a transcendent discernment of the spiritual truth about the Christian God, which was unveiled through continued study of the liberal arts.[107]

These twin processes of scientific soteriology and scientific sacralization are underappreciated key concepts pertaining to Carolingian thought about the liberal arts. The passage from Alcuin's letter to Charlemagne underscores how necessary arithmetic was considered to be for the exegesis of sacred texts. The significance of this passage extends beyond the role that the liberal arts played in revealing divine principles, for complicated arithmetical problems and astronomical dilemmas exercised problem-solving skills that are just as useful today as mastery of such techniques would have been for Carolingian students in the monastic and cathedral schools. For the Carolingians, the spiritual benefit of resolving computistical problems required knowledge about relationships, properties, and ratios that had been embedded in nature by the Creator. In addition, solving mathematical problems permitted the pupil to apply the principles to new questions. In a practical sense, the books in need of *emendatio* could be repaired only by clever readers capable of finding flaws. Similarly, biblical exegesis benefited from the critical reasoning skills necessary to break down scriptural passages into points that were hermeneutically coherent and that could stand on their own as a basis for orthodoxy.

Both mathematics and biblical exegesis demanded scrutiny of minutiae in order to arrive at a correct analysis, which in both disciplines achieved an incontrovertible result, in Alcuin's view. The crystalline truth of numbers and scripture relied on analogous operational rules and could be known by those who sought it. This was so because of the order established and placed within nature in the formative principles, which conveyed the neo-Platonic essences of things. Computistics, which was treated in detail by the *Handbook of 809,* conferred a dual reward. Intellectual pleasure resulted from both mathematical calculations and the observation of celestial bodies. In addition, the understanding of the organizational principles (*rationes*) of the heavens resulted in an appreciation of the workings of the cosmos. These explanations could only be ascertained through observation of heavenly movements. Humans alone held the privileged position of created beings capable of acquiring "perfect knowledge" through the seven columns of the liberal arts.[108]

At the other end of the Carolingian period, Remigius of Auxerre (d. 908) likewise underscored the belief in the spiritual benefits that resulted from study of the liberal arts. He is included here to demonstrate just how entrenched Alcuin's belief in spiritual renewal through study of the liberal arts became during the ninth century. In Remigius's case, however, this was because the liberal arts, such as astronomy, actually constituted an essential property of the human soul in need of cultivation:[109]

> Now every natural art has been placed in man's very nature and created with it; hence

> all men naturally possess the natural arts. But since, because of the sin of the first man they are hidden away in the minds of men and are plunged into profound ignorance, in learning we are doing nothing but recalling to the fore of the mind those same arts which have been hidden away in the depths of memory; and when we are occupied with other concerns, in neglecting the arts, we are doing nothing but dismissing them to return whence they had been recalled.[110]

In his *Commentum in Martianum Capellam* (Commentary on Martianus Capella, ca. 900), Remigius argued for an innate set of seven liberal arts that had been obscured within human souls by the biblical fall of man (Genesis 3). Remigius adapted John Scottus Eriugena's idea that all of the liberal arts were already stored within the human psyche at birth. Remigius's position, following Eriugena, made education a form of personal renewal that was particularly aligned with the Carolingian reform movements. Study restored access to the truths, and otherwise truth-yielding beliefs, intrinsic to the individual.[111] In other words, the pious cleric in training was equipped already with a comprehensive intellectual constitution capable of perfect discernment. Not only did the individual person contain a mental reservoir of topics of knowledge, but the linkages between conceptual structures could be unpacked through the soteriological effects of dialectic and the benefits of arithmetical study (or the pursuit of the *quadrivium*), as suggested likewise by Alcuin. The psycho-spiritual intellectual nexus of information obscured through the fall of man required learning and education in order to drag the salvific network of knowledge grounded in the liberal arts back from a postlapsarian morass.

It is important to reflect momentarily on where Remigius became aware of such an extraordinary thesis. Eriugena had espoused these views, for example, in his *De divisione naturae* (On the division of nature, otherwise known as the *Periphyseon*, ca. 865). Cora Lutz has noted that Eriugena suggests in that work that the soul is eternal precisely because "the arts are eternal and always cling to the soul unchangeably."[112] Lutz also notes Remigius's reliance in the passage from his commentary on the text of the anonymous, formerly labeled "Dunchad," commentary on Martianus Capella, the *Glossae in Martianum* (160.1).[113] Contreni has emphasized that although Remigius's commentary was typically derivative, it was his text that predominantly disseminated the ideas of his distinguished predecessors (such as Eriugena) to later medieval scholars. It therefore supplies a fitting summary of the Carolingian development of these conceits.[114]

Contreni argues, moreover, that Eriugena's influential view of the liberal arts, which was included in his own related commentary on Martianus Capella's *De nuptiis*, provided a spiritual impetus for their serious study. Through study, not only clerics but also lay pupils could restore their souls to their original state in the Garden of Eden prior to the fall. This extreme position made sacred study of the liberal arts a *sine qua non* for complete spiritual, Christian sanctification, or an integral component of the process of scientific soteriology.[115] Remigius added to the exegesis of Eriugena, according to Lutz. She explains that Remigius considered the arts to be liberal because "they feed the soul liberally with the fruits of truth, or because they seek a mind free and unencumbered by the confusion of the age."[116] The process of intellectual sanctification endorsed by Eriugena and Remigius restored the Christian soul to a purer state through education. For

Alcuin, the liberal arts had supplied the sustaining means to attain spiritual wisdom.[117] Remigius believed that spiritual renewal resulted from "perpetual feedings" at the table of truth, and that it was only through meditative release of anxiety and peaceful focus upon study that the soul could attain perfection. The inner dialectic between the individual soul, memory, and disciplined training resulted in the progressive work of scientific soteriology, effectively redeeming the mind of the individual one fact at a time (Romans 12:2). Alongside this exchange, a simultaneous sacralization of the *scientiae* contained within the liberal arts provided pathways for Christian redemption.

Both Eriugena and Remigius emphasized the role of wisdom in the process of the philosophical purification of the soul. This shared belief resulted in what is being called here the Carolingian doctrine of scientific soteriology. Lutz has underscored Remigius's mistrust of science for its own sake, because astronomy and the other liberal arts encouraged arrogance. Wisdom, however, could clear the mind with learning and peace. For this reason, the liberal arts led those who studied them to philosophy (the "love of wisdom") and the spiritual wisdom acquired through contemplation. For both Remigius and Eriugena, wisdom and spiritual understanding were the goals of ninth-century study of the liberal arts.[118] Their view, like the positions espoused by Alcuin and Hrabanus, was that the impetus for concerted study of the liberal arts during the Carolingian era was unequivocally and primarily the sanctification of the human soul. All intellectual interests remained ancillary to the Carolingian desire to impose a widespread plan for Frankish Christianization.

These discussions of the intellectual and spiritual benefits of the study of the liberal arts were complemented by lively debates interrogating the proper roles for imagery in Carolingian architecture and books. To fully appreciate the rationale for the joint presentation in text and image of the liberal art of astronomy within Drogo's copy of the *Handbook of 809*, it is necessary now to review Frankish considerations of aesthetics and classical forms.

Appropriation of the Classical Heritage and the *Opus Caroli*

All the considerations thus far have attempted to explain, through philosophical analysis and textual criticism, the grounds for the Carolingian adoption of various strategies of scientific sacralization and scientific soteriology. The study of astronomy played an important role in the personal sanctification of ninth-century students, who perfected their studies as they approached a more complete understanding of Christian orthodoxy. The discipline of astronomy also diverted the students' attention from earthly matters and fixed it on higher things, as Isidore had taught in the seventh century (*Etymologies*, III. lxxi.41).[119]

At this point, it is vital to reconsider what the readers of the *Handbook of 809* thought that their star pictures were doing. I contend that the imagery joined with the text in performing the explanatory and exegetical work of the star catalogue. Arguably, the crucial analytic discussion of imagery during the Carolingian era had its place in a complex polemic against the Byzantine worship of icons. In older literature, this text is known as the *Libri Carolini*, but today it is predominantly known by its original name, the *Opus Caroli regis contra synodum* (hereafter *Opus Caroli*). Perhaps the best record of the text is a copy now located in Rome, at the Biblioteca Apostolica

Vaticana (Vat. lat. 7207).[120] As Thomas F. X. Noble has argued, the manuscript itself is emblematic of the scenario described here for the creation of Drogo's copy of the *Handbook of 809*. As Noble independently labeled the production, it was a "'team' effort" with four scribes editing and reworking the *Opus Caroli*, not only recording an attempt at an official version of the text (probably read aloud to Charlemagne himself for approval) but also demonstrating aspects of the Frankish creative process of emendation. The *Opus Caroli*, and in particular the Vat. lat. 7207 manuscript, reveals a learned society eager to get things right. That motivation makes the information within the manuscript valuable for an analysis of the utility and status of Carolingian star pictures.[121]

Traditionally, the *Opus Caroli* has been discussed as a dismissal of themes inappropriate for church decoration, which has led some to disregard the utility of the polemic as an aesthetic analysis of pictorial representation. It is true that the *Opus Caroli* intends, first and foremost, to combat the excesses discerned in the Byzantine veneration of icons.[122] The book's themes address the misgivings Charlemagne allegedly had about large frescoes on church walls, or icons for an iconostasis. As a result, scholars such as C. R. Dodwell have established an inaccurate divide between a putative Carolingian moratorium on classical imagery in frescoes and the proliferation of related imagery in manuscripts made for more discerning intellectual elites—as the frescoes of Ulysses, Scylla, and a Siren within the second-tier sacred architectural confines of the Corvey westwork, *inter alia*, belie.[123]

Ann Freeman has identified Theodulf of Orléans as the author of the *Opus Caroli*, though he worked under the aegis of Charlemagne and discussed the content of the manuscript with the king.[124] Noble endorses this attribution in his encyclopedic reappraisal of Carolingian and Byzantine polemics prescribing the practical and liturgical role of imagery in sacred early medieval kingdoms. Given the erudition of Noble's tome, the issue can perhaps at long last be considered resolved.[125] The *Opus Caroli* was supposed to supply the official Carolingian response to what Charlemagne and the court in Aachen believed to be the apostasy of the Second Council of Nicaea (787). Nicaea II was a turning point for the history of Byzantium. The first historical period of Byzantine iconoclasm officially ended when the regent Irene and her son, Constantine VI, acquiesced to the council's decision to restore the veneration of icons.[126] Since the *Opus Caroli* is an unusual document with a complicated origin, I will offer a brief review of the impetus for its production.

Although the *Opus Caroli* (ca. 793) arose as the Carolingian response to what was believed to be the decision of Nicaea II, a lamentable "comedy of errors" had taken place. As Noble notes while introducing the timeline involved, "these speculations turn around who knew what, and when they knew it."[127] Members of the Carolingian court at Aachen failed to realize that the translation of the proceedings at Nicaea with which they were working was based on a quasi-official translation carried out by the curia in Rome. The solecisms in the clumsy translation resulted in a crabbed book and a flawed understanding of the intentions of Nicaea II overall. The translation appears to have suggested, erroneously, to certain members of the Carolingian court that icons were to be worshipped rather than venerated (celebrated through sincerely heartfelt *latreia* rather than physical gestures of *proskynesis*). This resulted in Frankish protest, since those writing on Charlemagne's behalf and at his behest believed him to be the defender and purveyor of Christian orthodoxy in Europe.

The *Opus Caroli* was supposed to demonstrate this point to Pope Hadrian I, who had (unbeknownst to those in Aachen) endorsed rather than decried the conclusion of the synod. Hadrian considered Nicaea II an opportunity for Christian reconciliation between the Eastern and Western churches. Charlemagne and Theodulf misunderstood the situation and attempted to assert the priority of Frankish orthodoxy. Their bungled effort to promote the intellectual superiority and purity of Christian faith throughout the Frankish lands only set the court at Aachen against the papal peacemaking agenda. This debacle nevertheless prompted the creation of the *Opus Caroli*, a comprehensive description of Carolingian aesthetic theory, since the courtiers in Aachen believed they were combating issues of heresy and idolatry that were intertwined with concerns about the proper use of votive imagery.

Theodulf and Charlemagne's plan to present the *Opus Caroli* to the Council of Frankfurt in 794 was derailed in 793 by the arrival of a text from Pope Hadrian I. The pope had critiqued a preliminary version of the polemic (drafted in 792), referred to as the *Capitulare adversus synodum*, and had declared pontifical support for the iconodules, rebuffing Theodulf and his collaborators. Given Charlemagne's embarrassment, the *Opus Caroli*, the foremost text on Carolingian beliefs about image-making, faded into obscurity for roughly half a century.[128] In one official record of the Council of Frankfurt, the *Annales regni Francorum* for 794, an even-handed historian looking back on the year's events commented, "The pseudosynod of the Greeks, which they held for the adoration of images and which they falsely called the seventh, was rejected by the bishops."[129] As Bernhard Scholz remarked in his popular translation of the *Annales regni Francorum*, the later revisions included anecdotal references, additional facts considered important but missing from earlier versions of the text, and at times critical reappraisals of Frankish political blunders or military maneuvers ending in defeat. The revised version of the annals thus provides a precious glimpse not only of the events during any given year but also of roughly contemporary criticism (ca. 815) of Frankish politics and the tenuous papal alliance.[130] Swept under the royal rug in its own day, the significance of the *Opus Caroli* has until recently been undervalued.[131]

The fundamental tenet of the image theory presented in the *Opus Caroli* is the ancillary status of pictures to words, which are more capable of communicating their intended meaning. According to Theodulf, truth is not conveyed through pictures but instead through holy writings, "non . . . in picturis, sed in divinis litteris."[132] Rosamond McKitterick has emphasized that for Carolingian prelates, this line of thinking became a standard component of ninth-century image theory.[133] McKitterick and David Ganz have drawn attention to a poetic epistle (ca. 835) that Hrabanus composed for his eventual successor as abbot of Fulda, Abbot Hatto (otherwise known as Bonosus, who served 842–46):

> The sign of writing is worth more than the form of an image and offers more beauty to the soul than the false picture with colours, which does not show the figures of things correctly. For script is the perfect and blessed norm of salvation and it is more important in all things and is more use to everyone. It is tasted more quickly, is more perfect in its meaning, and is more easily grasped by human senses. It serves ears, lips and eyes, while painting only offers some

> consolation to the eyes. It shows the truth by its form, its utterance, its meaning, and it is pleasant for a long time. Painting delights the gaze when it is new, but when it is old it is a burden, it vanishes fast and it is not a faithful transmitter of truth.[134]

This statement affirms the ephemeral nature of fading or flaking images on walls and in books, as opposed to the relative timelessness of a text. Hrabanus's claim does not merely concern the conservation of artifacts—frescoes and parchment leaves do blanch and mildew, after all—but rather draws attention to an epistemic problem: the text alone provided the reliable expression of content. This passage critiqued the capacity of paintings, such as the miniatures of the constellations in Drogo's copy of the *Handbook of 809*, to convey their contents adequately without text. In Drogo's book, the text alone supplied the informative catalogue of stars and was accompanied by the depictions of the constellations. The depictions themselves only facilitated the identification of a general outline or contour for a constellation in the night sky.

This assertion is arguably at odds with the famous iconodulistic position maintained by Gregory the Great (d. 604) in his second letter to Bishop Serenus of Marseille. Gregory had advocated for the evangelistic potential of images, which could relate biblical narratives to the illiterate. As Freeman notes, however, Theodulf drew upon Gregory's text in the *Opus Caroli* but elided the passages that were problematic for Theodulf's pessimistic view of the hermeneutic capacity of images.[135]

Theodulf also addressed this theme in book III.23 of the *Opus Caroli*: "Painters are thus able to commit events to the memory [that is, in pictorial form] but things which are perceptible only to the mind and expressible only in words cannot be grasped and shown by painters, but by writers."[136] Conceptual matters such as the nature of "holiness" were deemed solely expressible through words in an era long before Kandinsky and nonobjectivist, spiritual art.[137] For example, in book III.23 of the *Opus Caroli*, Theodulf rejected the possibility of depicting "the commands of Divine Law, given by God through Moses."[138] A useful indexical aspect of early medieval sign theory may actually have been familiar to astute students of Augustine, as discussed by Celia Chazelle. In *Christian Instruction*, Augustine had argued that "a sign . . . is a perceptible thing which makes 'something else come to mind, beyond the appearance that the sign itself bears to the senses,' as for example smoke recalls fire."[139] Images served a role as signs, facilitating the recollection of complicated bodies of information in the *Opus Caroli*.

This function of depictions such as the imagery of constellations in the *Handbook of 809* has been discussed in greater detail by Carruthers. The pictures of the constellations were intended to catalogue the relative positions of the stars in the night sky. Carruthers differs from those who emphasize the mythological narratives attached to the constellations as explanations for their origins. As already noted throughout this study, early medieval Carolingian teachers downplayed the pagan tales in their presentation of the liberal art of astronomy. Instead, the star pictures were needed as a way to recall the organization of the constellations, according to a memory practice that Carruthers labels a "stellar inventory." For early medieval teachers leery of pagan myths, the stories about the constellations (including the odd histories of their names in the Isidorean *Etymologies*, book III.lxxi) were "inventory fables," according to Carruthers's terminology. These narrative associations assisted the medieval student in locating a given constellation in the sky, thanks to a nexus of information linking the star pictures together. The patterns of the star clusters in the constellations were interlocked in a way that paralleled how the information was connected in the student's mind.[140]

A final point about Hrabanus's letter to Hatto merits mention. In an era when books were copied and illustrated by hand, the creation of text and pictures required enormous effort. The inks, pigments, parchment, and the manuscripts themselves involved great expense and the expenditure of skill and energy. McKitterick notes that Hrabanus's letter reminded Carolingian scribes of their hierarchically arranged grades of scripts, of which the Caroline minuscule used for Drogo's copy of the *Handbook of 809* was a refined and appropriate choice.[141]

Another concern for Theodulf, related to the inability of pictures to depict abstract themes, was how easily pictures could misconstrue reality. Chazelle has discussed the role that the concept of "falsity" played in Carolingian aesthetic theory. One of Theodulf's concerns had been to avoid having his theory lead to the erroneous conclusion that all images had actual referents. The image of Capricorn in Drogo's copy of the *Handbook of 809* (plate 11) could be problematic, since such hybrid aquatic goat-like creatures had not been created by God, nor were they exactly discussed by the holy texts. Such creatures were potentially idolatrous fabrications of perverse pagan imaginations.[142]

Moreover, pictures depicting key narrative events from the biblical story could be misidentified. In his critique of icons in the *Opus Caroli*, Theodulf permitted two general uses for medieval pictures. These uses remained distinct from the special uses for consecrated sacral objects like the Ark of the Covenant or liturgical accoutrements, as discussed by Chazelle:[143] "We do not speak against images for the memory of past deeds and the beauty of churches, since we know that they were made thus by Moses and Solomon, although as type figures, but we reject their most insolent or rather most superstitious adoration which we cannot discover to have been instituted by the patriarchs, prophets, apostles, or apostolic men."[144]

The emphasis placed here on typologies is significant for the present discussion. Not only do the iconographic visual histories inform the normative and acceptable modalities of artistic production, but the physical manufacture of an artwork remains within the pale of acceptability precisely because it adheres to standard formal types. The syntactical formal parameters or criteria with which any Carolingian image had to comply simultaneously established the legitimacy of that form for Frankish use and were bound eternally with the benefit of the image for liturgical or educational implementation. The Carolingian pedagogical reliance on the star catalogues was an extension of this overarching theological-aesthetic justification for images of the classical constellations in manuscripts such as Drogo's copy of the *Handbook of 809*.

In addition, the semantic import of the celestial models of the heavens construed and studied by eighth- and ninth-century Frankish painters and prelates emerges as a series of considerations that range from (a) the lack of an actual reference in the world (entailing a charge of *falsity*) to (b) the suitable compliance with a formal visual model (embodying a legitimate iconographic *prototype* or *archetype*) and even to (c) an adequate reference to established astronomical knowledge and tradition through the conventional referent in the star picture.[145] The images of the heavens in the *Handbook of 809* supplied more than Carruthers's stellar inventory. They situated Frankish readers within a nexus of cultural traditions and confluences which were themselves being worked out in the social arena and monastic schools of Charlemagne's kingdom.

Frankish painters asserted the formal normativity and sacred suitability of their images of the constellations. They also situated the readers of the handbooks upon the same precarious precipice they themselves occupied while working for their monastic and episcopal patrons throughout the Frankish lands. Regardless of whether each eighth- or ninth-century Frank with stylus in hand recognized his individual role to play within the codification process that fixed in time the early medieval beliefs about the use of imagery, Carolingian manuscript illuminators reified the Christian appropriation of the cultural wealth of classical antiquity and established the ground rules for the reinsertion of such visual knowledge within the Frankish schools. The collection of copies of the *Handbook of 809*, the revised type-C edition of 818 (according to Borst's nomenclature from chapter 1), and other documents of Carolingian astronomical appropriation such as the Leiden *Aratea* demand that the creative work of these painters be celebrated and linked to the potential for sociocultural formation that remains foundational to understanding the Carolingian renewal. Such an approach examines every copy of the *Handbook of 809* and its corollaries as one stratigraphic layer of a sociocultural palimpsest built up and refined through the energetic efforts of the copyists who interpreted the visual models before them and laid down the precedents for the ubiquitous future reinterpretations of the star pictures that were to become an enduring celestial backdrop to the Western pictorial tradition.

Insofar as the star pictures fulfilled their scientific and pedagogical roles within treatises such as the *Handbook of 809*, these images supplemented, complemented, and illustrated their texts through diverse visual strategies. The pictures also, however, participated in the liturgical and scholastic reform programs under way throughout the Frankish lands. Consequently, images of the heavens such as that of Hercules (plate 2) performed an integral, spiritual role in the personal renewal of the minds and hearts of Frankish pupils. For a few intellectual appreciators of Carolingian celestial art, the miniatures may even relate to the liturgical vessels that Theodulf considered a vital class of objects within his Christian society. Although secondary to vessels for the Mass, such as chalice or paten, the Carolingian books about astronomy exemplified the spiritual processes of individual renewal that they enabled, providing a complementary set of textual and visual stepping stones that Frankish students could follow to that personal end. In Theodulf's words, chalice and paten make possible "the holy mysteries of our redemption."[146] The twin processes of scientific sacralization and scientific soteriology advanced the pedagogical corollaries of such arcane dogma.

The discussion of pictorial falsity also necessarily touches upon the ontological status of depictions (signifiers) that refer to that which they signify (the signified, which may or may not be an actual referent), according to an Augustinian theory of representation. Chazelle has outlined the implication of Theodulf's theory of the sign from book I of the *Opus Caroli*. There, Theodulf inserted text from *De diversis quaestionibus* (On various questions) by Augustine, addressing the meaning of several terms pertaining to an early Christian and Carolingian semiotic theory of signification. Briefly stated, an *image* emanates from and embodies, materializes, or otherwise reinstates the visual form of its "prototype" in a direct relationship to it. This could be the mirror reversal of a coffee cup on the desk, or the physical creation of a cup, molded, cut, wrapped, and printed according to factory standards for that cup. In other words, the distinction can work for either two-dimensional or three-dimensional Carolingian artistic creation.

In this respect, Theodulf's Augustine is quite interesting concerning the act of creation: an image is a materialized embodiment of a formal idea (so far this is nicely Platonic) that denotes a class of objects likewise derived from the abstract, formal archetype or prototype. A condition of *likeness* emerges when two independent objects, which need not be identified with any prototype in particular, nevertheless display accidental features that they share. An *equality* is entailed when two or more objects share a short list of essential or accidental properties but remain independent entities. The idea of *equivalence* (between two images in an equality relation) ensured that every picture of Sagittarius, which conformed *in every essential and meaningful way* to a particular prototype such as the proposed lost archetypal *Handbook of 809* from Aachen, was a Sagittarius-picture.[147]

Theodulf attempted to make this recondite distinction more accessible in book I.8 of the *Opus Caroli*: "Every egg is similar to every other egg, in so far as it is an egg. But a partridge's egg, in so far as it is an egg, is similar to a hen's egg, but is nevertheless not its image because it is not expressed by it, and it is not equal because it is smaller and comes from a different kind of animal."[148] The use of physical objects and organic forms (such as an egg) is intriguing. The egg is both protected by its shell and comestible, both wild and domesticated, and although ubiquitous, its specific, distinct features depend on the species that laid it. Overall, the partridge and hen lay eggs that display a *likeness* as eggs but do not share enough properties to manifest an *equality* relation. Both eggs might refer to a formal idea of *eggness*, which they then share, to be likenesses of one another, but neither is a direct *image* of the other. Still, an egg is created through a natural series of events that result in its formation. Inherent in this rhetoric is a latent event-based theory of action expressing manifest formal properties in created things and permitting comparisons between taxonomic classes or categories of them.

At this juncture, a slippery and tortuous path around a charge of visual *falsity* was permitted to those artists who accurately adhered to archetypal *artistic* models. This is an important and routinely overlooked—albeit fundamental—aspect of the Carolingian creative process, and one design strategy in the ever-energetic Frankish manuscript practice of *compilatio*. A more pessimistic way to cast the same information would be to emphasize that a hard interpretation of the *falsity* proviso precludes the ability of Capricorn to denote anything whatsoever for a good Christian audience, since it is a fictive creature. This inserts a hard and soft use of the *falsity* codicil into the discussion. The hard reading asserts the preeminence of essentially Platonic ideal forms for an aesthetics of Christian art ultimately derived from Augustinian metaphysical views. A soft reading, which would have been more palatable to Theodulf the poet than Theodulf the voice of Charlemagne in the *Opus Caroli*, permits the development of an intriguing aesthetic theory of early medieval scientific illustration.

In a more general way this was also true, however, for a *likeness*, which denotes the more typical bond between copied star pictures. Examples could proliferate in the handbooks for teaching purposes. Every instance that conformed *for the most part* to a particular prototype (whether the prototype/archetype was a presentation copy of a book or an ontological reality about the cosmos) was an effective *likeness*. The status of *likeness* made the image worthy of pedagogical use, just like the original (plate 11).[149] This is, of course, a *sine qua non* for illustrated scientific books to the present day. Whereas the *image* maintains a tight grasp on the actual historical and scientific cosmos through its relationship to the physical universe, the *likeness* permits a more fluid confluence of shared characteristics.

And this is where an aesthetics of scientific manuscript illumination becomes even more complicated. On the one hand, the star pictures do embody certain formal properties that relate directly to the organization of physical materials in the cosmos; that is, after all, what the constellations purport to do for stargazers, sailors, farmers, and so on. On the other hand, the constellations are themselves, as Carruthers has argued, human abstractions clustered into discernible patterns as mnemotechnical tools. The *falsity* proviso becomes more difficult to apply when the star picture refers to a prototype that is a physical abstraction invented by human ingenuity (like a constellation) but believed to actually relate to the organizational pattern of fixed stars in the nested system of heavenly spheres surrounding the earth (ignoring for the moment any problems deriving from precession).[150] In this respect, a well-constructed star picture of the constellation Hercules would perhaps be an *image* of that to which it refers, a zone of space defined by the formal unification of stars into a pattern for human observers. As Chazelle has rightly argued, Augustine and Theodulf do not permit this, since "such a link with the prototype did not occur in the unconsecrated artistic image."[151] The formal circularity of the conceptual apparatus and language for talking about art offered by Theodulf and Augustine reveals its own intrinsic inadequacies. The jargon nevertheless simultaneously supplies a rich linguistic set of tools that, when used with care, provide the art historian with a meaningful, early medieval way to think about and discuss images of the heavens.

In another passage from the *Opus Caroli*, Theodulf argues that even appropriately rendered forms of unconsecrated images required accompanying texts, or *tituli*. This was true even for a great biblical figure such as Mary, depicted in a standard iconic pose that recalls the Byzantine Virgin Hodegetria:

> Even if we suppose that an image of the Holy Mother of God ought to be adored, how are we to identify her image, or differentiate it from other images? When we see a beautiful woman depicted with a child in her arms, with no inscription provided, how are we to know whether it is Sarah holding Isaac, or Rebecca with Jacob, or Bathsheba with Solomon, or Elizabeth with John, or indeed any other woman with her child? Or if we consider the fables of the gentiles, which are often depicted, how are we to know whether it is Venus with Aeneas, or Alcmene with Hercules, or Andromache with Astyanax? For if one is adored [by mistake] for another, that is delusion, and if one is adored that should never have been adored, that is madness.[152]

Here Theodulf makes two points that are important for a study of how Carolingians understood their star pictures. First, the *tituli* clarified the intended meanings of the images. In princely copies of the *Handbook of 809*, such as Drogo's, the texts from the *De ordine* adduced the relative placements of the stars and drew significantly upon the memory-enhancing backdrop of the traditional figural forms of the constellations. These texts operated like extended *tituli*, ensuring that teachers and students would correctly use such books. In the *Handbook of 809*, the text accompanying the image of Taurus (as in Drogo's copy; plates 6–7) correctly identifies the animal as the constellation rather than as a bovine character in a bucolic scene (fig. 65), or just an arbitrary reclining bull without specific referent. Similarly, the textual or scriptural explanation accompanying an icon transformed the vision of woman and child into a meaningful object of veneration. (Although philosophically rigorous in his analysis, Theodulf's criticism here lacks enough faith in the common sense of individual people.)

Second, Theodulf asserts that the pagan worship of celestial bodies,[153] which was always a concern vis-à-vis the star pictures, was unequivocally a form of "madness." Using Isidore's terminology discussed above, natural astrology permitted the tracking of celestial phenomena for monastic purposes. Superstitious astrology, which could lead to the adoration of pagan deities and their actual physical counterparts in the heavens, was unthinkable. The deviant excess of such astrological practices lay in the realm of the insane because of the failure to comply with normative Christianity, or in other words, orthodoxy.[154]

Misgivings in the *Opus Caroli* about such pictures found their way into a lengthy invective in book III.23: "And is it not obvious that it is contrary to Scripture if one depicts rivers and streams and their confluence as men pouring water out of urns [like Heridanus (plate 15)]? And if the sun and the moon and the other adornments of the sky are depicted in human form [like Orion (plate 13) or Hercules (plate 2)], their heads crowned with rays [figs. 3 and 52], does not all of this run quite contrary to Holy Scripture?"[155] As Freeman has explained, this passage drew attention to the divide between the truth of the referents found in scriptural texts and the fictitiousness found in portrayals of classical narratives. In this context, it is hard to believe that Theodulf only had wall paintings in mind. Presumably, he also recalled illuminated manuscripts such as the *Calendar of 354* or illustrated copies of Vergil (fig. 65) and the *Iliad* while drafting his critique: "In Scripture there is nothing depraved, nothing unseemly, nothing impure, nothing false . . . in pictures many falsities are found, many depravities, many things contrary to logic and decency."[156]

The physicality of common pictures, such as the miniatures and diagrams in Drogo's copy of the *Handbook of 809*, required that any referents also be material, rather than spiritual, for both Augustine and Theodulf. This was true because a *likeness* did not relate directly and immediately to any prototype; according to the definitions, only an *image* laid claim to such a strong relation with the physical world and its putative Platonic ideal forms. In addition to the distinctions and possibilities adduced above, it is worth noting that an *image* of the stars (which did or did not constitute a constellation) could have been provided by a reflection in the pool of an atrium or a fountain in a monastic cloister. As mentioned above, it was in fact a *likeness* of the constellation that was presented in paint in Drogo's copy of the *Handbook of 809* with the form appropriate to the figure and as little mythological content as possible.

The *likenesses* of human figures such as the depiction of Hercules (plate 2), however, lacked any actual referent in reality. Unlike the depiction of a saintly legend in a fresco cycle, the miniature in Drogo's book offered the viewer a mythological "falsity" birthed in pagan "depravity." In the Hercules picture, there were therefore two falsities: first, there could never be a man named Hercules who walked the earth as a soulless material form, and second, there was never any demigod named Hercules, period.[157] The case of Capricorn was naturally even more problematic under this analysis, since it was a hybrid creature.

Carolingian ideas about the liberal arts were necessarily interconnected with ninth-century concerns about the forms of artistic illustration appropriate for teaching about the liberal art of astronomy. Consequently, it is unsurprising that Isidore's discussion of images from his *Etymologies* XIX.xvi.1 (*De pictura*) influenced both Hrabanus in his *De rerum naturis* (*De universo*) XXI.9 (ca. 842–46, as discussed in chapter 3)[158] and Theodulf in the composition of his *Opus Caroli*:[159] "A picture is an image representing the form of something which, when seen, prompts the mind to recall it. A picture may be called a kind of fiction; it is a fictitious image, not the truth. . . . Thus there are certain pictures which, using color, exceed [the limits of] corporeal actuality and, while seeking to promote our confidence, proceed to a lie—like those which depict a Chimaera with three heads, or Scylla in human form in her upper part, but girt about with dogs' heads below."[160] This definition of a picture as a *likeness* assisting in the processes of recollection reinforces the other ideas already discussed above. It is important to underscore that the idea of "falsity" applied here related particularly well to the imaginary hybrid forms like Capricorn. But convincing depictions of fictitious characters such as Hercules would have also been considered deceitful, given the odd moralizing gloss on representation. This posed a final problem for the Carolingian understanding of the star pictures.

A homily by Origen offered Hrabanus a sure way to identify the subtle difference between a likeness and an idol. Hrabanus's exegetical commentary on *Exodus* (ca. 822–29) argues,

> If someone gives gold, silver, wood or stone the shape of a quadruped, a snake or a bird, and sets it up to be adored, he makes a likeness and not an idol, nevertheless if he sets up a picture for this purpose it is a likeness. Someone makes an idol when he [according to the Apostle's teaching "An idol is nothing" (1 Corinthians 8:4)] makes what does not exist. What is it that does not exist? An image which no eye has seen, but which the spirit has imagined. For example, if you imagine a dog's or a ram's head on human limbs, or if you put two faces on one man or join a man's torso to the lower parts of a horse [plates 12 and 16] or a fish [fig. 69].[161]

The miniature of the constellation Capricorn (plate 11) in Drogo's copy of the *Handbook of 809* ran the risk at this point of being labeled an idol, since it was an imaginary hybrid creature that did not exist, and that could be shown excessive attentions, or even *adored*, by a pagan or a proponent of superstitious astrology.

Actually, the definition of *likeness* used in the *Opus Caroli*, together with the two acceptable uses for pictures (i.e., that they decorate and recall important events), offers a way to resolve the ontological dilemmas that the constellation pictures faced. A miniature in Drogo's book resembled the essential features of the arranged cluster of stars composing the constellation that could be used to identify the patterned arrangement of the constellation in the sky. In tandem with the textual description, functioning as an extended *titulus* (from the *De ordine* in the *Handbook of 809*), the form of the constellation aided the student in recalling the relative placement of the stars. This was so even without the representations of the actual stars, as in Drogo's miniatures.

The stars described in the text, according to the ninth-century theory of images in the *Opus Caroli*, need not have been displayed as points of light and would have been redundant for a *likeness*, agreeing in general form and characteristics with the actual arrangement of stars in the sky and the archetypal version of the *Handbook of 809*. It is important to remember that the three-part handbooks such as Munich 210 (see chapter 3) quickly reverted to the Aratean practice of designating stylized, inaccurate distributions of the stars for the reader's sake. The shape mattered most, whether it was represented in a finished form or awaited the viewer's attempt to connect the dots. The fact that the constellations did exist in the sky precluded them from being considered idols, as long as they remained the objects of study of natural astrology. Each *likeness* of the constellations recalled important events in a Carolingian view of history: their divine creation, classical Greco-Roman storytelling and literary traditions, and the original human ingenuity of clustering the endless heavens into discrete, cognizable patterned zones of space through the mediation of the constellations locked within the realm of fixed stars.

Orienting the Heavens Within: The Microcosm, the Macrocosm, and the Carolingian Idea of Order

Those star pictures and their texts in the *De ordine* worked together to create within the *Handbook of 809* a cohesive, ordered presentation of the cosmos and a Christian outlook on the heavens. The intersection of the sacred and the secular, according to the Carolingian cosmological view of the heavens described here, converged in the soul of the individual. As discussed throughout this book, the Carolingian explanations of the utility of the liberal arts emphasized the role that education played in the sanctification of the cleric in training and, by extension, all good Franks.[162] Platonic and Pythagorean cosmological ideas about order also established the human being as a representative site of convergence for the celestial principles and ratios manifested by the heavens (fig. 1).

In the diagram of "The Pythagorean Theory of the Harmonic Intervals" in Drogo's copy (Madrid 3307; fig. 4), the geocentric model of the solar system displays the planetary names symmetrically arrayed from the zodiac to the earth at 12:00, as discussed in chapter 3. Below, the tonal values provide the distances between the celestial bodies. At 90 degrees to the left of the centrally located column of texts is a horizontal arm, designating the wandering stars, or planets.[163] Bianca Kühnel has argued compellingly that medieval artists often displayed the information contained in their diagrams using a cross-like organizational scheme.[164]

According to Kühnel's convincing iconography of medieval diagrams, the Pythagorean diagram presented an established musical and mathematical order. But in addition, the diagram symbolically alluded to the sanctified order of spiritual renewal made possible throughout Creation by Christ's

Crucifixion. In Kühnel's words, "In order to be able to stop measuring Carolingian diagrams according to scientific standards alone and to start looking for their *interpretatio christiana*, we have to keep in mind that . . . medieval diagrams never attempted to make faithful copies of ancient sources but were rather intent on improving, correcting, and modifying them, with one aspiration in mind: to transcend classical culture and to maintain, disseminate, and purify the Christian religion, as the work of salvation in all its aspects."[165]

Isidore of Seville had explained that the ordered arrangement of the solar system in the diagram of intervals reflected the ordered constitution of the individual, according to the medieval conceit of the microcosm. His description of this relationship supplied the connection between his discussion of the liberal arts of music and astronomy in the *Etymologies*: "But just as this proportion in the universe derives from the revolution of the spheres, so even in the microcosm it has such power beyond mere voice that no one exists without its perfection and lacking harmony. Indeed, by the perfection of this same art of Music, meters are composed of arsis and thesis, that is, by rising up and setting down."[166] The essential proportional symmetry, ordering the arrangement of the planets, also provided a necessary condition for the existence of human beings. Music supplied a craft capable of human mastery through the ordered manipulation of octaval ratios. The Creator had similarly imbued people with ratios that governed their constitutions and revealed the ordered individual universe of a human body, known as a microcosm. Apparently, this is the earliest known reference to the microcosm in Western Latin literature. Fontaine has called this aspect of Isidore's philosophy a "religious anthropology."[167] Isidore's view established a series of well-known correspondences that rely on the interconnectedness of the so-called quaternities, or fourfold divisions of nature into discrete properties. The human body, for example, is divided into the four elements—earth, air, water, and fire—that correspond to the flesh, breath, blood, and vital heat.[168]

Isidore's classic expression of the correspondences between the various quaternities came from his original diagram in chapter 11 of his *De natura rerum*, in which he discussed the fundamental components of the universe (fig. 1).[169] A faithful late eighth-century example from Salzburg of the Isidorean diagram displays the information coherently (Munich, Bayerische Staatsbibliothek, Clm 16128, folio 16r).[170] The diagram includes a structural schema of interlocking circles that draws attention to the cross-like presentation of the information. This reinforces Kühnel's supposition that diagrams such as the Isidorean representation of the correspondences between the microcosm and macrocosm included an idea of sanctifying spiritual renewal made possible through the Crucifixion. This salvific work in nature took place in the "mundus, annus, homo" listed in the center of the diagram. The universe, temporality, and human beings were all composed of a related, ordered configuration of parts that displayed the created symmetry of God's handiwork in need of renewal.[171]

In the apology for arithmetic from book III of Isidore's *Etymologies*, numbers assume an important role in the expression of natural order:

> The reckoning of numbers ought not to be despised, for in many passages of sacred writings it elucidates how great a mystery they hold. Not for nothing it is said in praise of God (Wisdom 11:21), "Thou hast ordered all things in measure, and number

> [*numerus*], and weight." . . . It is even our lot to depend on the discipline of numbers to some extent when through it we name the hours, when we dispute about the course of the months, and when we recognize the duration of the turning year. Indeed, through numbers, we are provided with the means to avoid confusion. Remove numbers from all things, and everything perishes. Take away the computation of time, and blind ignorance embraces all things; those who are ignorant of the method of calculation cannot be differentiated from the other animals.[172]

Faith Wallis has indicated the significance of this passage for medieval computistics. Isidore emphasized the need for numbers in calendrical reckoning and therefore revealed his esteem for reckoning moments of the year with accuracy. As noted above, human beings could not exist without the ordered arrangement of the body and soul, reflected in the ordered arrangement of the macrocosmic solar system. The mathematical ratios uniting the microcosm and the macrocosm could be expressed through numbers. In his defense of arithmetic, Isidore clearly stated that, without numbers, not only the ordered system of existence but also existence itself ceased to be possible. Such dissolution and disarray would be contrary to the natural order established by God. One of the important ways in which human beings demonstrated their superiority over the other animals was by performing computistical calculations using tables like those in Drogo's copy of the *Handbook of 809* for fixing feasts such as Easter, as discussed in chapter 1.[173]

Without their adroitness at time reckoning—derived from natural astrology—recommended by Hrabanus Maurus for ninth-century Christian clerics, such monastics could not generate the yearly church calendar. The celebration of feasts and festivals was only possible because of the artificial conceptual grid of temporal units imposed onto the endless stream of observed astronomical and historical events. These divisions of time were the *sine qua non* for political and spiritual order in the early Middle Ages.[174] The *Handbook of 809*, and particularly the special copy of that manuscript made in medieval Metz for Drogo by four peripatetic painters, permitted clerics to adhere to daily liturgical rites. At the center of this initiative were the prelates entrusted with the task of creating the sequence of holy days and events for their parishioners to follow. The order that men such as Drogo and important ecclesiastical women imposed on the Frankish lands and throughout the Carolingian empire unified a Christian kingdom. Through sacred study of the liberal arts in the schools and the salvific work of the ninth-century sciences, Carolingian teachers attempted to restore the original perfect order that had been placed within individual Franks by their Creator. The twin processes of scientific soteriology and the sacralization of science made the *Handbook of 809* possible. Princely copies of the handbook enabled local prelates, such as Bishop Drogo at Metz, to harness the scientific practice of renewal contained therein, consolidating constellations of control and imposing a simulacrum of Christian order beneath a Frankish dome of heaven beyond which, Drogo believed, Charlemagne and his concubine Regina awaited.

NOTES

INTRODUCTION

1. Obrist, "Le diagramme Isidorien," 95–100, 113–14; Fontaine, *Isidore de Séville: Genèse et originalité*, 297–310. The cultural and historical scientific precursors to these diagrams have been thoroughly assessed in Obrist, *La cosmologie médiévale*.
2. Borst, "Das Aachener Verhör von 809" and "Die Aachener Enzyklopädie von 809," both in Borst, *Schriften zur Komputistik*, 3:1034–1334; Borst, *Buch der Naturgeschichte*, 156–65. Borst's sedulous studies must, however, be read in conjunction with interventions that examine the precursors to the Lorsch calendar (789), as in Englisch, *Zeiterfassung*, esp. 9–40, 101–2; at 17 she writes, "in fact, it appears that the calendar is considered as a product and component of the conceptual consideration of a problem from the sector fixing the feast of Easter, which the Frankish kingdom of the eighth and early ninth centuries moved and that through the computus of the Venerable Bede was insufficiently answered." Englisch's important observation is that the Frankish study of computus was an active, creative enterprise and inaccurately believed to be grounded in Bedan or external authority. On the contrary, as will be shown throughout this book, the Carolingian engagement with astronomical and calendrical science attests to an innovative, solution-oriented, and extremely creative effort at grappling with textual and visual precedents as prelates, sages, and painters fashioned an appropriately Frankish outlook on the heavens. Please note that throughout this book all translations are my own unless otherwise indicated.
3. Borst, "Das Aachener Verhör von 809," 1034–53; at 1035, Borst sensibly argues that Archbishop Arn of Salzburg and Abbot Fridugis of Tours could not participate in the proceedings on computus because of their integral involvement in the doctrinal aspects of the synod convened at Aachen in November 809 and lasting into the New Year. Most importantly, at the synod the Frankish prelates stipulated the papally sanctioned view that the Holy Spirit emanates from both the Father and the Son (*filioque*). For more on the synod, see the critical edition of key texts in Willjung, *Konzil von Aachen 809*, esp. 1–4. For a succinct summary of the impetus for and outcome of the synod, see McKitterick, *Charlemagne*, 311–15.
4. Borst, "Das Aachener Verhör von 809," 1040–44. Also see C. Jones, "Early Medieval Licensing Examination," 19–29, esp. 26. On the date for the vernal equinox, it is necessary to consult Springsfeld, *Alkuins Einfluß*, 105n70.
5. Borst, *Schriften zur Komputistik*, 1:63–96; Borst, *Karolingische Kalenderreform*, esp. 30–99, 157–68.
6. Englisch, *Zeiterfassung*, 102.

7. Eastwood, *Ordering the Heavens,* 10–11, 313–72. Also see McCluskey, *Astronomies and Cultures,* 20–38.
8. The outdated thesis concerning an Anglo-Saxon origin in Alcuin's York for the Plinian excerpts was offered by Rück, *Auszüge aus der Naturgeschichte,* 87–88. Instead, creative compilers on Frankish soil established the canonical set of excerpts, as discovered by V. King, "Investigation," 1–80. And, finally, in two volumes the diagrams accompanying the Carolingian cycle of Plinian texts were also shown to be original: Eastwood and Graßhoff, *Planetary Diagrams,* 14–15; Eastwood, *Ordering the Heavens,* 95–158. An important recent dissertation must be taken into account for any future serious engagement with early medieval astronomical art from either the East or the West: Anderson, "World Image," 207–349 for the present study.
9. Nees, *Early Medieval Art,* 172–84; Beckwith, *Early Medieval Art,* 11–14; Stalley, *Early Medieval Architecture,* 71–74; Reudenbach, *Godescalc-Evangelistar.*
10. McKitterick, *Charlemagne,* 94–96, 314–15.
11. Ibid., 314.
12. Dutton, "Of Carolingian Kings," 94.
13. Ibid., 93–94, 234 nn1–2, with the cited translations; for the original, see Dümmler, *Epistolae karolini aevi,* 4.2:176 l. 32–4.2.177 l. 3.
14. Kühnel, *End of Time,* 13–24, 65–159; McGurk, "Carolingian Astrological Manuscripts"; Plotzek et al., *Glaube und Wissen,* 15–64, along with the numerous catalogue entries assessing the scientifically and spiritually significant codices. For the most nuanced account of Carolingian engagement with classical tradition to date, see Nees, *Tainted Mantle,* 3–109.
15. Borst, *Buch der Naturgeschichte,* 161.
16. For example, see Borst, "Die Aachener Enzyklopädie von 809," 1054–65; Eastwood, *Ordering the Heavens,* 99.
17. For example, see Blume, "Sichtbares Bild," 292–301; Carruthers, *Craft of Thought,* 24–32. The earliest studies of the imagery contained within the *Handbook of 809* fall under this category: Neuß, "Meisterwerk"; Neuß, "Karolingische Kopie"; and Koehler and Mütherich, *Karolingischen Miniaturen,* 3.2:119–27. And not surprisingly, the text volume accompanying the facsimile edition likewise adopts a philological and iconographic approach: Mariana, *Códice de Metz.*
18. There is, however, a rich literature on the Leiden *Aratea,* including a comprehensive text volume that accompanied the facsimile edition: Bischoff et al., *Aratea;* also see Katzenstein and Savage-Smith, *Leiden Aratea;* Dekker, "Provenance"; C. Verkerk, "*Aratea,*" 245–87; and Dekker, "Carolingian Planetary Observations."
19. Blume, Haffner, and Metzger, *Sternbilder,* 1:43–51, 66–75; this exhaustive cataloguing enterprise has recently attempted to situate astronomical imagery within a meaningful conceptual and historical framework from the early to high Middle Ages, paying close attention to the philological significance of Aratean textual transmission and the careful negotiation of Christian artists with their classically inspired star pictures. Although this undertaking is useful for gaining a sense of the broad thematic brushstrokes, the fine points of the arguments (when there are arguments rather than bald assertions) do not always reflect an accurate assessment of the copiously presented data. For example, the authors present the implausible scenario that Drogo's copy of the *Handbook of 809* was made ca. 820 in Aachen for Louis the Pious, relying upon a cycle of illustrations that was created *sui generis* at the time of the manufacture of Madrid 3307, and not at the time of the creation of the *Handbook of 809* following the synod of that year. The only reasons offered for this are the dearth of comparative material, a situation evidently rectified after the creation of the Leiden *Aratea* in ca. 816 or while court artists were compiling Louis's pictorial program (using what precisely is unclear), and the changing opinions of monks about the suitability of star pictures. The former is art-historically inaccurate, and the latter fails to take into account the spiritually beneficial aspects of study of the liberal art of astronomy, as argued in this book. On a related note, the manuscript Cologne, Dombibliothek, MS 83-II was, according to Blume, Haffner, and Metzger, completed without illustration in 805. Allegedly, the dearth of visual resources required that empty spaces be left in the Cologne codex for about thirteen years, until suitable imagery could be found in a revised edition of the Carolingian handbook such as Vienna, Österreichische Nationalbibliothek, Cod. 387, created ca. 818. All of these unjustified statements require

that the original *Handbook of 809* prepared after Charlemagne's synod lacked a cycle of illustration on no grounds other than the circular argument presupposing the dearth of extant imagery available for consultation and an endemic mistrust of the pagan roots of the star pictures—in order to give artificial primacy to the Leiden *Aratea*. Such special pleading is not isolated in *Sternbilder.* The diverse forms of imagery remaining from the early medieval period attest instead to the richness of early medieval imaging of the heavens, and the creative ways that Carolingian painters achieved this goal should be celebrated.
In fact, the present volume attempts to correct this methodological problem. Speculation without argumentation should not skew the biographies of individual books; on the contrary, the factual stories of individual codices, drawing upon all art-historical and palaeographic tools available to researchers, should be used to clarify aspects of cultural production within specific spatiotemporal parameters, avoiding overgeneralization.

20. The missing portions are books I.1–2, I.5–9, VI.1–7, and most of VII.1.
21. Borst, "Die Aachener Enzyklopädie von 809," 1054–68.
22. A similar approach was applied to the study of the cathedral school of Laon by Contreni, *Cathedral School,* 29–77.
23. Borst, "Die Aachener Enzyklopädie von 809," 1063–64; Borst, *Schriften zur Komputistik,* 1:90–91.
24. Isidore, *Etymologies,* trans. Barney et al., III.lxxi.41, 107.
25. Ganz, *Corbie.*
26. The best resource on celestial globes and their significance for early medieval astronomy and astronomical illustration in the West is, now, Dekker, *Illustrating the Phaenomena,* 84–103, 111–230; also see Savage-Smith, *Islamicate Celestial Globes,* 3–16, and Künzl, "Globus/Globe," 81–153 for the English version of the text. An important dissertation in this respect is also Dolan, "Role of Illustrated *Aratea* Manuscripts."
27. Corradini, "Rhetoric of Crisis," with the citation at 320.
28. Eastwood, "Astronomy in Christian Latin Europe," 240, and the other collected essays in this volume all attest to Eastwood's suitable methodological approach, celebrating intelligent strategies for presenting Carolingian portrayals of the stars and planets. These texts should be considered in conjunction with Eastwood, *Ordering the Heavens,* 406–25.
29. Blume, Haffner, and Metzger, *Sternbilder,* 1:66.
30. Consider, for example, Chance, *Medieval Mythography,* 57–60, 220–60, 302–3, 347–54; Le Bourdellès, *Aratus Latinus;* Maass, *Commentariorum in Aratum Reliquiae;* Martin, *Histoire du texte.*
31. Haffner, *Sternbilderzyklus;* Mütherich, "Bilder."
32. Although outdated in places, the most useful resource to consult remains Gombrich, *Aby Warburg.*
33. Rampley, "Symbol to Allegory," 51. This perspective needs to be considered in conjunction with Forster, "Introduction," with earlier bibliography; the many helpful notes and addenda in this volume supplement Gombrich's comments in *Aby Warburg.*
34. Rampley, "Symbol to Allegory," 41–55.
35. Ibid., 51.
36. Forster, "Introduction," 6–38; as Forster explains, Warburg "was prepared to treat the work of art as something historically transmitted, and often discontinuous and lacking in unity" (38).
37. Forster, "Introduction," 28–35; also see Gombrich, *Aby Warburg,* 1–18, 310–11, and, in the same volume, Saxl, "The History of Warburg's Library (1886–1944)," 325–38.
38. Forster, "Introduction," 28.
39. Boll, *Sphaera.*
40. Saxl, *Verzeichnis,* vols. 1–2.
41. Saxl and Meier, *Verzeichnis,* vol. 3.
42. McGurk, *Catalogue,* vol. 4.
43. Panofsky and Saxl, "Classical Mythology," 228–29.
44. Ibid., 236.
45. Ibid., 228.
46. Panofsky, *Renaissance and Renascences,* 82–100, 104–8, at 106 for the citation.
47. Borst, *Karolingische Kalenderreform,* 319.
48. Borst, *Buch der Naturgeschichte,* 164; Borst, "Die Aachener Enzyklopädie von 809," 1054.
49. Borst, "Die Aachener Enzyklopädie von 809," 1054. Also see Bede, *Reckoning,* intro. Wallis, xv–xxviii.
50. Eastwood, *Ordering the Heavens,* 95–126.
51. Springsfeld, *Alkuins Einfluß,* 105–18.
52. Ibid., 105–15, 384–90.

53. In her introduction to Bede's *Reckoning*, Wallis poignantly remarks, "What is striking is the substitution of a cosmological for a computistical framework: while the *Reckoning of Time* is ordered by the units of time-reckoning, as they appear in the fundamental documents of computus, the great Carolingian encyclopaedias reconfigure the material in terms of the categories of natural science" (xcii).
54. Englisch, *Artes Liberales,* 58–67, 108–26.
55. Ibid., 339–49. Wallis's introduction to *Reckoning* emphatically endorses the significance of Englisch's work: "This in turn changed the definition of science itself. She sees Bede's way of combining elements from the traditional *artes liberales* with the technical, problem-oriented literature of the Christian calendar as essentially scientific in that it is structured, empirical, and rational" (xxvii).
56. Blume, Haffner, and Metzger, *Sternbilder,* 1:47, 354 refers to Drogo's copy of the *Handbook of 809* as "the *Libri computi,*" "Aachener Enzyklopädie von 809," and "a handbook," supporting the present nomenclature.
57. Saxl, "Reg. lat. 309," and "Vat. lat. 645," in *Verzeichnis,* 1:59–62, 71–73.
58. Borst, "Die Aachener Enzyklopädie von 809," 1057–60.
59. Englisch, *Artes Liberales,* 339–49.
60. Borst, "Die Aachener Enzyklopädie von 809," 1060–64.
61. For the definitive critical edition with a new German translation and a helpful introduction to all the debates surrounding this vital document, see Mordek, Zechiel-Eckes, and Glatthaar, *Admonitio generalis,* 1–25, with the appropriate section listed as chap. 70 (olim chap. 72 in earlier literature) on 222–25.
62. Ibid., 222. For the older edition, also see Boretius, *Capitularia regum francorum,* MGH, Leg. II, Cap. I, 59–60. Readily accessible translations can be found in Dutton, *Carolingian Civilization,* 80–81, and Munro, *Laws,* 15, who supplied the translation used here.
63. Contreni, "Carolingian Renaissance," 716–18, emphasizes, however, that there is ample evidence of learned Carolingian women. Hincmar of Reims, for example, made recommendations for the instruction of boys in areas without girls. This of course made explicit the male-centered exclusion of female students while perhaps inadvertently indicating the presence of girls, and, apparently, all-girls schools. Certainly the prohibitions against the fraternization between priests and the nuns they assisted pointed to the instruction of girls by female teachers within monastic sites. Consequently, the avowed interest in the education of boys and the principles used to guide the reforms of education for men should probably be extrapolated to women as well.
64. For a concise history of Carolingian liturgical music and its relationship to Roman chant, see Crocker, "Carolingian Chant," 142–70.
65. Munro, *Laws,* 15, provided all the translations for this document unless otherwise noted; this passage was slightly adapted here. Boretius, *Capitularia regum francorum,* MGH, Leg. II, Cap. I, 59–60; Mordek, Zechiel-Eckes, and Glatthaar, *Admonitio generalis,* 224, ll. 320–26.
66. Munro, *Laws,* 15: "and let them join and associate to themselves not only children of servile condition, but also sons of free men." *Ad. Gen.,* MGH, Leg. II, Cap. I, 60; Mordek, Zechiel-Eckes, and Glatthaar, *Admonitio generalis,* 222, l. 318–224, ll. 320.
67. Consider, for example, McKitterick, *Written Word,* 1–22. For a concise history of grammar and its instruction in the Frankish lands, supplying a foil to McKitterick's understanding of written Latin as a pervasive governmental official language and instead presenting a picture of practical missionaries struggling to convert and teach the Franks, see Law, "Study of Grammar," 88–110. The flourishing literary culture of Carolingian society should also be remembered in this context, since it was the men, not the boys, who were assigned the mature task of *emendatio;* see Garrison, "Emergence of Carolingian Latin Literature," 111–40. For more on the roles of Alcuin and monasticism in the Carolingian intellectual traditions that were considered manifestations of suitable spiritual service for a limited elite, also see Diem, "Monastic Schools," 27–44. McKitterick, *Charlemagne,* 318 argues that Charlemagne's native tongue was a Rhenish Franconian type of Old High German.
68. Contreni, "Carolingian Renaissance," 747–48.
69. Borst, *Streit,* 90–93.
70. Ibid., 14–17.
71. Borst, *Karolingische Kalenderreform,* 231–311, 341–43.

72. Englisch, *Zeiterfassung,* 85–100. See Borst's rejoinders, occasionally reduced to special pleading, in Borst, *Streit,* 70–89.
73. See Palmer, "Paschal Controversy," 240 for the discussion of computistical "creativity."
74. On the key players in the educational and spiritual reforms, see Eastwood, *Ordering the Heavens,* 8–15, 126–27; Dutton, introduction to *Charlemagne's Courtier,* xi–xiv, but that anthology must also be considered alongside Noble, introduction to *Charlemagne and Louis the Pious,* 7–11; Theodulf of Orléans, *Theodulf,* trans. Andersson, 1–4; Haarländer, *Rabanus Maurus,* 29–57. (I follow Lynda Coon's spelling of Hrabanus Maurus in this book; see Coon, *Dark Age Bodies,* 1–16.) Any discussion of a putative univocal Frankish court structure must be read in conjunction with McKitterick, *Charlemagne,* 137–213, 345–50.
75. McKitterick, *Charlemagne,* 212–22.
76. Fried, "Netzen der Wissensgesellschaft," 141–47. For a completely different exegesis, see Anderson, "World Image," 324–25, 336–38.
77. McKitterick, *Charlemagne,* 237–43 and 292–380, with the citations at 240 and 294.
78. See Anderson's helpful discussion of variability in "World Image," 324–25.
79. Müller, "Irritierende Variabilität."
80. Nees, "On Carolingian Book Painters," at 209 for the citation.
81. Ibid., 219–34, esp. 226–27, adds that under Nees's model an itinerant painter such as the Transfiguration Master was a collaborator with site-specific scribes or artists who coalesced at a work site.
82. Ibid., 227. Nees has also, for example, attempted to re-create an art-historical biography for the early medieval artist Godescalc, in "Godescalc's Career." In a recent lecture accessible online, Nees has also argued for the utility of mapping artistic networks of exchange and influence using art-historical methods and textual documentation; consider his "Networks or Schools? The Production of Illuminated Manuscripts and Ivories During the Reign of Charlemagne," paper presented at "Charlemagne: Les temps, les espaces, les hommes—Construction et déconstruction d'un règne," Deutsches Historisches Institut Paris, 26–28 March 2014, http://charlemagne.hypotheses.org/tag/lawrence-nees.
83. Koehler and Mütherich, *Karolingischen Miniaturen,* 8 vols.
84. This is the criticism often levied against Weitzmann, *Illustrations in Roll and Codex;* also see Dolezal, "Elusive Quest." In what follows, the useful aspects of Weitzmann's archaeological methodology for manuscript study will be elucidated.
85. Isidore, *Etymologies,* trans. Barney et al., III.lxxi.41, 107.

CHAPTER 1

1. For the early Christian roots of the *mundus/saeculum* distinction, see P. Brown, *Authority and the Sacred,* 12–26.
2. Edgar, ed., *Vulgate Bible,* 1:4–7, 10–13.
3. Eastwood, *Ordering the Heavens,* 31–36; also see McCluskey, *Astronomies and Cultures,* 17–25.
4. Evans, *History and Practice,* 39–42.
5. Augustine, *City of God,* 435–36, at 436.
6. McCluskey, *Astronomies and Cultures,* 99–110.
7. Dutton, "Of Carolingian Kings," 100–27.
8. Ibid., 95–127.
9. Kühnel, *End of Time,* 170–71. For the commentary of Augustine, see Augustine, *Iohannis Evangelium,* x, 12, 108.
10. Borst, "Alkuin und die Enzyklopädie von 809," 68–73; Borst, "Die Aachener Enzyklopädie von 809," in Borst, *Schriften zur Komputistik,* 3:1054–1334, with a helpful review of the literature on 1054–70. And any discussion of Carolingian computistics, calendrical science, or astronomy must pay homage to Borst, *Karolingische Kalenderreform,* xxi–xxii, xxvii–xxviii, 45–69, 319–24.
11. Bordona, *Manuscritos,* 1:269, cat. 568, 1:270, ill. 240.
12. Neuß, "Meisterwerk"; Neuß, "Karolingische Kopie."
13. For more biographical information, see the historical notes provided by *Kolumba* in Cologne, s.v. "Wilhelm Neuß," http://www.kolumba.de (accessed 14 September 2006); *Biographisch-bibliographisches Kirchenlexicon,* s.v. "Wilhelm Neuss," cols. 1128–30, http://www.bautz.de/bbkl/n/neuss_w.shtml (accessed 20 September 2006).
14. Neuß, "Meisterwerk," 37–38; Neuß, "Karolingische Kopie," 115.

15. Saxl, *Verzeichnis*, vols. 1 and 2; Saxl and Meier, *Verzeichnis*, vol. 3; McGurk, *Catalogue*, vol. 4.
16. For example, see Mütherich and Gaehde, *Carolingian Painting*, 26, cat. 13.
17. Nordenfalk, "Buchmalerei," 296–97, cat. no. 479.
18. Van der Horst, Noel, and Wüstefeld, *Utrecht Psalter*, 198–99, cat. no. 12.
19. Camps i Sòria, *Catalunya*, 337, cat. no. 70.
20. Mariana, *Códice de Metz*, supplied the text that introduces the facsimile. Also see the helpful Spanish-language edition of the text in Cartelle, *Códice de Metz*.
21. Neuß, "Karolingische Kopie."
22. Neuß, "Meisterwerk," 38–39.
23. Koehler and Mütherich, *Karolingischen Miniaturen*, 3:120–21. See also McGurk, *Catalogue*, 4:52–61.
24. Neuß, "Karolingische Kopie," 119; Borst, "Aachener Enzyklopädie von 809," 1084; also see Saxl, *Verzeichnis*, 1:59–66.
25. Saxl, *Verzeichnis*, 1:71–76.
26. Neuß, "Meisterwerk," 38–39.
27. Bede, *Reckoning*, 425–26.
28. Boretius, *Capitularia regum francorum*, MGH, Leg. II, Cap. I, 235, no. 117 , "Quae a presbyteris discenda sint," found in Munich, Bayerische Staatsbibliothek, Clm 6325, fol. 132v.
29. Schieffer, *Karolinger*, 121.
30. Neuß, "Karolingische Kopie," 119; Boschen, *Annales Prumienses*, 22–23, 59–61 argued for a *terminus post quem* of 824 for the manuscript, because his reproduction of the leaves from the handbook did not permit him to corroborate simply the more accurate dating set forth by Neuß.
31. Barbero, *Charlemagne*, 138–39. See also Boschen, *Annales Prumienses*, 59n50. P. King, *Charlemagne*, 149 offers the following translation from the entry for 813 from the *Moissac Chronicle*: "and he conferred the authority of rulership upon [Louis the Pious] and commended his sons Drogo, Theoderic and Hugo to him." Also see Abel, von Jasmund, and Rau, *Karolingischen Reichsgeschichte*, 1:232. In the *Vita Hludowici* of Thegan (who died mid-ninth century), particular attention is paid to emphasize Louis the Pious's wishes for his half-brothers Drogo and Hugo to study and prepare for a life of religious service (which of course eliminated their threat as potential political rivals): "24. In those days he ordered that his brothers, Drogo, Hugh, and Theoderic receive the tonsure, in order to alleviate any conflict, and ordered they be taught the liberal arts. Later, he established them honorably; he gave Drogo a see, and he gave Hugo several monasteries."
32. Boschen, *Annales Prumienses*, 21–22, 98–103 also provides a good overview of the so-called *Annales Laubacenses*, which were included within the nineteen-year tables and provide the best reason to associate the manufacture of the manuscript with the Hainault.
33. Boschen, *Annales Prumienses*, 98–103 adds that the odd date for the death of Hugo is a standard feature of the *Annales Laubacenses*. In the *Annales San Quintini* the date is 844, but the date 845 also appears in the *Annales San Dionysii*. The discrepancies reveal that there was a lack of consensus about the date of Hugo's death, and this confusion, although troubling, is not overly significant.
34. H. Seibert, "Lobbes, St-Pierre de," in Auty et al., *Lexikon des Mittelalters*, vol. 5, cols. 2061–62.
35. Baroffio, "Biblioteca Capitolare," 181–83.
36. McGurk, "Carolingian Astrological Manuscripts," 321. This reverses McGurk's earlier assessment that the manuscript had a probable origin after 867; see McGurk, *Catalogue*, 4:52–61. McGurk roughly sketches the odd history of the manuscript, which was confiscated from the library in *Verona* during the Napoleonic Italian campaigns (ca. 1796–97). Baroffio, "Biblioteca Capitolare," 181 explains that the shipment of 115 books, including Monza f–9/176, was repatriated to Italy after the Treaty of Vienna, but Monza f–9/176 was accidentally sent to *Monza*, not back to the Biblioteca Capitolare in Verona in 1817. See, too, Belloni and Ferrari, *Monza*, 106–7.
37. Koehler and Mütherich, *Karolingischen Miniaturen*, 3:121 add that the annals were written down progressively beginning with the year 851.
38. Belloni and Ferrari, *Monza*, 106.
39. Crosby, *Saint-Denis*, 94 notes the relative obscurity of Abbot Louis, who eventually was replaced by none other than Charles the Bald himself. After Abbot Louis died, Charles the Bald assumed control of the royal abbey to ensure ongoing and easy access to its prosperous resources. Later, Charles the Bald was buried there; also see Schieffer, *Karolinger*, 169.
40. Vieillard-Troiekouroff, "Art carolingien," 77–80 explains that the manuscript includes additions of

importance for the Royal Abbey of Saint-Denis; this must be consulted in tandem with Borst, "Aachener Enzyklopädie von 809," 1084.

41. Boschen, *Annales Prumienses*, 21.
42. Anderson, "World Image," 302–25 has proposed an alternative solution, linking luxury copies of the *Handbook of 809* to various royal Frankish chaplains and the chancellery. This is a compelling thesis, not to be dismissed lightly. In its favor is the localization of authoritative documents in court centers renowned for superlative scribal practice. On the other hand, this view favors the centralization of models and types that is belied by the iconographic programs of the copies of the *Handbook of 809* that appear to be locally interpreted, ad hoc constructions with imperial benefactors. Although Anderson notes the lack of uniformity revealed by the contents of these books, the link to the Carolingian schools and scriptoria advanced by Drogo, Hugo, Hubert, and Louis is more important than a putative connection to *capellae*. For more on the royal princes who also held positions in the court chapel system, also see Fleckenstein, *Hofkapelle*.
43. See note 10.
44. Dutton, *Charlemagne's Courtier*, xix–xx argues for the composition of the *Vita Karoli* in the mid-820s, against Rosamond McKitterick, who considers Einhard's text laden with propagandistic and eulogistic tendencies appropriate to roughly the year 818, during the incipient years of rule by Louis the Pious, and against Noble, *Charlemagne and Louis the Pious*, 13, who prefers a nominally later date of 828–29, recasting the *Vita* as a *bonum exemplum* of Frankish rulership for a foundering nexus of Carolingian communities. In any case, it is vital to underscore that all dates in the second half of the 820s point toward a common interest in the cultivation of learning for personal and more broadly Frankish benefit.
45. Dutton, *Charlemagne's Courtier*, xviii–xx, xxiii–xxiv, 32 explains that there was a division of labor at court when it came to the education of the emperor. Alcuin was responsible for the general pedagogical training of the king, including astronomy (encompassing calculation and the observation of the stars), rhetoric, and dialectic. Arithmetic and astronomy commonly belonged to the quadrivium in the Middle Ages (along with geometry, which could also fall under the purview of calculations, and, finally, music). Peter of Pisa, another deacon, taught the king grammar. See also Borst, "Alkuin und die Enzyklopädie von 809," 58.
46. Fleckenstein, "Alcuin im Kreis der Hofgelehrten," esp. 3–7 provides an excellent survey of the role played by Alcuin at the Frankish court as instructor of the liberal arts and special consultant to Charlemagne. Evidently, Fleckenstein does not see the meeting in Parma as the first encounter between Alcuin and Charlemagne. He believes that Alcuin's teacher, Ælberht (d. 780), had already managed to gain Alcuin an audience with the king while Alcuin traveled about on church business.
47. McKitterick, *Charlemagne*, 348; Bullough, *Alcuin*, 337.
48. McKitterick, *Charlemagne*, 345–50.
49. Dutton, *Charlemagne's Courtier*, 32: "[Charlemagne] also attempted to [learn how to] write and, for this reason, used to place wax-tablets and notebooks under the pillows on his bed, so that, if he had any free time, he might accustom his hand to forming letters. But his effort came too late in life and achieved little success."
50. Fleckenstein, "Alcuin im Kreis der Hofgelehrten," 4–5.
51. Borst, "Alkuin und die Enzyklopädie von 809," 59–60; Springsfeld, "Karl der Große," 57. For the text, see Lohrmann, "Alcuins Korrespondenz," 104–13. For additional commentary and background, also see Springsfeld, *Alkuins Einfluß*, 62–90.
52. Warntjes, *Munich Computus*, 349–52.
53. Borst, "Alkuin und die Enzyklopädie von 809," 60. For an explanation of the *saltus lunae*, see the helpful survey of Alcuin's astronomical correspondence with Charlemagne in Lohrmann, "Alcuins Korrespondenz," 88–89. An ever-useful resource is Wallis's introduction to Bede, *Reckoning*, here at xliii.
54. Springsfeld, "Karl der Große," 58–63. The *De saltu lunae* text is edited at letter 126 in Dümmler, *Epistolae karolini aevi*, MGH, Epistolarum IV, Karolini Aevi II, 185–87.
55. Borst, "Alkuin und die Enzyklopädie von 809," 60–61.
56. Ibid., 61–63.
57. Cavallo, "Biblioteca longobarda"; Nees, "Godescalc's Career," 36.
58. Borst, "Alkuin und die Enzyklopädie von 809," 61–62.

59. Bullough, *Alcuin*, 336–37. Also see Higham, *Northumbria*, 144–49.
60. Bullough, *Alcuin*, 339.
61. Ibid., 336–39; Borst, "Alkuin und die Enzyklopädie von 809," 59–62.
62. A careful and well-articulated expression of the significance of Alcuin for Carolingian computistical study is Springsfeld, *Alkuins Einfluß*, 20–26; she argues for the traditional position that Alcuin arrived at court in 782.
63. Springsfeld, "Karl der Große," 53–66.
64. Von Euw, "Zeitrechnung," 3; Springsfeld, *Alkuins Einfluß*, 20–26.
65. Bullough, *Alcuin*, 359–61.
66. Borst, "Alkuin und die Enzyklopädie von 809," 62–63.
67. Borst, *Karolingische Kalenderreform*, 317–18.
68. Neuss, "Meisterwerk," 39. The reconstruction of Madrid 3307 remains indebted to two critical studies: Mariana, *Códice de Metz*, 21–31; Cartelle, *Códice de Metz*.
69. Saxl, "Reg. lat. 309" and "Vat. lat. 645," in *Verzeichnis*, 1:59–66, 71–76.
70. Springsfeld, "Karl der Große," 63 moots the possibility that Hildebald was one of the "Egyptian Boys."
71. Borst, "Alkuin und die Enzyklopädie von 809," 71–72; for more on the Lorsch calendar, also see Borst, *Karolingische Kalenderreform*, 241–98, esp. 242 for the intertwining pragmatic and personal motivations of Charlemagne's spiritual and administrative effort to reform the calendar. Also see Borst, *Buch der Naturgeschichte*, 126–38. These sources must be considered in conjunction with Englisch, *Zeiterfassung*.
72. Mariana, *Códice de Metz*, 23–24.
73. Evans, *History and Practice*, 20; Bede, *Reckoning*, intro. Wallis, xlv–xlvi; Von Euw, "Zeitrechnung," 11.
74. Warntjes, *Munich Computus*, 346–48; Declercq, *Anno Domini*, 60–65.
75. Declercq, *Anno Domini*, 60–63; Warntjes, *Munich Computus*, 346–48.
76. Von Euw, "Zeitrechnung," 11–12; Evans, *History and Practice*, 20. See the excellent history of the nineteen-year tables in Bede, *Reckoning*, intro. Wallis, xxxiv–lxiii, 352–53.
77. Declercq, *Anno Domini*, 66–67, 80–81, 97–112; Bede, *Reckoning*, intro. Wallis, xlviii–lxiii.
78. Bede, *Reckoning*, intro. Wallis, l–li.
79. Palmer, "Paschal Controversy," 213–16. Also see Borst, *Schriften zur Komputistik*, 1:462–508.
80. Bede, *Reckoning*, intro. Wallis, li–lii, chaps. 47 and 53, 126, 136–37.
81. Von Euw, "Zeitrechnung," 5–12 argues that the annotation adjacent to the line of data for the year 768 in the thirteenth table of the nineteen-year tables on fol. 15r provided the date for the manuscript. A scribe made additions to the nineteen-year tables shortly after the completion of the manuscript. The number twenty-seven ("XXVII") was set alongside "Carlus" at 768, designating the number of years that had lapsed since Charlemagne's coronation. This yields the date of the manuscript, or 795. MS 103 is a compilation of texts by Bede including, *inter alia*, a calendar (fols. 1r–8v); nineteen-year tables (9r–22v); *De natura rerum* by Bede (23v–35r); *De temporibus* by Bede and *Ex Bedae Computo* (35r–45v); and *De temporum ratione* by Bede (52v–184r). For a more complete list of the contents of the manuscript from which these comments were culled, also see Von Euw, "Beda Venerabilis," cat. 23.
82. Von Euw, "Zeitrechnung," 12. For more on the later tradition of illustrated manuscripts, including texts by Bede, see Bober, "Illustrated Medieval Schoolbook," 65–97.
83. An excellent introduction to the information contained within columns of the nineteen-year tables can be found in Declercq, *Anno Domini*, 101–5.
84. Von Euw, "Zeitrechnung," 1–3, 11–12. Warntjes, *Munich Computus*, 341–42; Wallis in the introduction to Bede, *Reckoning*, explains that the bissextile day derived its name from the double occurrence of the vi kalends of March, or 24 February, during a leap year (xli).
85. Augustine, *City of God*, 432–89; McKitterick, *History and Memory*, 245–59; Kühnel, *End of Time*, 65–115.
86. Declercq, *Anno Domini*, 112–17.
87. Ibid., 138–47.
88. McKitterick, *History and Memory*, 86–87; Bede, *Reckoning*, chap. 48, 130, 339–40; also see A. Jones, *Later Roman Empire*, 448–69.
89. Von Euw, "Zeitrechnung," 12. Also see Mariana, *Códice de Metz*, 15.
90. Von Euw, "Zeitrechnung," 12.

91. Evans, *History and Practice*, 325–26; Cartelle, *Códice de Metz*, 21–22; Declercq, *Anno Domini*, 67–68, 102–3; Warntjes, *Munich Computus*, 348.
92. Evans, *History and Practice*, 185–86; Bede, *Reckoning*, xli.
93. Mariana, *Códice de Metz*, 15.
94. Cartelle, *Códice de Metz*, 22.
95. Ibid., 22.
96. Jansen and Pohle, *Künste am Hofe Karls*, 40–41; Mariana, *Códice de Metz*, 16; Cartelle, *Códice de Metz*, 24; Bede, *Reckoning*, intro. Wallis, xviii–xix, 296–97; Declercq, *Anno Domini*, 69–71, 104; Warntjes, *Munich Computus*, 343.
97. Mariana, *Códice de Metz*, 16; Cartelle, *Códice de Metz*, 24.
98. Declercq, *Anno Domini*, 104.
99. Von Euw, "Zeitrechnung," 12.
100. Evans, *History and Practice*, 188–90; Van de Vyver, "Nombre d'or."
101. Von Euw, "Zeitrechnung," 12; the Roman system of dating vexes modern readers, unless it is recalled that the Romans reckoned their calendar by looking ahead to one of three identifiable days each month. The first important day, termed the *kalendae* or the *kalends* of each month, denoted day one. The second important day was the *nonae* or *nones*, which commonly fell on the fifth. The third important day of the month was the *idus* or *ides*, which commonly fell on the thirteenth. In four exceptional months, which broke these rules, the seventh of the month was instead reserved for *nones*, as was the fifteenth for the *ides*; they are March, May, July, and October, easily recalled with the mnemonic "bad MMoJO." This is all nicely explained, along with notes on the Julian calendar, in Evans, *History and Practice*, 163–66.
102. Von Euw, "Zeitrechnung," 11–12.
103. Bede, *Reckoning*, intro. Wallis, xviii–xix, chap. 6, 24–28, 273–75; Springsfeld, *Alkuins Einfluß*, 105.
104. Cartelle, *Códice de Metz*, 20–21; Von Euw, "Zeitrechnung," 12.
105. Von Euw, "Zeitrechnung," 12.
106. Mariana, *Códice de Metz*, 15; Bede, *Reckoning*, intro. Wallis, xliii–xlvi; Cartelle, *Códice de Metz*, 21–22.
107. Boschen, *Annales Prumienses*, 9–10, 90–95, 227–28.
108. Höhl, *Ottonische Buchmalerei*, 11–12.
109. Parisse, "Metz, capitale d'Austrasie," 84 lists the death of Lothar I as 20 September 855, although the 29 September 855 date is more commonly accepted. Büchler and Dümge, *Archiv der Gesellschaft*, 22–23 transcribes the entry for the year 855, announcing the date for the death of Lothar I in the *Necrologium Prumiense* as "III. Kal. Octobris," or 29 September 855. A Carolingian record of these events can also be found in the entry from the Annals of Saint-Bertin for 855; see Grat, Vielliard, and Clémencet, *Annales de Saint-Bertin*, 71:

    > Lotharius imperator, morbo correptus vitamque desperans, monasterium Proneae in Ardvenna constitutum adiit, seculoque et regno penitus abrenuntians, tonsus est, vitam habitumque monachi humiliter sumens. Dispositoque inter filios qui secum morabantur regno, ita ut Lotharius cognomen eius Franciam, Karlus vero Provintiam optinerent, intra sex dies vita decessit tertio kalendarum octobrium, atque in eodem monasterio sepulturam, ut desideraverat, consecutus est.
    >
    > The emperor Lothar, being consumed by illness and despairing his life, requested that he enter the monastery of Prüm in the Ardennes, renouncing completely the world and the kingdom, was tonsured, and humbly assumed the life and character of a monk. Having assigned his kingdom amongst his sons who were staying with him, with the result therefore that Lothar his namesake obtained Francia, but Charles obtained Provence, within six days he departed life the third kalends of October [29 September 855], and in the same monastery he attained the grave, as he had desired.

110. Koehler and Mütherich, *Karolingischen Miniaturen*, 3:146; Fried, "Frankish Kingdoms," 146. It is routinely yet inaccurately reported that Drogo became an archbishop, which is impossible since Metz never attained archiepiscopal status and Drogo never assumed pastoral control of another see; also see M. Parisse, "Metz," in Auty et al., *Lexikon des Mittelalters*, vol. 6, cols. 585–89.
111. Höhl, *Ottonische Buchmalerei*, 11–12.
112. Boschen, *Annales Prumienses*, 75–77.
113. Koehler and Mütherich, *Karolingischen Miniaturen*, 3:120; Neuß, "Karolingische Kopie," 119n20.

114. Koehler and Mütherich, *Karolingischen Miniaturen*, 3:120; Boschen, *Annales Prumienses*, 75–78; Neuß, "Karolingische Kopie," 119.
115. Koehler and Mütherich, *Karolingischen Miniaturen*, 3:120.
116. Boschen, *Annales Prumienses*, 73–74 presupposes that Drogo's copy of the *Handbook of 809* had come to Prüm by 848, when Wandalbert consulted the manuscript in the creation of his famous calendar. This position fails to recognize the fact that the revisions within the Madrid calendar and the calendar of Wandalbert of Prüm are independent and occasionally overlapping expressions of mid-ninth-century developments. The priority of neither can be affirmed without additional evidence. Borst, *Karolingische Kalenderreform*, 245–301 argues that the first imperial Carolingian, so-called Lorsch, calendar type was constructed in 789 as a synthesis and expansion of the earlier calendars created by Willibrord in 705 and Godescalc in 781–82. By 840, one of the important copies of this calendar was in Prüm (Berlin, Preußischer Staatsbibliothek zu Berlin, Phillipps 1869, fols. 1r–11v). Calendar reforms and alternative forms of the calendars were prevalent in the Carolingian heartland. There is no logical way to affirm or confirm a necessary relationship between the Wandalbert verses and the calendar in the Madrid copy of the *Handbook of 809*. A far more plausible scenario respects the authority of the *Handbook of 809* and the hegemonic status of the lost archetype, which could have been consulted by Wandalbert using extant or lost Carolingian copies, texts, notes, or excerpts.
117. Janini and Serrano, *Manuscritos litúrgicos*, 73–76.
118. Boschen, *Annales Prumienses*, 81, 180 agrees that the work of Hand B has 855 as a *terminus post quem*.
119. Koehler and Mütherich, *Karolingischen Miniaturen*, 3:144–46, 156.
120. Fleckenstein, *Hofkapelle*, 1:55–56, 83–84, 118–21; Koehler, *Manuscrit latin 9428*, 9.
121. Koehler, *Manuscrit latin 9428*, 9–10 adds that the *Drogo Sacramentary* was removed from the cathedral of Metz, came to Paris, and was accessioned in 1802 into the collection of the Bibliothèque nationale de France.
122. Koehler and Mütherich, *Karolingischen Miniaturen*, 3:143–46.
123. Boschen, *Annales Prumienses*, 75–81.
124. Ibid., 77.
125. Ibid., 75–82, at 82 for the transcription.
126. Ibid., 26, 82, 183, 195–227.
127. Mariana, *Códice de Metz*, 38.
128. Boschen, *Annales Prumienses*, 75–84, 195–228.
129. Mariana, *Códice de Metz*, 34–38, at 34 for the citation.
130. J. Laudage, "Regino," in Auty et al., *Lexikon des Mittelalters*, vol. 7, cols. 579–80.
131. Boschen, *Annales Prumienses*, 77–78, 183–210.
132. See, for example, Borst, *Karolingische Kalenderreform*, 319.
133. Boschen, *Annales Prumienses*, 19.
134. Borst, "Alkuin und die Enzyklopädie von 809," 72; Bede, *Reckoning*, 9–13, 254–63; Cartelle, *Códice de Metz*, 111–12.
135. Borst, "Alkuin und die Enzyklopädie von 809," 72.
136. Consider, for example, Obrist, *La cosmologie médiévale*, 147–94.
137. Boschen, *Annales Prumienses*, 18.
138. Borst, "Alkuin und die Enzyklopädie von 809," 72.
139. Koehler and Mütherich, *Karolingischen Miniaturen*, 3:122, especially 122n90.
140. Eastwood and Graßhoff, *Planetary Diagrams*, 4–5.
141. Le Bourdellès, *Aratus Latinus*, 85–89, 99–107; Haffner, *Sternbilderzyklus*, 15.
142. Le Bourdellès, *Aratus Latinus*, 262. For more on Abbot Grimo, see Ganz, *Corbie*, 19–20.
143. Le Bourdellès, *Aratus Latinus*, 260–63.
144. Kaczynski, *Greek*, 5–25, 115–16.
145. Le Bourdellès, *Aratus Latinus*, 85. Also see Hasper, *Hyginus, philosophus*.
146. Neuß, "Karolingische Kopie," 120.
147. Le Bourdellès, *Aratus Latinus*, 85–86. Recently, this result has been confirmed by Dekker, *Illustrating the Phaenomena*, 116–55.
148. Le Bourdellès, *Aratus Latinus*, 86. The two important excerpts are found on Madrid 3307, fol. 53r, lines 25–26 for Gemini, "in commissura zodiaci atque lactei circulorum Gemini sunt locati," and for Sagittarius, fol. 53v, lines 25–27, "habens post se Sagittarium in commissura circulorum zodiaci atque lactei sub delphino constitutum."
149. Le Bourdellès, *Aratus Latinus*, 72–87, 102; see also Maass, *Commentariorum in Aratum Reliquiae*, 180, 309–12.

150. Le Bourdellès, *Aratus Latinus*, 71–76. Evans, *History and Practice*, 41 explains that the *Catasterisms* of Eratosthenes (third century B.C.E.) provided a list of the constellations complementary to the one Aratus offered in the *Phaenomena*. These descriptions of the constellations supplemented the Aratean ones, because they related the mythological narratives attached to the constellations and explained the distribution of the stars within particular constellations. Also see Dekker, *Illustrating the Phaenomena*, 192.

151. Le Bourdellès, *Aratus Latinus*, 74–81; Martin, *Histoire du texte*, 38–51.

152. Vieillard-Troiekouroff, "Art carolingien."

153. Katzenstein and Savage-Smith, *Leiden Aratea*, 5–36.

154. Le Bourdellès, *Aratus Latinus*, 102.

155. Van der Horst, Noel, and Wüstefeld, *Utrecht Psalter*, 200–201, cat. 13.

156. Le Bourdellès, *Aratus Latinus*, 99–107.

157. Haffner, *Sternbilderzyklus*, 75–79.

158. Borst, "Alkuin und die Enzyklopädie von 809," 72–73.

159. Mariana, *Códice de Metz*, 31; Cartelle, *Códice de Metz*, 211–22.

160. Borst, "Alkuin und die Enzyklopädie von 809," 73–74. A complete critical edition is introduced with earlier bibliography by Borst, "Die Salzburger Enzyklopädie von 818 (Lib. calc.)," in Borst, *Schriften zur Komputistik*, 3:1367–451. For notes on Munich 210, which Borst considers the best witness to the redaction, see also Bierbrauer, *Karolingischen Handschriften*, 73–75.

161. Neuß, "Karolingische Kopie," 115–19.

162. Boschen, *Annales Prumienses*, 56–57.

163. Parisse, "Metz, capitale d'Austrasie," 70.

164. Diebold, *Word and Image*, 82–84 believes, for example, that the scene depicting the presentation of the Vivian Bible to Charles the Bald (Paris, Bibliothèque nationale de France, MS lat. 1, fol. 423r, ca. 845) pictorially relates Charles the Bald to the biblical king David in a representation of David on fol. 215v, which supplies a frontispiece to the Psalms; Parisse, "Metz, capitale d'Austrasie," 70.

165. Contreni, "Carolingian Renaissance," 710.

166. Parisse, "Metz, capitale d'Austrasie," 67–68; Parisse, "Remarques sur le destin," esp. 73–75; and alternatively, Desportes, "La capitale cléricale," esp. 59.

167. Parisse, "Metz, capitale d'Austrasie," 70.

168. Schieffer, *Karolinger*, 11–12.

169. Parisse, "Metz, capitale d'Austrasie," 70–71.

170. Ibid., 70–72 adds that the disciplined monastic community at St. Arnulf's could not have been installed prior to ca. 660. It is easy to see how the installation of the saintly relics could encourage monastic reform throughout Metz and on site. Abel, von Jasmund, and Rau, *Karolingischen Reichsgeschichte*, 216 provides a related note from the *Vita Hludowici* of Thegan: "Saint Arnulf, while he was a duke in his youth, begat Duke Ansgisus; Duke Ansgisus begat Duke Pippin the Elder; Duke Pippin the Elder begat Duke Charles the Elder; Duke Charles the Elder begat the Pippin, whom the Roman Pope Stephen consecrated and anointed as King; King Pippin the Elder begat the Charles, whom the Roman Pope Leo consecrated and anointed as emperor in the church where the most blessed body of Peter, first of the apostles, lay, on the day of the Nativity of our Lord Jesus Christ." For more on Saint Arnulf and Metz, see Oexle, "Hl. Arnulf." Also see E. Hlawitschka, "Arnulf," in Auty et al., *Lexikon des Mittelalters*, vol. 1, cols. 1018–19; Schieffer, *Karolinger*, 12–16.

171. Koehler and Mütherich, *Karolingischen Miniaturen*, 3:120.

172. Boschen, *Annales Prumienses*, 46–47.

173. McKitterick, *Written Word*, 192–95.

174. Boschen, *Annales Prumienses*, 25n80. For more on the library at Murbach, including its extant ninth-century inventory (copied in 1464 and stored at the Archives départementales du Haut-Rhin in Colmar), see Milde, *Klosters Murbach*, 8 passim.

175. Höhl, *Ottonische Buchmalerei*, 34n208.

176. Borst, *Karolingische Kalenderreform*, 319; Kühnel, *End of Time*, 109–10.

177. Bischoff, "Paläographische Fragen," 97, reproducing the essay from *Frühmittelalterliche Studien* 5 (1971): 101–34.

178. McKitterick, *Written Word*, 192–93; Milde, *Klosters Murbach*, 8–16, 36–61.

179. McKitterick, *Written Word*, 192.

180. For a helpful introduction with examples of the relevant scripts, see Clemens and Graham, *Manuscript Studies*, 140–44.

181. Bischoff, *Latin Palaeography*, 112–18.

182. Lowe, *Codices Latini Antiquiores*, 9:x.

183. Clemens and Graham, *Manuscript Studies*, 141.

184. Lowe, *Codices Latini Antiquiores*, 8:48, no. 1193.

185. Ibid., 6:16, no. 753.
186. Ibid., 6:15, no. 751; Milde, *Klosters Murbach*, 42, 55, item 169; lastly, Lowe, *Codices Latini Antiquiores*, 9:x.
187. Koehler and Mütherich, *Karolingischen Miniaturen*, 3:122, esp. 122n90; Nordenfalk, "Buchmalerei," 296–97, cat. 479; Janini and Serrano, *Manuscritos litúrgicos*, 73–76; Mütherich and Gaehde, *Carolingian Painting*, 26, cat. 13; McGurk, "Carolingian Astrological Manuscripts," 321; Mariana, *Códice de Metz*, 32–33; Höhl, *Ottonische Buchmalerei*, 34; Van der Horst, Noel, and Wüstefeld, *Utrecht Psalter*, 198–99, cat. 12; Laffitte and Denoël, *Trésors carolingiens*, 191.
188. Koehler and Mütherich, *Karolingischen Miniaturen*, 3:120–22.
189. Ibid., 3:100–105. This discussion must be read in conjunction with Bischoff, "Court Library," 76–85.
190. Laffitte and Denoël, *Trésors carolingiens*, 203–5, cat. 55.
191. Koehler and Mütherich, *Karolingischen Miniaturen*, 3:122.
192. Nees, "On Carolingian Book Painters," 209.
193. Boschen, *Annales Prumienses*, 75–82, 195–210.
194. Mariana, *Códice de Metz*, 35; J. Deeters, "Maastricht," in Auty et al., *Lexikon des Mittelalters*, vol. 6, cols. 53–54.
195. Janini and Serrano, *Manuscritos litúrgicos*, 74, reported again in Mariana, *Códice de Metz*, 35; J.-L. Kupper, "Notker," in Auty, *Lexikon des Mittelalters*, vol. 6, cols. 1288–89.
196. Neuß, "Meisterwerk," 63–64; Mariana, *Códice de Metz*, at 35 for the citation.
197. Mariana, *Códice de Metz*, 35–36.
198. Neuß, "Meisterwerk," 63–64; Janini and Serrano, *Manuscritos litúrgicos*, 74; Mariana, *Códice de Metz*, 35–36.
199. Mariana, *Códice de Metz*, 36.
200. Ibid.
201. Ibid., 34–37.
202. Ibid., 36–37; Janini and Serrano, *Manuscritos litúrgicos*, 73–74.

CHAPTER 2

1. Bede, *Reckoning*, intro. Wallis, lxxxix–xcii, 243–46; all translations from Bede's *Reckoning* are from the edition cited here. As Wallis explains, we cannot assume that all Carolingian clerics knew about Bede's book. Even so, a high-ranking Frankish prelate like Drogo would more likely than not have been aware of the ideas and principles contained within the *Reckoning*, given the textual transmission of the treatise during the Carolingian era.
2. Ibid., 370–73. This remains an important concern of the Carolingian prelates supervising the creation of the astronomical encyclopedias, including Archbishop Arn of Salzburg; on this and related matters, see Borst, *Schriften zur Komputistik*, 2:820–32, and 3:1367–79, esp. 1372.
3. McKitterick, *History and Memory*, 99–100.
4. Dutton, "Of Carolingian Kings," esp. 118–20.
5. Sedulius, *Christian Rulers*, 9–11; Dutton, "Of Carolingian Kings," 120. Also see Hellman, *Sedulius Scottus*, 1–18.
6. Sedulius, *Christian Rulers*, 15.
7. Ibid., 100, poem 2. All translations from Sedulius are taken from the edition cited here, translated by Doyle. Also see Dutton, "Of Carolingian Kings," 120.
8. Sedulius, *Christian Rulers*, 99, poem 1.
9. Ibid.
10. McCluskey, "Astronomies in the Latin West," 153; also see his related treatment in *Astronomies and Cultures*, 131–64, esp. 135.
11. Dutton, "Of Carolingian Kings," 120–21.
12. Aratus, *Phaenomena*, trans. Kidd, 3–5; Katzenstein and Savage-Smith, *Leiden Aratea*, 5, 11; Haffner, *Sternbilderzyklus*, 15. All translations from the original Aratus will be from the Kidd edition cited here. Also see Dolan, "Illustrated *Aratea* Manuscripts."
13. Aratus, *Phaenomena*, 4–23.
14. Katzenstein and Savage-Smith, *Leiden Aratea*, 12 supplied the translation. Aratus, *Phaenomena*, 4–16 argues that the *locus classicus* for such disregard is to be found in *De orat.* I.69 with additional support from *Rep.* I.22. The Latin text of *De orat.* I.69 reported by Kidd follows: "si constat inter doctos hominem ignarum astrologiae ornatissimis atque optimis versibus Aratum de caelo stellisque dixisse," which corresponds in my literal translation to "if only it is agreed by learned folk that a certain man Aratus who was ignorant of astrology [and astronomy] spoke with the most distinguished and noble verses about the sky and the [fixed or wandering] stars." Aratus,

Phaenomena, 16–21 further traces the development of the unversed Aratus thesis back to Hipparchus (fl. ca. 147–127 B.C.E.).

15. Aratus, *Phaenomena,* 16.
16. Ibid., 73.
17. Ibid., 14–18, 237.
18. Madrid 3307, fol. 54r. The translations from the *Handbook of 809* are my own unless otherwise specified. As Borst noted in his critical edition of the *Handbook of 809,* the copy from Madrid is the most reliable witness to the textual tradition. For that reason, Drogo's copy of the book has been utilized and significant textual discrepancies noted by Borst in his textual apparatus have been included for the reader's benefit in this chapter; for the definitive critical edition of the text, see Borst, *Schriften zur Komputistik,* 3:1054–1334. A convenient facsimile is offered here in plates 1–17 and figs. 3–6.
19. Mariana, *Códice de Metz,* 45.
20. Haffner, *Sternbilderzyklus,* 31–73 reconstructed the original pictorial program of Aratean manuscripts, which relate to the Latin translation of the *Phaenomena* by Germanicus. This reconstruction provides important comparisons to the cycle of imagery in Madrid. See also the original descriptions of the "Visible Signs" in Aratus, *Phaenomena,* 72–105.
21. Lindberg, *Beginnings,* 41–42; McCluskey, *Astronomies and Cultures,* 20–24; Beck, *Ancient Astrology,* 20–23, 26–28; Evans, *History and Practice,* 54–58.
22. Introductory treatments include Ridpath, *Stars and Planets,* 16; McCluskey, *Astronomies and Cultures,* 15–24; Beck, *Ancient Astrology,* 20–37; Eastwood, *Ordering the Heavens,* 95–126; Eastwood and Graßhoff, *Planetary Diagrams,* 12–14, 35–40.
23. Gundel, *Zodiakos,* 15–22 provides an excellent introduction to the astronomical terminology appropriate to an understanding of the zodiac. Also see Von Euw, *Leidener Aratus,* 17–19, Evans, *History and Practice,* 54–55.
24. Evans, *History and Practice,* 245–47; Beck, *Ancient Astrology,* 20–21.
25. Von Euw, *Leidener Aratus,* 17; Beck, *Ancient Astrology,* 26–28; McCluskey, *Astronomies and Cultures,* 17–20.
26. McCluskey, *Astronomies and Cultures,* 20–28.
27. Aratus, *Phaenomena,* 73–75.
28. Le Bourdellès, *Aratus Latinus,* 102–5; Cordoliani, "Les traités de comput," 52–53, cat. 4.
29. Le Bourdellès, *Aratus Latinus,* 252–63; Aratus, *Phaenomena,* 52–55.
30. Le Bourdellès, *Aratus Latinus,* 71–76, 269–75.
31. Borst, "Alkuin und die Enzyklopädie von 809," 60–62.
32. Le Bourdellès, *Aratus Latinus,* 52–60.
33. Ibid., 80–81 postulated the need for two missing copies of the *Revised Aratus* that could account for the two discrete recensions of the manuscript tradition. The so-called manuscript *y* was used to create the manuscript in Paris, MS lat. 12957 and the manuscript in Cologne, Dombibliothek, MS 83 (II). For more on the latter, see Plotzek et al., *Glaube und Wissen,* 136–56; cat. 24.
34. Plotzek et al., *Glaube und Wissen,* 136; Von Euw, "Zeitrechnung," 1–2, 12–13.
35. Von Euw, "Zeitrechnung," 13–14. An illustrated astronomical-computistical manuscript prior to the reign of Louis the Pious suggests a longer scientific manuscript tradition. Haffner, *Sternbilderzyklus,* 89 argues that a renewal of interest in illustrated scientific manuscripts began during the reign of Louis the Pious. There is no reason to set the beginning so late.
36. Von Euw, "Zeitrechnung," 1–2.
37. Koehler, *Manuscrit latin 9428,* 9; see also Koehler and Mütherich, *Karolingischen Miniaturen,* 3:144–46, 156.
38. Le Bourdellès, *Aratus Latinus,* 272.
39. Consult a facsimile with notes at "St. Gallen, Stiftsbibliothek, Cod. 902," *Codices Electronici Sangallenses,* ed. Christoph Flüeler and Ernst Tremp, 22 March 2007, http://www.cesg.unifr.ch/virt_bib/handschriften.htm. See, too, Scherrer, *Verzeichniss,* 316–18; Tremp, Schmuki, and Flury, *Benediktinisches Mönchtum,* 128. For the edited text of the *Recensio interpolata,* see Maass, *Commentariorum in Aratum Reliquiae,* 180. A brief codicological note is helpful here. The term *Recensio interpolata,* used by Maass, refers to the revised translation of the *Aratus Latinus,* including emendations from other texts. This anthology of altered texts, which seamlessly recombined into a new version of the Aratean description of the heavens, constituted the so-called *Scholia Sangermanensia,* according to Alfred Breysig. Le Bourdellès prefers this nomenclature for one component of his complete reconstruction of the *Revised Aratus,* which also included texts by Pliny, Hyginus, Fulgentius, Isidore, and arguably Bede. This is probably right, but any manuscript belonging to the

so-called *Recensio interpolata* is a manuscript containing a portion of the *Revised Aratus* text, according to the nomenclature employed by Le Bourdellès. For more on the history of this text and its transmission, see Le Bourdellès, *Aratus Latinus*, 72–74. Although somewhat outdated now, O'Connor, "Star Mantle," 62–69, includes useful summaries of the complicated transmission of early medieval astronomical texts. A more recent introduction can be found in Dekker, *Illustrating the Phaenomena*, 116–20, complemented by her more thorough treatment in "Provenance," esp. 5–8.

40. Haffner, *Sternbilderzyklus*, 15, 75–79. See also Mütherich, "Bilder," 32; Martin, *Histoire du texte*.
41. Roberts and Skeat, *Birth of the Codex*, 36–37. Gamble, "Bible and Book," 22–26 corroborates their result. During the early Christian period, the codex replaced the rotulus as the primary vehicle of textual transmission by the fourth century.
42. Haffner, *Sternbilderzyklus*, 77; Le Bourdellès, *Aratus Latinus*, 12, 23–26.
43. Weitzmann, *Illustrations in Roll and Codex*, 71–73 calls an illustration like those in Madrid a "column picture" in the "papyrus style"; Weitzmann, *Book Illumination*.
44. "St. Gallen, Stiftsbibliothek, Cod. 250," *Codices Electronici Sangallenses*, ed. Christoph Flüeler and Ernst Tremp, 22 March 2007, http://www.cesg.unifr.ch/virt_bib/handschriften.htm.
45. Weitzmann, *Illustrations in Roll and Codex*, 89–103.
46. Le Bourdellès, *Aratus Latinus*, 102–7.
47. "Figulus, Publius Nigidius," *Encyclopedia Britannica*, 2007, Encyclopedia Britannica Online, 9 April 2007, http://www.britannica.com/eb/article-9034238; Le Bourdellès, *Aratus Latinus*, 28.
48. Le Bourdellès, *Aratus Latinus*, 26–29; "Lactantius," *Encyclopedia Britannica*, 2007, Encyclopedia Britannica Online, 9 April 2007, http://www.britannica.com/eb/article-9046767.
49. Le Bourdellès, *Aratus Latinus*, 26–29.
50. Thiele, *Antike Himmelsbilder*, 71–72, 157–58 offers an alternate hypothesis that the satyr's form could have derived from a celestial globe, as argued here. In addition, Hyginus dubbed the satyr Krotos of Sositheos, according to *Astronomica*, Book II, 27. However, it is somewhat suspect to assert dogmatically that the satyr form of Sagittarius entered into the archetypal *Handbook of 809* or Drogo's copy from any particular sources. In fact, these texts and objects displayed an underappreciated amount of textual and visual exchange during the late antique and early medieval periods.
51. The total number of constellations is erroneously listed as 41 in the catalogue entry from the exhibition: Van der Horst, Noel, and Wüstefeld, *Utrecht Psalter*, 198–99, cat. 12. See also Ridpath, *Stars and Planets*, pp. 92-93, 138.
52. Salzmann, *On Roman Time*, 253–56.
53. Early descriptions of the pictorial cycle can be found in the following older publications: Neuß, "Meisterwerk," 48–50; Neuß, "Karolingische Kopie," 122–36; Koehler and Mütherich, *Karolingischen Miniaturen*, 3:123–27. For an example of a Carolingian encyclopedia derived from polycyclic sources, see Le Berrurier, *Pictorial Sources*, esp. 116–17.
54. Dekker, "Carolingian Planetary Observations," with earlier bibliography; Mütherich, "Book Illumination."
55. Madrid 3307, fol. 54v.
56. Martin, *Histoire du texte*, 103.
57. Ibid., 38–40; Le Bourdellès, *Aratus Latinus*, 106–7.
58. Le Bourdellès, *Aratus Latinus*, 72–74, 252–63.
59. Unruh, "Esus und Jesus," examines the intersection of autochthonous German deities in Trier, their Roman counterparts, and their later Christian replacements. For a discussion of this religious aspect of the *mundus*, see P. Brown, *Authority and the Sacred*, 1–26.
60. Le Bourdellès, *Aratus Latinus*, 71–81, 254.
61. McCluskey, *Astronomies and Cultures*, 112, 171–80. An example of a star clock remains in a manuscript from Saint-Gall; also see Schmuki, Ochsenbein, and Dora, *Cimelia Sangallensia*, 120–21, with illustration.
62. Blume, Haffner, and Metzger, *Sternbilder*, 1:67 argue that faint traces of gilding ensure that there were once stars included. Even if this were so, the images of the constellations would still not have been "accurate" by modern standards and would have been ineffective for stargazing without the additional consultation of parallel implements, especially a celestial globe.
63. Carruthers, *Craft of Thought*, 24–29.
64. Koehler and Mütherich, *Karolingischen Miniaturen*, 4:79–83, 108–116; on Latin translations, see Le Bourdellès, *Aratus Latinus*, 23–26; Mütherich, "Bilder."
65. Mütherich, "Bilder," 62.

66. For an introduction to parataxis and hypotaxis, see Moreland and Fleischer, *Latin Literature and Style*, 60, 87–88.
67. Dutton, "Of Carolingian Kings," 108 argues that some Frankish histories also juxtaposed celestial phenomena with various occurrences throughout the realm, although the texts treated them as discrete and disjointed events, leaving the reader to appreciate an implied but unstated notion of causality. He considers this a paratactic structure, because there is no explicitly stated relationship between the otherwise notable happenings.
68. Mütherich, "Bilder," 62.
69. Stückelberger, *Bild und Wort*, 45–46, esp. fig. 21.
70. Dekker, "Provenance," 16–20.
71. Von Euw, *St. Galler Buchkunst*, 446–49.
72. Dekker, "Provenance," 6; a convenient French translation of the text that probably reflects the original by Eratosthenes is Charvet and Zucker, *Le ciel;* for Ursa Minor, see 33.
73. Madrid 3307, fol. 54v.
74. Aratus, *Phaenomena*, 75, lines 30–44.
75. There is a relationship, for example, between the miniatures of the constellations in Drogo's copy of the *Handbook of 809* and the painterly modeled forms of the figures in such frescoes as the *Erotic Pair with Silver Sphere*, which decorated a ceiling in a Roman palace found under the cathedral of Trier. The frescoes and the palatial program have been linked to relatives of Constantine the Great and dated to the 330s. Two coins commemorating the construction of a palace in Constantinople confirmed this connection and supported a date ca. 333–34. On these points, see Weber, *Constantinische Deckengemälde*, 42–45.
76. Madrid 3307, fol. 54v.
77. Madrid 3307, fol. 55r.
78. Aratus, *Phaenomena*, 77, lines 64–70.
79. Mütherich, "Bilder," 37; Ridpath, *Stars and Planets*, 92–93.
80. Madrid 3307, fol. 55r.
81. Mütherich, "Bilder," 62.
82. Madrid 3307, fol. 55r.
83. Madrid 3307, fol. 55v.
84. Gundel, *Zodiakos*, 12–17 explains that by the first millennium B.C.E., the basic set of zodiac signs and their corresponding constellations had been formalized. The eleven-part zodiac cycle remained nonetheless a historical counterpart to the standard twelve-part cycle. The consistency of zodiac signs across cultures and disparate time periods, albeit with certain fluctuations or deviations such as the eleven-sign zodiac, was the direct result of commercial interaction across the Mediterranean. Individual iconographic peculiarities did arise, however, as a result of the "corrupting sea"; see Horden and Purcell, *Corrupting Sea*.
85. Gundel, *Zodiakos*, 17.
86. Salzmann, *On Roman Time;* Stern, *Calendrier de 354*. Hourihane, *Time*, liv–lix argues that precursors to the medieval pictorial cycles of the labors/occupations of the months derived from classical agrarian traditions, and therefore later examples could also include iconographically embedded allusions to their pagan roots. Generally speaking, the presentation of the months moved away from isolated Roman personifications toward more complicated vignettes and even genre scenes celebrating medieval daily activities. Also see Henisch, *Medieval Calendar*.
87. Aurigemma, *Scorpion*, traces the development of the iconography and significance of Scorpio over the *longue durée*.
88. Madrid 3307, fol. 55v.
89. Madrid 3307, fol. 56r.
90. Aratus, *Phaenomena*, 29, 79–83, lines 96–146.
91. Mütherich, "Bilder," 39.
92. Le Bourdellès, *Aratus Latinus*, 102–7 suggests that Adalhard played an integral role in the creation of the *De ordine*. His involvement could hardly have been restricted to the text alone. Adalhard probably participated in the managerial process of selecting images for inclusion in his prototypical copy of the *De ordine* brought from Corbie to Aachen, but perhaps, too, over the painted forms agreed upon in Aachen.
93. Madrid 3307, fol. 56r.
94. Hamilton, *Mythology*, 249.
95. Gundel, *Zodiakos*, 70.
96. Katzenstein and Savage-Smith, *Leiden Aratea*, 23.
97. Cologne, Dombibliothek, MS 83 II, fol. 158v: "Castor interea et Pollux quos nos Geminus vocamus" (Meanwhile Castor and Pollux, whom we call the Gemini).
98. Mütherich, "Bilder," 39–40; P. Brown, *Authority and the Sacred*, 25 notes that similarly, in late antique Ephesus, Christian Romans chiseled crosses into the

foreheads of figures erected near the Prytaneion in honor of Augustus and Livia.

99. G. Henderson, "Emulation," esp. 267.
100. Madrid 3307, fol. 56v.
101. Aratus, *Phaenomena,* 138–39; lines 892–904.
102. Ibid., lines 905–908.
103. There were no depictions of donkeys in the Leiden *Aratea,* fol. 18v, and the *Recensio Interpolata* pictures like Saint-Gall 902, page 86. Cologne, *Revised Aratus,* fol. 158v lacked an illustration altogether.
104. Le Bourdellès, *Aratus Latinus,* 106 explains that a copy of Pliny's *Naturalis historia* belonged to the monastery of Corbie in the ninth century, but is now mutilated and housed in three separate collections: Paris, Bibliothèque nationale de France, MS lat. 6796; Rome, Biblioteca Apostolica Vaticana, Vat. lat. 3861; Leiden, Bibliotheek der Rijksuniversiteit, Voss. lat. F 61.
105. P. D. A. Harvey, "Columella im Mittelalter," in Auty et al., *Lexikon des Mittelalters,* vol. 3, cols. 67–68, http://www.brepolis.net/bme, in *Brepolis Medieval Encyclopaedias—Lexikon des Mittelalters Online,* accessed 24 May 2007, notes that there were two copies of this agricultural treatise available in the early Middle Ages. Two ninth- or tenth-century copies have survived from Corbie and Fulda.
106. Eastwood, "Astronomy in Christian Latin Europe," 235–36.
107. Eastwood (ibid., 236) added that this pragmatic form of Christian astronomy deviated from the mathematical astronomy that had preceded it during the Hellenistic era.
108. Madrid 3307, fol. 56v. The more accurate version of the text from Rome, Biblioteca Apostolica Vaticana, Reg. Lat. 309, fol. 59r reads: "in p- podo claram unam." The description of the constellation Leo is possibly corrupted; the total should perhaps be nineteen stars. The debate turns on whether there are one or two stars in the felid's forepaws. In other words, the text is either redundant for emphasis, or there is a mistake in the textual transmission. Borst, *Schriften zur Komputistik,* 3:1254–55 identified one copy of the *Handbook of 809* that reported nineteen (novemdecim) as the total number of the stars: Bamberg, Staatsbibliothek, Classicus 55, made in Reims, before 900. In any case, these findings suggest that Reg. lat. 309 derived from the archetype along a different branch from Drogo's copy of the *Handbook of 809.* Furthermore, on a more astronomical note, the total number of stars commonly accepted to belong to the modern constellation of Leo is only sixteen. The bright star in the paws is probably derived from observation of Zeta (ζ) Leonis; also see Ridpath, *Stars and Planets,* 103.
109. Mütherich, "Bilder," 40–41 described the mythological associations linked to Leo.
110. Madrid 3307, fol. 57r.
111. Katzenstein and Savage-Smith, *Leiden Aratea,* 23.
112. Aratus, *Phaenomena,* 84–85, lines 156–59: "and rumour has reached you of the Goat herself and the Kids, who have often looked down on men being tossed upon the heaving sea."
113. Mütherich, "Bilder," 42 adds that the full-length Taurus was associated with zodiac wheels, even if the individual representations of the constellations rarely displayed the complete bovine form.
114. Madrid 3307, fol. 57r.
115. Katzenstein and Savage-Smith, *Leiden Aratea,* 24.
116. Ibid., 27.
117. Madrid 3307, fol. 57v.
118. Madrid 3307, fol. 58r.
119. Ibid.
120. Mütherich, "Bilder," 43–44: for example, the image in Madrid recalled the pictures of Pegasus in Saint-Gall 902 on page 89, and on fol. 32v in the Leiden *Aratea.*
121. Mütherich, "Bilder," 44; Gundel, *Zodiakos,* 18, 69.
122. Madrid 3307, fol. 58r.
123. Madrid 3307, fol. 58v.
124. Ibid.
125. Gundel, *Zodiakos,* 74; Katzenstein and Savage-Smith, *Leiden Aratea,* 27; Mütherich, "Bilder," 44. On fol. 39r, lines 244–45, the text of the Leiden *Aratea* explained, "They cannot travel freely, instead chains conjoined with a knot hold each one of them at the tail." A helpful transcription and German translation were prepared by Bernhard Bischoff in Bischoff et al., *Aratea,* 114–15.
126. Gundel, *Zodiakos,* 74.
127. Ibid., 18, 69; Mütherich, "Bilder," 44.
128. Madrid 3307, fols. 58v–59r.
129. Madrid 3307, fol. 59r.
130. Ibid.

131. Madrid 3307, fol. 59v. Reg. lat. 309, fol. 96v supplied a correct form of the text, "ceterae." Also see Borst, *Schriften zur Komputistik*, 3:1258.
132. Mütherich, "Bilder," 46–47. See also Katzenstein and Savage-Smith, *Leiden Aratea*, 28–29.
133. Himmelmann, *Antike Götter*, 4–5 emphasizes that nudity could be associated with idols, rather than antiquity, during the medieval period.
134. Dutton, *Charlemagne's Courtier*, 31: "[Charlemagne] was fond of the books of Saint Augustine, particularly the one called the *City of God*."
135. Augustine, *City of God*, book XXII, chap. 17, 1057. Also see Bynum, *Resurrection*, 98.
136. Augustine, *City of God*, book XXII, chap. 17, 1057.
137. Madrid 3307, fol. 59v.
138. Madrid 3307, fol. 60r.
139. Madrid 3307, fol. 60v. Cartelle, *Códice de Metz*, 181 proposed in his Spanish translation of this text that 'enchiridion' should be interpreted as a 'handhold,' or 'hilt,' of a sharp weapon, as opposed to the more common late Latin use of the term to identify a 'dagger.'
140. Madrid 3307, fol. 60v.
141. Ridpath, *Stars and Planets*, 114.
142. Aratus, *Phaenomena*, 96–97, lines 319–21; Mariana, *Códice de Metz*, 45–46.
143. See Mariana, *Códice de Metz*, 45–52 for more information concerning the style of painterly execution.
144. Mütherich, "Bilder," 47; Katzenstein and Savage-Smith, *Leiden Aratea*, 28.
145. Aratus, *Phaenomena*, 94–95, at 102–3 for the citation, with accompanying commentary at p. 295 for lines 300–310, at p. 325 for lines 399–401.
146. Gundel, *Zodiakos*, 72–73.
147. Aratus, *Phaenomena*, 94–95, lines 287–95; Katzenstein and Savage-Smith, *Leiden Aratea*, 28.
148. Aratus, *Phaenomena*, 94–95, lines 300–307.
149. On polycyclic strategies of Carolingian creation, see Le Berrurier, *Pictorial Sources*, esp. 117.
150. Aratus, *Phaenomena*, 112–13, lines 536–56. Aratus simply refers to Sagittarius as the Archer.
151. Ibid., 94–97, 295, lines 300–315, at 314–15 for the citation.
152. McCluskey, *Astronomies and Cultures*, 171–80 notes that it is unclear when the astrolabe was earliest in use throughout western Europe. For the history of the astrolabe, also see North, "Astrolabe," 211–20.
153. Aratus, *Phaenomena*, 295. Also see Charvet and Zucker, *Le ciel*, 133–35; Mütherich, "Bilder," 47.
154. Thiele, *Antike Himmelsbilder*, 71–72, 157–58; Mütherich, "Bilder," 47.
155. Madrid 3307, fol. 60v.
156. Madrid 3307, fol. 61r.
157. Katzenstein and Savage-Smith, *Leiden Aratea*, 32; Mütherich, "Bilder," 49–50.
158. Katzenstein and Savage-Smith, *Leiden Aratea*, 24–32.
159. Madrid 3307, fol. 61v.
160. Madrid 3307, fol. 62r.
161. Gameson, *Early Medieval Bible*, 53–77.
162. Katzenstein and Savage-Smith, *Leiden Aratea*, 32.
163. Charalampos Tsochos, "Thyrsos," in Cancik and Schneider, *Neue Pauly*, vol. 12.1, col. 526.
164. Also consider Vat. lat. 645, fol. 65r.
165. Mütherich, "Bilder," 51.
166. Isidore, *Etymologies*, trans. Barney et al., 114; Isidore, *Etymologiarum*, IV.ix.12: "A certain Greek, Chiron, invented medical practice for draft animals. For this reason he is pictured as half man, half horse." See Fritz Graf, "Chiron," in Salazar, *Brill's New Pauly*, vol. 3, cols. 233–34.
167. Mütherich, "Bilder," 51.
168. The total, which should be twenty-two, was recorded incorrectly as twenty-six in all copies of the *Handbook of 809* examined.
169. Madrid 3307, fol. 62r.
170. Madrid 3307, fol. 62v.
171. Mütherich, "Bilder," 62.
172. Le Bourdellès, *Aratus Latinus*, 71–74, 80.
173. Isidore, *Etymologies*, trans. Barney et al., 10–17.
174. Fontaine, *Isidore de Séville: Genèse et originalité*, 297–310, esp. 299–301; Kühnel, *End of Time*, 123–33.
175. Isidore, *Etymologies*, trans. Barney et al., 106 (III. lxxi.38).
176. Borst, *Schriften zur Komputistik*, 3:1054–55.
177. Mütherich, "Bilder," 37; for an image of the text, see Haffner, *Sternbilderzyklus*, 216, illustration 15.
178. Mütherich, "Bilder," 37, 60–61; Hamilton, *Mythology*, 172–74; Charvet and Zucker, *Le ciel*, 41–43.
179. Panofsky and Saxl, "Classical Mythology," esp. 235–38.
180. Ibid., 235.
181. Ibid., 236, 230.
182. Panofsky, *Renaissance and Renascences*, 42–113.

183. See, for example, the excellent discussion in Sauerländer, "'Barbari,'" 123–37.
184. Panofsky, *Renaissance and Renascences,* 50.
185. Ibid., 82.
186. Ibid., 83.
187. Ibid., 84.
188. See Ullman, *Idea of Kingship,* 1–20 for the classic statement of ecclesiastic responsibility affecting Frankish administrative policy. This position must be considered now in conjunction with the contributions by Mayke de Jong and Donald Bullough in Story, *Charlemagne,* 103–50.

CHAPTER 3

1. The most comprehensive treatment at present of the Frankish debate concerning images is Noble, *Images,* 215–30. A powerful corrective to a textually reductionist perspective can be found in Kessler, *Seeing,* 19–43, 87–105, 131–50, 165–79. A concise history of Carolingian aesthetic interpretation is Appleby, "Instruction and Inspiration."
2. Nelson, "Kingship and Royal Government," 398–406.
3. Calkins, *Illuminated Books,* 49 notes that the physical healing power of the written word extended to animal science as well: the submerged Book of Durrow allegedly supplied a veterinary remedy for diseased seventeenth-century livestock.
4. Kessler, *Seeing,* 89.
5. McCormick, "Birth of the Codex"; also see Roberts and Skeat, *Birth of the Codex,* 1–75.
6. The key resources treating the scriptorium at Metz include Koehler and Mütherich, *Karolingischen Miniaturen,* 3:97–167; Laffitte and Denoël, *Trésors carolingiens,* 188–205.
7. Laffitte and Denoël, *Trésors carolingiens,* 189–91; Koehler and Mütherich, *Karolingischen Miniaturen,* 3:98–100.
8. Ibid., 3:97–100.
9. Scholz, *Carolingian Chronicles,* 113.
10. Koehler and Mütherich, *Karolingischen Miniaturen,* 3:98–99; McKitterick, *Charlemagne,* 146–47; Fleckenstein, *Hofkapelle,* 1:48–49.
11. Koehler and Mütherich, *Karolingischen Miniaturen,* 3:97–98; Laffitte and Denoël, *Trésors carolingiens,* 189–92; also see Contreni, "Carolingian School."
12. Scholz, *Carolingian Chronicles,* 114–15; Kessler, *Seeing,* 175.
13. Scholz, *Carolingian Chronicles,* 114–15.
14. Kessler, *Seeing,* 175.
15. Rabanus, "*De Universo,*" book XXI.9, col. 563C; my translations use this edition of the text. Also see Haarländer, *Rabanus Maurus,* 14–60, 168, and Coon, *Dark Age Bodies,* 13–41.
16. Isidore (Isidoro de Sevilla), *Etimologías,* XIX.xvi.1, 121; also see Isidore, *Etymologies,* trans. Barney et al., 3–28 and XIX.xvi.1–2, 380.
17. Carruthers, *Craft of Thought,* 200–201.
18. Isidore, *Etimologías,* XIX.xvii.1–2, 123–25. For an English translation, see Isidore, *Etymologies,* trans. Barney et al., 380.
19. Heyse, *Untersuchungen zu den Quellen,* 1, 65, 146.
20. Rabanus, "*De Universo,*" book XXI.10, col. 563D.
21. Kessler, *Seeing,* 22.
22. Eastwood, "Plinian Astronomy in the Middle Ages and Renaissance," 204; Eastwood, "Plinian Astronomical Diagrams," 143–44. Eastwood has altered the details of his opinions about Carolingian and Capellan diagrams. It is therefore helpful to read his texts in this order: Eastwood and Graßhoff, *Planetary Diagrams—Descriptions, Models, Theories,* 1–29; Eastwood and Graßhoff, *Planetary Diagrams,* 1–48; Eastwood, *Ordering the Heavens,* 95–178, esp. at 99 for the Carolingian originality thesis. For a complementary review of related themes, see Obrist, *La cosmologie médiévale,* esp. 171–94, 311–19.
23. Eastwood, "Plinian Astronomy in the Middle Ages and Renaissance," 203–4 formerly argued that this diagram of the apsides alone could have preceded the creation of the other diagrams and the compilation of the four Plinian excerpts into the *Handbook of 809.* Eastwood had argued that the presentation of the apsides in the planetary configuration in the Leiden *Aratea,* fol. 93v, supplied the model for the diagram of the apsides in the *Handbook of 809.* Eastwood had dated that planetary configuration in Leiden to 28 March 579, given the arrangement of the planets relative to the signs of the zodiac in an earlier article: Eastwood, "Origins and Contents of

the Leiden Planetary Configuration," 2. He rescinded this opinion, however, and later advocated a date of 18 March 816; see Eastwood, "Astronomy in Christian Latin Europe," 247. As a result, Eastwood came to believe instead that the apsidal diagram in the handbook, made ca. 811, later influenced the construction of the planetary configuration on fol. 93v of the Leiden *Aratea;* see Eastwood and Graßhoff, *Planetary Diagrams,* 4–5. These diagrams appeared *ex nihilo* and were copied repeatedly. Their inclusion in the archetypal *Handbook of 809* would best explain their rapid and widespread diffusion. On an 816 date for the planetary configuration in the Leiden *Aratea,* also see Dekker, "Carolingian Planetary Observations."

24. Eastwood, "Astronomy in Christian Latin Europe," 235–38 emphasized that certain aspects of the Christianization of medieval Europe can be traced through the history of late antique and early medieval astronomy. The findings of this chapter support his conclusion. Examination of the pictorial cycle of the *De ordine,* however, reveals that the pictures sometimes demonstrated the Christianization of astronomy better than the texts. For an older treatment and critical edition of the Plinian excerpts, see Rück, *Auszüge aus der Naturgeschichte,* 34–36; this edition was updated by V. King, "Investigation," 137–69. For an English translation presented in conjunction with the Latin Plinian text from the *Handbook of 809,* also see Eastwood and Graßhoff, *Planetary Diagrams,* 25–27, alongside the critical edition in Borst, *Schriften zur Komputistik,* 3:1260–76.

25. Haffner, *Sternbilderzyklus,* 165–70 concluded her reconstruction of the cycle of illustrations in Aratean manuscripts. Certain images were commonly associated with the Aratean manuscripts but missing from the archetypal *Handbook of 809.* These included the depiction of the five planets, the representation of the Milky Way, an occasionally or optionally inserted image of the seasons, a depiction of the zodiac, a personification of the sun, and a personification of the moon. Also see Mütherich, "Bilder," 52–58.

26. Eastwood and Graßhoff, *Planetary Diagrams,* 12. The classic introduction to the relevance of Plinian astronomical texts for early medieval scientific literature remains Rück, *Auszüge aus der Naturgeschichte,* 1–95, especially 81–89. This essay is even more important for an understanding of the relationship between the Plinian excerpts and the third redaction of the astronomical handbooks (that is, version C, according to Borst's nomenclature, found in such books as Munich 210). Eastwood, "Plinian Astronomy in the Middle Ages and Renaissance," 200–203, 225n40 critiqued Rück's belief that the Plinian excerpts from the *Handbook of 809* originated in Bede's Britain. The compilation of Plinian excerpts from books II and XVIII of the *Naturalis historia* probably originated instead on the European mainland without relying upon a preexisting Hiberno-Saxon compilation during the ninth century. In any case, Rück's hypothesis about an English origin resulted in the alternate name for the excerpts, sometimes referred to as the Plinian "York Excerpts." For the different roles the diagrams played in the *Handbook of 809,* see Eastwood and Graßhoff, *Planetary Diagrams,* 12–14.

27. Eastwood and Graßhoff, *Planetary Diagrams,* 23–27 provided the text and the translation; also see Eastwood, *Ordering the Heavens,* 95–107.

28. Eastwood, *Ordering the Heavens,* 31–37, and Obrist, *La cosmologie médiévale,* 171–75, introduce the most important resources treating Macrobius at present with copious earlier bibliography.

29. Obrist, *La cosmologie médiévale,* 172; Eastwood, "Astronomy in Christian Latin Europe," 245; Eastwood, *Ordering the Heavens,* 31–32; also see Macrobius, *Commentary on the Dream of Scipio,* trans. William Harris Stahl (1952; New York: Columbia University Press, 1990), 1–70.

30. Eastwood, *Ordering the Heavens,* 37.

31. Ibid., 156–57, 179–80, 194–96, 238–44; Martianus Capella, *Seven Liberal Arts,* VIII.857, 2:333; also see Shanzer, *Philosophical and Literary Commentary,* 8–28, although Shanzer's remarks have received neither suitable nor widespread acceptance, leaving 410–39 C.E. a viable range of dates for the *De nuptiis,* as recounted nicely in Teeuwen, *Harmony,* 15–20.

32. Eastwood, "Astronomy in Christian Latin Europe," 245–47; Eastwood and Graßhoff, *Planetary Diagrams,* 129–30; Eastwood, *Ordering the Heavens,* 107, 179–260.

33. Fontaine, *Isidore de Séville: Géenèse et originalité,* 299, and Obrist, "Le diagramme Isidorien," 95–164.

34. Eastwood, "Astronomy in Christian Latin Europe," 246.

35. Martianus Capella, *Seven Liberal Arts,* VIII.857, 333.
36. Eastwood, "Astronomy in Christian Latin Europe," 246; Eastwood, *Ordering the Heavens,* 238–59, esp. 257.
37. Eastwood, "Astronomy in Christian Latin Europe," 245–47; Eastwood and Graßhoff, *Planetary Diagrams,* 129–30; Eastwood, *Ordering the Heavens,* 238–59.
38. Eastwood and Graßhoff, *Planetary Diagrams,* 12; consider Contreni, "John Scottus."
39. Rück, *Auszüge aus der Naturgeschichte,* 36–37; Eastwood and Graßhoff, *Planetary Diagrams,* 29 provided the text and the translation.
40. Eastwood and Graßhoff, *Planetary Diagrams,* 12–14.
41. Ibid., 27–29; Eastwood, *Ordering the Heavens,* 107–9.
42. Kühnel, "Carolingian Diagrams," 365–75.
43. Kühnel, *End of Time,* 116–59, esp. 139–42.
44. Rück, *Auszüge aus der Naturgeschichte,* 37–40; Eastwood and Graßhoff, *Planetary Diagrams,* 33–35; Eastwood, *Ordering the Heavens,* 109–19.
45. Eastwood and Graßhoff, *Planetary Diagrams,* 30–35, at 34–35 for both the supplied text and its translation: "ut subiecta figura demonstrat."
46. On the originality of the diagrams, which might have been created especially for the archetypal *Handbook of 809,* see Eastwood and Graßhoff, *Planetary Diagrams,* 4–5.
47. Ibid., 30–35, at 33–35 for the text and translation; Eastwood, *Ordering the Heavens,* 109–19.
48. Eastwood and Graßhoff, *Planetary Diagrams,* 30–35.
49. Eastwood, *Ordering the Heavens,* 109–18.
50. Eastwood and Graßhoff, *Planetary Diagrams,* 30–35. See also Eastwood, "Plinian Astronomical Diagrams," 154–56; Eastwood, *Ordering the Heavens,* 117.
51. Eastwood and Graßhoff, *Planetary Diagrams,* 12, 32; Eastwood, *Ordering the Heavens,* 109–17.
52. Eastwood, "Plinian Astronomical Diagrams," 154; Eastwood, *Ordering the Heavens,* 110.
53. Eastwood and Graßhoff, *Planetary Diagrams,* 38–40 provided the text and translation; Rück, *Auszüge aus der Naturgeschichte,* 40–43; Eastwood, *Ordering the Heavens,* 119–26.
54. Eastwood and Graßhoff, *Planetary Diagrams,* 12–13, 35–37; Eastwood, "Plinian Astronomical Diagrams," 157–59; Eastwood, *Ordering the Heavens,* 120–24. Azimuth defines a celestial location in degrees relative to cardinal compass points, whereas altitude assesses angular elevation relative to the horizon, as explained in Evans, *History and Practice,* 28, 100.
55. Eastwood and Graßhoff, *Planetary Diagrams,* 12–13, 35–39, at 39 for the citation; Eastwood, *Ordering the Heavens,* 119–26.
56. Eastwood and Graßhoff, *Planetary Diagrams,* 12–13, 35–39, at 39 for the citation; Eastwood, *Ordering the Heavens,* 119–26.
57. Eastwood and Graßhoff, *Planetary Diagrams,* 12–13, 35–39, at 39 for the citation; Eastwood, *Ordering the Heavens,* 119–26.
58. Eastwood and Graßhoff, *Planetary Diagrams,* 12–13.
59. Ibid., 39; Eastwood, *Ordering the Heavens,* 119–20.
60. Eastwood and Graßhoff, *Planetary Diagrams,* 12–14; Contreni, "Carolingian School," 86.
61. Koehler and Mütherich, *Karolingischen Miniaturen.*
62. De Jong, *Penitential State.*
63. Schieffer, *Karolinger,* 87–89 notes that Adalhard returned to the court in Aachen in 790 to become an advisor to Charlemagne. It is important to underline the (somewhat circumstantial) connections that linked the kingdom of Italy, Adalhard of Corbie, and the Carolingian court in Aachen. Ganz, *Corbie,* 23–24 prefers the outdated view that Adalhard's sister Theodrada was the abbess of Soissons. Ganz correctly notes, however, that a regular correspondence was exchanged between Adalhard and Alcuin. Letters by Alcuin to Adalhard remain from 790, 799, and 801. This demonstrates that Adalhard was truly regent in name alone and traveled back and forth from Corbie. Since Adalhard replaced Alcuin as the spiritual father of Carolingian astronomical studies after the earlier master's death, this is an important relationship to emphasize in a history of Carolingian star pictures. Also see Mütherich, "Erneuerung," 560–61, 568–69.
64. Nees, "Art and Architecture," 841. For a brief discussion of the artist (or at least scribe) named Thomas, see Nees, *Early Medieval Art,* 168. For more on Thomas and his participation in Echternach on the gospel book, see Netzer, *Cultural Interplay,* 62, 112–21.
65. Nees, *Early Medieval Art,* 157–59 notes the traditional connection between the translation of Cuthbert's relics to Lindisfarne in 698 and the manufacture of the book. Since the manuscript took at least a generation to complete, this tradition is thoroughly compatible with the later addition into the manuscript of a note

suggesting the completion of the manuscript by 721. The note designated Bishop Eadfrith (d. 721) as patron. Also see M. Brown, *Lindisfarne Gospels,* 13–149.

66. Nees, *Early Medieval Art,* 164–66; Bruce-Mitford, "Codex Amiatinus," 1–25, esp. 11–17.
67. Nees, *Early Medieval Art,* 153–55. Lowden, "Biblical Illustration," 43–45 calls into question any direct link between the gospel book and Augustine of Canterbury; also see Wormald, *Miniatures.*
68. Nees, *Early Medieval Art,* 150–66 is a readily available resource. For a monographic survey with illustrations of this manuscript evincing an interest in exegetical manuscript illumination and a strong text-image correlation, see Nees, *Gundohinus Gospels.*
69. Caillet, *L'art carolingien,* 198–99.
70. Panofsky, *Renaissance and Renascences,* 42–54.
71. Porcher, "Book Painting," in Malraux and Parrot, *Europe of the Invasions,* 121–28. More recent treatments include Pace, *Arte medievale,* 1:3–14; Pace, *VIII secolo,* 21–30, 305–12, 326–32.
72. Porcher, "Book Painting," in Malraux and Parrot, *Europe of the Invasions,* 122–28, fig. 138 emphasizes the related figural style that appeared in Italian manuscripts of the same period, such as the late sixth- or early seventh-century Ashburnham Pentateuch, which had already attained Frankish territory by roughly 750 and was housed in Tours during the ninth century. For more on the Ashburnham Pentateuch, which was probably made in Rome during the tenure of Gregory the Great (d. 604), see D. Verkerk, "Ashburnham Pentateuch." Also see the twin studies by Narkiss, "Further Study" and "Reconstruction." This classic treatment must now be considered alongside Mitchell, "Display of Script." Also see J. Jarnut, "Desiderius," in Auty et al., *Lexikon des Mittelalters,* vol. 3, cols. 724–25.
73. Bischoff, "Court Library," 76–92.
74. For more on the links connecting Charlemagne, Lombardy, and Byzantium, also see Delogu, "Lombard," 290–310.
75. Porcher, "Book Painting," in Malraux and Parrot, *Europe of the Invasions,* 128–47 notes the use of the term *ubi* as a sign indexing temporal and spatial transformation within the compilation. Crivello, *"Omelie sui Vangeli" di Gregorio Magno,* 82–86 provides a complete description and catalogue of the miniatures in the Vercelli manuscript.
76. Belting, "Probleme," 104–5, 125–43 stresses that Godescalc did not import his Italo-Byzantinizing, painterly style after his journey to Rome. Presumably, Godescalc traveled to Rome in order to attend the baptism of Charlemagne's sons at the Lateran Baptistery in 781. Rather than looking to frescoes and murals, the precursors of the new Carolingian style of painting are to be found in manuscripts, according to Belting. For an equally important survey of pre-Carolingian book decoration and its historical sources, up to the decisive turn toward the refined show script known as Caroline minuscule, see Holter, "Buchschmuck." Also see Mitchell, "Display of Script," 2:937–39.
77. Porcher, "Book Painting," in Malraux and Parrot, *Carolingian Renaissance,* 75–78; also see Underwood, "Fountain of Life."
78. Belting, "Probleme," 125–42; Laffitte and Denoël, *Trésors carolingiens,* 93–94. Also see Camille, "Word, Text, Image."
79. Nees, "On Carolingian Book Painters," 212–13, 235n24, 236n50; as reported in Nees, a transcription is available in Reudenbach, *Godescalc-Evangelistar,* 5–26, 31, 98–101. Nees, "Godescalc's Career," 24–27 believes the Egino Codex to be the work of Godescalc as well.
80. Nees, "On Carolingian Book Painters," 213.
81. Porcher, "La peinture provinciale," notes that artists in Corbie, Laon, and northern French scriptoria added the Lombardic (Italian) convention of a frontispiece, depicting a cross under a portico supported by two columns, to their books. This convention was introduced in the mid-eighth century; noteworthy examples include Paris, Bibliothèque nationale de France, MS lat. 12168 from Laon, and Rome, Biblioteca Apostolica Vaticana, Reg. lat. 316, possibly from Corbie.
82. Nees, "Godescalc's Career," 24.
83. Ibid., 21–43. Also see Rosenthal, *Illuminations,* 95–105; this book reveals the specifically Italo-Byzantine features of the Roman Vergil.
84. Mütherich and Gaehde, *Carolingian Painting,* cat. XV, 26–27, 93–95, plates 30–31 supplies an example of a figure whose facial features were rendered using this stylistic convention in the second half of the ninth

century at Saint-Amand: the gospel-author portrait of John, Gospels of Francis II, Paris, Bibliothèque nationale de France, MS lat. 257, fols. 147v–148r.

85. Pächt, *Buchmalerei*, 17–25; Haffner, *Sternbilderzyklus*, 78 significantly updates the problematic treatment in Bethe, *Buch und Bild*, 50–55.
86. Pächt, *Buchmalerei*, 24.
87. Borst, "Alkuin und die Enzyklopädie von 809," 73–75; Bierbrauer, *Karolingischen Handschriften*, 73–75.
88. Weitzmann, *Illustrations in Roll and Codex*, 84–89.
89. Ibid., 86–112.
90. Dolezal, "Elusive Quest," provides the best critique of the Weitzmannian analytic method. Her study evaluates the equally problematic model of the quest for an original gospel text, which was supposed to be revealed by examinations of the Byzantine lectionary at the University of Chicago. In both cases the problematic underlying assumption was that changes and emendations to earlier texts were always deleterious obfuscations.
91. Melzak, "Antiquarianism," 636.
92. Schieffer, *Karolinger*, 109–11 underlines that Louis the Pious, king of Aquitaine, had actually been in *de facto* control of the empire since 811.
93. Laffitte and Denoël, *Trésors carolingiens*, 167–70 now provides the best succinct introduction to the school of Reims. The older treatment in Mütherich, "Carolingian Manuscript Illumination," remains helpful. Also see Van der Horst, Noel, and Wüstefeld, *Utrecht Psalter*, 184–85; Koehler and Mütherich, *Karolingischen Miniaturen*, vol. 6, parts 1–2 with a succinct account of the fall of Ebo at 6.1:12; Duckett, *Carolingian Portraits*, 20–57, esp. 42–51 for the historic events of 833–35; De Jong, *Penitential State*, 254–55.
94. Duckett, *Carolingian Portraits*, 44, 51; Schieffer, *Karolinger*, 131.
95. Koehler and Mütherich, *Karolingischen Miniaturen*, 6.1:32–36 explains that Paris, Bibliothèque nationale de France, MS lat. 265 (after 825) provided a verifiable link between Vienna and the development of book painting in the various foundations of Reims.
96. Laffitte and Denoël, *Trésors carolingiens*, 167. In this, Denoël follows the lead of Mütherich, who argued equally for the localization of specific manuscripts made by Saint-Remi, Saint-Thierry, or naturally at the cathedral; also see Koehler and Mütherich, *Karolingischen Miniaturen*, 6.1:10–54, esp. 49–51, and 6.2:8–65, esp. 61.
97. Mütherich, "Carolingian Manuscript Illumination," 104–19; also see Van der Horst, Noel, and Wüstefeld, *Utrecht Psalter*,168–70.
98. Chazelle, "Archbishops Ebo and Hincmar," esp. the concluding remarks on 117–19.
99. Homburger, *Illustrierten Handschriften*, 101–17. Van der Horst, Noel, and Wüstefeld, *Utrecht Psalter*, 190–91.
100. Von Steiger and Homburger, *Physiologus bernensis*, 19–31; Woodruff, "Physiologus of Bern," 226–30.
101. Koehler and Mütherich, *Karolingischen Miniaturen*, 6.1:38–39, 172–82.
102. Henkel, *Studien zum Physiologus*, 1–23; von Steiger and Homburger, *Physiologus bernensis*, 9–12; Van der Horst, Noel, and Wüstefeld, *Utrecht Psalter*, 190.
103. Henkel, *Studien zum Physiologus*, 1–4; von Steiger and Homburger, *Physiologus bernensis*, 9–11.
104. Van der Horst, Noel, and Wüstefeld, *Utrecht Psalter*, 190.
105. Homburger, *Illustrierten Handschriften*, 101; von Steiger and Homburger, *Physiologus bernensis*, 11.
106. Woodruff, "Physiologus of Bern," 237–46; Haffner, *Sternbilderzyklus*, 78.
107. Weitzmann, *Book Illumination*, 32–33; Homburger, *Illustrierten Handschriften*, 103.
108. Woodruff, "Physiologus of Bern," 238–41, 246; Homburger, *Illustrierten Handschriften*, 102–3.
109. Homburger, *Illustrierten Handschriften*, 103–4.
110. Ibid., 101, 115; Woodruff, "Physiologus of Bern," 253; Isidore, *Etymologies*, trans. Barney et al., 249; von Steiger and Homburger, *Physiologus bernensis*, 11, 44.
111. Mütherich and Gaehde, *Carolingian Painting*, 63–65.
112. Koehler and Mütherich, *Karolingischen Miniaturen*, 3:101 with updates from Laffitte and Denoël, *Trésors carolingiens*, 188–205.
113. Koehler and Mütherich, *Karolingischen Miniaturen*, 3:122n90.
114. Laffitte and Denoël, *Trésors carolingiens*, 173–74; Koehler and Mütherich, *Karolingischen Miniaturen*, 6.1:13–17, 150–58.
115. Laffitte and Denoël, *Trésors carolingiens*, 173; Koehler and Mütherich, *Karolingischen Miniaturen*, 6.1:13, 24, 150–52; Mütherich, "Carolingian Manuscript Illumination," 108–11.

116. Mütherich, "Carolingian Manuscript Illumination," 111.
117. Ibid., 108; Van der Horst, Noel, and Wüstefeld, *Utrecht Psalter*, 184–85; Mütherich and Gaehde, *Carolingian Painting*, 56–62; Koehler and Mütherich, *Karolingischen Miniaturen*, 6.1:73–84.
118. Koehler and Mütherich, *Karolingischen Miniaturen*, 6.1:22–24; Laffitte and Denoël, *Trésors carolingiens*, 173.
119. Laffitte and Denoël, *Trésors carolingiens*, 203–4.
120. Nees, "On Carolingian Book Painters," notes the links between the artists who worked in the Reims and Metz workshops and attempts to reconstruct the artistic biography of his proposed Transfiguration Master. He suggests that one of the painters of the Utrecht Psalter might have contributed to the Drogo Sacramentary as well.
121. Schieffer, *Karolinger*, 131.
122. Chazelle, *Crucified God*, 254–58 with earlier bibliography.
123. Ibid., 254–66, 283–85.
124. Ibid., 256–63.
125. Ibid., 257–85.
126. Ibid., 259–60.
127. For more on the complicated interrelationships among these symbolic interpretations of globes, see Schramm, *Sphaira*, 55–70.
128. For a late third-century example of Urania teaching Aratus with a celestial globe, see the image from the famous "signed" Monnus mosaic in Trier: Hoffman, Hupe, and Goethert, *Mosaike aus Trier*, 138–41, plate 67. On celestial globes, consult Künzl, "Globus/Globe," and Dekker, *Illustrating the Phaenomena*, 116–256.
129. McGurk, "Carolingian Astrological Manuscripts," 320. For one of the best essays on Carolingian astronomy with passing references to the Corbie manuscript in Paris (MS lat. 12957), see Obrist, "La représentation carolingienne," 23–24. Mütherich, "Bilder," 35, 35n26 seems uncertain about the date of the manuscript, which she variously ascribes to the first and second quarters of the ninth century. Perhaps the manufacture of the book commenced in the first quarter, only to conclude in the second? Ganz, *Corbie*, 49 sets the date for a back flyleaf of Paris, MS lat. 12957 in the early ninth century, but that says nothing about the date of the manuscript itself. So a date ca. 810–40 is offered here as a guarded proposal. The preponderance of the evidence argues for a date that is closer to the time of manufacture of the copy of the *Revised Aratus* in Cologne, Dombibliothek, MS 83-II, however, and therefore suggests a date closer to 810 than 840.
130. Dekker, *Illustrating the Phaenomena*, 192, 254–55.
131. Ibid., 194–202.
132. Ptolemy, *Almagest*, trans. Toomer, vii–x, 404–6; see the helpful foreword by Owen Gingerich.
133. Murdoch, *Album of Science*, 73, cat. 68, and 244–45, cat. 220; Stückelberger, *Bild und Wort*, 21–23.
134. Murdoch, *Album of Science*, 246–47, cats. 221–23.
135. Dekker, *Illustrating the Phaenomena*, 101 leaves open this possibility.
136. Savage-Smith, *Islamicate Celestial Globes*, 11; Gundel, *Zodiakos*, 204, cat. 8; Dekker, *Illustrating the Phaenomena*, 84–102, 111–13.
137. Dekker, *Illustrating the Phaenomena*, 84–88; Thiele, *Antike Himmelsbilder*, 19–42.
138. Künzl, "Globus/Globe," 7–11, 30, 71 disagrees with this assessment of the images of Sagittarius. Instead, Künzl argues that the damaged form of Sagittarius on the Farnese Globe represents a dorsal view of a rearing Centaur, according to a more typical iconography for Sagittarius. Since the hand of Atlas has obstructed the critical spot on the globe, to a certain extent this must remain an open question. A chased globe in Mainz, made ca. 150–220 C.E. in the eastern Roman Empire, displays an image of Sagittarius as a centaur. The key distinction between the depictions of Sagittarius on the Mainz globe and on the Farnese Globe remains that in the Mainz globe, the shooting centaur representing Sagittarius elevates its front right leg, drawing attention to its anatomical form as a quadruped. Without the back legs to support the weight of the centaur, the hybrid would collapse. The images of Sagittarius resembling a satyr in the *Handbook of 809* and the depiction of Sagittarius on the Farnese Globe differed from the form on the Mainz globe. In both the Farnese Globe and Drogo's copy of the *Handbook of 809*, the satyr plants both its legs and hooves necessarily upon the "ground." Otherwise, the representation of Sagittarius from the Farnese Globe would have displayed an elevated right leg projecting out beyond the stable left leg, as in the globe in Mainz. This point

was evident in the print by Folkes, although on the statue these details are difficult to discern, to say the least. Dekker, *Illustrating the Phaenomena*, 100, 114 prefers to consider simply the illegible form of Sagittarius as what remains of a Centaur, as one would typically expect.

139. Gundel, *Zodiakos*, 72–73.
140. The best summary of celestial hemispheres is now Dekker, *Illustrating the Phaenomena*, 118–42, 207–27; also see Von Euw, *St. Galler Buchkunst*, 446–49.
141. Dekker, *Illustrating the Phaenomena*, 10–35.
142. Ibid., 35.
143. Eastwood and Graßhoff, *Planetary Diagrams*, 27–35.
144. Kirchner, *Phillipps-Handschriften*, 30 dates both manuscripts to the tenth century. Rose, *Meerman-Handschriften*, 287–93, cat. 129 (MS 1830) and cat. 130 (MS 1832), rightly prefers to date MS 1830 to 874, as recorded in the manuscript on fol. 8v, through a year reported in the book "ab incarnatione domini." Rose dates MS 1832 to ca. 873, because a scribe entered a date "ab incarnatione domini" of 873 on fol. 46r, although he notes that there was also an "ab incarnatione domini" date of 878 supplied on fol. 10r; on this, also see Eastwood, *Ordering the Heavens*, 259.
145. Contreni, *Cathedral School*, 2–3, 124–25, 129.

CHAPTER 4

1. Dutton, *Charlemagne's Courtier*, 32. For more on Charlemagne's cultivation of Latin and the seven liberal arts at Aachen, also see Fried, "Karl der Große," 1:25–44.
2. Dutton, *Charlemagne's Courtier*, 31.
3. Ibid., 32.
4. Borst, "Alkuin und die Enzyklopädie von 809," 57–73. Alcuin drew upon three of his earlier writings (composed 782–96) in his letters about astronomy to Charlemagne (798–99). For more on this erudite exchange, see Lohrmann, "Alcuins Korrespondenz."
5. Isidore, *Etymologies*, trans. Barney et al., 99; Isidore, *Etymologiarum*, 1:III.xxv.1. Subsequent references are to these editions. Also see Fontaine, "Isidore de Séville et l'astrologie," 275, 290–91.
6. See Appendix VI, "From Isidore to my lord and son Sisebut (i.e. the Visigoth King of Spain; probably 620)," in Isidore, *Etymologies*, trans. Barney et al., 7–11, at 413 for the citation. Also see Fontaine, *Isidore de Séville et la culture classique*, 2:453–57.
7. Nelson, "Voice of Charlemagne," 76–88.
8. Ibid., 80–85; Ganz, "Theology and the Organisation of Thought," 758–62.
9. Ullman, *Idea of Kingship*, 1–29, 135–46, at 6 for the citation; this text must now be read alongside Nelson, "Kingship, Law, and Liturgy." At the intersection of conciliar debate, intellectual theological interrogation of ideas, and spiritual authority, also see Marenbon, "Alcuin."
10. Nelson, "Kingship, Law, and Liturgy," 243; De Jong, *Penitential State*, 29, 65, 67–69, 77, 80, 156, 231–56.
11. Contreni, "John Scottus," at 4 for the citation; also see Contreni, "Carolingian Renaissance."
12. Ramírez-Weaver, "Classical Constellations."
13. Contreni, "Carolingian Renaissance," 709–19 explains the Carolingian educational system, including pedagogical training for young women.
14. Hanfmann, "Continuity of Classical Art"; Lindberg, "Fate of Science."
15. Haffner, *Sternbilderzyklus*, 22, 75–79; Le Bourdellès, *Aratus Latinus*, 102–4; also see Katzenstein and Savage-Smith, *Leiden Aratea*.
16. Madrid 3307, fol. 54r; Borst, *Schriften zur Komputistik*, 3:1251. The translations from this manuscript are my own, unless otherwise specified.
17. Madrid 3307, fol. 54r; Borst, *Schriften zur Komputistik*, 3:1251.
18. Carruthers, *Craft of Thought*, 24–29.
19. Munro, *Laws of Charles*, 15; Mordek, Zechiel-Eckes, and Glatthaar, *Admonitio generalis*.
20. Garrison, "Emergence of Carolingian Latin Literature," 111–19.
21. Nees, *Tainted Mantle*, 3–17, esp. 9–11.
22. On the Christian interpretation of the liberal arts in Frankish society, see also Marenbon, "Carolingian Thought," 171–74. On the specific relationship between the liberal art of astronomy and the cultivation of religious instruction, see McCluskey, "Astronomies in the Latin West," 152–56.
23. Garrison, "Emergence of Carolingian Latin Literature," 119.
24. Lemay, "True Place of Astrology," 63–64.
25. McCluskey, *Astronomies and Cultures*, 29–48.

26. Ibid., 99–113 argues for the composition of the *De cursu stellarum* following Gregory's ascension to the see of Tours in 573. For the critical edition of the text, also see Gregory of Tours, *De cursu stellarum ratio*, in *Miracula et opera minora*, MGH, Scriptorum rerum Merovingicarum, 1.2:404–22.
27. Flint, *Rise of Magic*, 97–99, 137–38; McCluskey, *Astronomies and Cultures*, 101. Also see McCluskey, "Gregory of Tours."
28. McCluskey, *Astronomies and Cultures*, 104–10; Gregory of Tours, *De cursu stellarum ratio*, chap. 23, 415.
29. Burckhardt, *Ausgewählte Miniaturen*, 9 with earlier bibliography.
30. Le Bourdellès, *Aratus Latinus*, 26–29.
31. Hübner, *Begriffe*, 5–35.
32. Le Bœuffle, *Astronomie/Astrologie*, 56.
33. Charvet and Zucker, *Le ciel*, 41.
34. Blume, "Sichtbares Bild," 294–300.
35. Theodulf, *Verse*, poem 46, 142–46, with an introduction and additional bibliography; all translations are from this edition. For the critical edition, consult Dümmler, *Poetae latini aevi carolini*, MGH, Poetarum Latinorum Medii Aevi, 1:544–47. Also see Ommundsen, "Liberal Arts," 188–92. These volumes of the *Poetae latini* can be consulted online courtesy of the HathiTrust.
36. Theodulf, *Verse*, 145; Dümmler, *Poetae latini aevi carolini*, 1:546.
37. Theodulf, *Verse*, 145; Dümmler, *Poetae latini aevi carolini*, 1:546–47.
38. Carruthers, *Craft of Thought*, 209–13, at 212 for the citation.
39. Nees, *Tainted Mantle*, 147–301 has assessed this theme with great rigor, focusing upon the related problem of the Hercules ivories in the Cathedra Petri. Nees insightfully explains the cultural appropriation of classical motifs—coupled with a pervasive anxiety about their forms and connotations—for specifically Carolingian Christian audiences.
40. For a critical edition, see Hrabanus, *De institutione clericorum*; subsequent references will be to this edition, unless otherwise specified. There is also a corrupted edition of the *De clericorum institutione* readily available in Migne, *Patrologia Latina (PL)* 107, cols. 291–420a. Detlev Zimpel identified the Migne edition as a reprint of volume 6 from the complete works of Hrabanus prepared by George Colvener in 1626–27 (150, 234).
41. Haarländer, *Rabanus Maurus*, 27–30. For a concise biography of Hrabanus's life, see Coon, *Dark Age Bodies*, 13–41.
42. Contreni, *Cathedral School*, 29–58 supplies relevant information about the creation of a cathedral library for ecclesiastic and educational purposes and the active engagement with its books by a learned community of scholars.
43. Hrabanus, *De institutione clericorum*, 44–48.
44. Haarländer, *Rabanus Maurus*, 130–35.
45. Marenbon, "Carolingian Thought," 173.
46. Ribémont, *Les origines des encyclopédies*, 27–33. See also Marrou, "Arts libéraux," 5–27, esp. 16.
47. Marenbon, "Carolingian Thought," 172–73 supplies the discussion and translation. For a complete English translation, see Augustine, *Divine Providence*, trans. Russell, esp. 144–47; also see Augustine, *De ordine*, 129.
48. Shanzer, *Philosophical and Literary Commentary*, 13–15. Also see Augustine, *Retractions*, 21–22; Augustine, *Retractionum*, 17.
49. Vitruvius, *Ten Books on Architecture*, 199.
50. Martianus Capella, *Seven Liberal Arts*, 2:346, IX.891. Shanzer, *Philosophical and Literary Commentary*, 14–15.
51. Shanzer, *Philosophical and Literary Commentary*, 14–16, 26–28, 42–43. Grant, *Foundations*, 9–17 adds, however, that adequate astronomical education during the late antique period and the early Middle Ages was naturally limited by the capabilities of well-meaning but poorly trained teachers.
52. Hrabanus, *De institutione clericorum*, 483: "ut quidam dixit, dignum est religiosis argumentum magnumque curiosis tormentum." Compare this with Augustine, *De ordine*, 130, lines 15–16: "magnum religiosis argumentum tormentumque curiosis." The translations are mine, unless otherwise noted.
53. Fontaine, *Isidore de Séville et la culture classique*, 2:459–60; Carruthers, *Craft of Thought*, 24–28.
54. Cassiodorus, *"Institutions,"* trans. and ed. Halporn, 225; Cassiodorus, *Institutiones*, ed. Mynors, 152–53. Subsequent references will be to these editions, unless otherwise specified.
55. Fontaine, "Isidore de Séville et l'astrologie," concludes his essay with a methodologically sound

recommendation for early medieval investigation that merits report in summary (300n1). Fontaine argues that an appreciation for an early medieval author's (or, by extension, painter's) oeuvre can only be assessed by (1) analyzing original sources, (2) clarifying all instances of emendation, deletion, and insertion, and (3) making appeals to the cultural context surrounding the work in keeping with a sound cultural historical methodology.

56. Cassiodorus, "*Institutions*," 225; Cassiodorus, *Institutiones*, 153.
57. Cassiodorus, "*Institutions*," 225; Cassiodorus, *Institutiones*, 153. Also see Hrabanus, *De institutione clericorum*, 483.
58. Cassiodorus, "*Institutions*," 225; Cassiodorus, *Institutiones*, 153. Also see Hrabanus, *De institutione clericorum*, 483.
59. McCluskey, *Astronomies and Cultures*, 32–36.
60. Hrabanus, *De institutione clericorum*, 483, lines 14–15.
61. McCluskey, *Astronomies and Cultures*, 35n28 disagrees.
62. Augustine, *City of God*, 979–81. Subsequent references are to this edition.
63. Ibid., book XXI.9, 984–85: "For in any case I have sufficiently argued that it is possible for living creatures to remain alive in the fire, being burnt without being consumed, feeling pain without incurring death; and thus by means of a miracle of the omnipotent Creator."
64. Hrabanus Maurus, *De institutione clericorum*, 484; Isidore, *Etymologies*, trans. Barney et al., 99; Isidore, *Etymologiarum*, 1:III.xxvii.1.
65. Hrabanus, *De institutione clericorum*, 484, line 23: "licet ad unam disciplinam ambae pertineant."
66. Isidore, *Etymologies*, trans. Barney et al., 99; Isidore, *Etymologiarum*, 1:III.xxvii.1–2. Hrabanus, *De institutione clericorum* includes insubstantial changes to Isidore's text, attributable to the textual transmission of the *Etymologiarum* (484).
67. Hübner, *Begriffe*, 27, 33–34. Also see Fontaine, *Isidore de Séville et la culture classique*, 2:467n1: "in astrologia caeli conversio, ortus, obitus motusque siderum (Cicero, *De orat.* I.187)"; see also the more complete analysis by Fontaine, "Isidore de Séville et l'astrologie," in which Fontaine notes Isidore's alternate use of the term "potestates" to describe the "powers" of *astrologia* in his *Differentiae* II.39.152 (274–75). Otherwise, Isidore's definition of *astrologia* from the *Differentiae* closely recalls his definition of *astronomia* from the *Etymologiarum*; see also Thorndike, *History of Magic*, 1:632–33. The fact that Isidore vacillated in his personal denotation of *astrologia* makes Hrabanus's efforts at clarification through compilation all the more complex and interpretative.
68. Hübner, *Begriffe*, 31–33.
69. Thorndike, *History of Magic*, 1:632. Also see Hübner, *Begriffe*, 33n78.
70. Flint, *Rise of Magic*, 92–101, 128–31 explains that the Council of Braga (560–65) proscribed the practices of linking zodiac signs to biblical patriarchs, souls, and/or physical body parts; these practices derived from the heretical beliefs espoused by Priscillian (d. 385/86). Also see Hübner, *Begriffe*, 34.
71. Lejbowicz, "Théorie et pratique," 625.
72. Isidore, *Etymologies*, trans. Barney et al., 99; Isidore, *Etymologiarum*, 1:III.xxv.2.
73. Lejbowicz, "Théorie et pratique," 623–30. Also see Fontaine, *Isidore de Séville et la culture classique*, 3:1102–4.
74. Hrabanus, *De institutione clericorum*, 484. Zimpel's edited version of the passage used here differs slightly in text but significantly in meaning from Migne, *PL* 107, cols. 403d–404a. I wish to thank Gregory Hays for assistance with this translation; any inaccuracies remain mine.
75. Flint, *Rise of Magic*, 98.
76. Lejbowicz, "Théorie et pratique," 628–30.
77. See chapter 2 above.
78. Aratus, *Phaenomena*, 138–39, lines 892–908; Eastwood, *Ordering the Heavans*, 170–72.
79. Madrid 3307, fol. 70r–v; see also Borst, *Schriften zur Komputistik*, 3:1287.
80. Other examples include Monza f–9/176, fol. 63v, ca. 864 (fig. 36), and Vat. lat. 645, fol. 58v, ca. 827 (fig. 38).
81. Eastwood, *Ordering the Heavens*, 170–77 considered book V of the *Handbook of 809* a utilitarian "*vade mecum*."
82. See note 66 above.
83. Barton, *Ancient Astrology*, 86–89.
84. Sharpe, "Isidore of Seville," 19.
85. Isidore, *Etymologies*, trans. Barney et al., 115. For a concise introduction to Galenic humoral theory and its influence upon early medieval medical texts,

including those by Isidore, see Sharpe, "Isidore of Seville," 22–26.
86. Flint, *Rise of Magic*, 129–31.
87. Fontaine, *Isidore de Séville: Genèse et originalité*, 307.
88. Haffner, *Sternbilderzyklus*, figs. 70 and 72.
89. Isidore, *Etymologies*, trans. Barney et al., 107.
90. Ribémont, *Les origines des encyclopédies*, 301–2.
91. Ibid., 305–10.
92. Carruthers, *Craft of Thought*, 124–30 provides the translation (at 128) of Hrabanus's text and emphasizes Hrabanus's interest in reclaiming the decorative presentation of language rather than the ideas. In the case of the illustrated Aratean texts preceding the *Handbook of 809*, Hrabanus would have probably intended his bibliophilic recycling project to apply to the text, the pictures, and the manner in which the text had been enlivened through wordplay. Also see Migne, *PL* 107, col. 396a–b; G. Brown, "Introduction: The Carolingian Renaissance," 38–40; Hrabanus, *De institutione clericorum*, 470.
93. Migne, *PL* 107, col. 396b–c: "ne pereat qui infirmus est in scientia nostra frater, propter quem Christus mortuus est, si viderit in idolio nos recumbentes." Also see Hrabanus, *De institutione clericorum*, 470.
94. Augustine, *Christian Instruction*, vol. 2 of *Fathers of the Church*, 112–13, 2.40.60. Also see Lutz, "Remigius' Ideas," 69.
95. Coon, *Dark Age Bodies*, 16–29, 134–44 discusses Hrabanus's influential ideas of spiritual renewal through meditation and knowledge acquisition that also find a corollary in the astronomical imagery of the Leiden *Aratea* and the *Handbook of 809*.
96. Von Euw, "Alkuin als Lehrer," 251–62. For more on Alcuin's impact upon the history of Carolingian manuscript illumination, see Nees, "Alcuin and Manuscript Illumination."
97. Jansen and Pohle, *Künste am Hofe Karls*, 38–41. Also see Fleckenstein, "Alcuin im Kreis der Hofgelehrten"; Borst, "Alkuin und die Enzyklopädie von 809," 53–78; Lohrmann, "Alcuins Korrespondenz."
98. Dutton, *Charlemagne's Courtier*, 32, par. 25.
99. Ibid., 32.
100. For more on the classical uses of the scientific terms employed here, see Evans, *History and Practice*, 75–79, 289–99.
101. Marenbon, "Carolingian Thought," 172. Also see Holtz, "L'œuvre grammaticale d'Alcuin."
102. Marenbon, "Carolingian Thought," 172–73 for both commentary and the translation used here; Migne, *PL* 101, cols. 853b–c. Also see Lejbowicz, "Computus," 170–72, and Matter, "Alcuin's Theology," 91–105.
103. Alberi, "'Mystery of the Incarnation.'"
104. Dutton, *Carolingian Civilization*, 79–80.
105. Alberi, "'Mystery of the Incarnation,'" 509; for the Latin text, see Dümmler, *Epistolae karolini aevi*, 239, no. 148; for this specific text, the translations are my own, unless otherwise noted. Also see Lohrmann, "Alcuins Korrespondenz," 82–86, and Bullough, *Alcuin*, 236–38.
106. Alberi, "'Mystery of the Incarnation,'" 509.
107. Marenbon, "Carolingian Thought," 173. For more on Alcuin and Carolingian spiritual renewal, see Schmitz, "Bonifatius und Alkuin."
108. Alberi, "'Mystery of the Incarnation,'" 506–9; see note 102 above.
109. Marenbon, "Carolingian Thought," 173–74.
110. Lutz, "Remigius' Ideas," 66–68 provides beneficial commentary and the translation of the text used here. For the original, see Remigius, *Commentum in Martianum Capellam*, ed. Lutz, 2:26, 160.1, lines 12–19.
111. Beaumont, "Latin Tradition," 285–87.
112. Lutz, "Remigius' Ideas," 70–71, esp. note 29; Eriugena, *De div. nat.*, I.44 from Migne, *PL* 122, col. 486c: "artes esse aeternas, et semper immutabiliter animae adhaerere." For a concise synopsis of the philosophical development of Eriugena's thinking on the liberal arts during the ninth century, see Mathon, "Formes et la signification," 47–64.
113. Lutz, "Remigius' Ideas," 67–68, esp. note 10. Contreni, "Three Carolingian Texts," 808–13 argues against any certainty concerning the historic attributions of the Dunchad commentary to either Duncaht of Reims or Martin Hiberniensis of Laon (d. 875), and prefers to leave the matter an open question, positing additional alternatives for further consideration elsewhere. Teeuwen, "Study of Martianus," 185–93 argues instead that the anonymous Dunchad commentary was the coordinated, collaborative result of courtly savants at the court of either Louis the Pious or Charles the Bald. Also see Teeuwen, *Harmony*, 33–150.

114. Contreni, "John Scottus," 4–5. Also see Jeauneau, *Études Érigéniennes*, 23–28, 82–84, 154–55.
115. Contreni, "John Scottus," 3–6.
116. Lutz, "Remigius' Ideas," 77n74 notes that this citation cannot be attributed to Eriugena. Whether this idea was an original contribution by Remigius, or derived from another source, is unclear. Remigius, *Commentum in Martianum Capellam*, ed. Lutz, 112–13, 25.14: "quae idcirco liberales dicuntur quia liberaliter fruge veritatis animam pascunt, vel quod liberam et expeditam mentem a tumultibus saeculi requirant."
117. Contreni, "John Scottus," 5.
118. Lutz, "Remigius' Ideas," 79–81.
119. Isidore, *Etymologies*, trans. Barney et al., 107.
120. Chazelle, "Matter, Spirit, and Image." The title *Opus Caroli* became standardized with Freeman's 1998 edition.
121. Noble, *Images*, 166–68.
122. Brubaker and Haldon, *Byzantium*, 9–68.
123. Coon, *Dark Age Bodies*, 134–44. For more on the Corvey westwork, see Conant, *Carolingian and Romanesque Architecture*, 63–64; Dodwell, *Painting in Europe*, 22.
124. Freeman, "Introduction," 17–33.
125. Noble, *Images*, 183–84.
126. Freeman, "Introduction," 1–2.
127. Noble, *Images*, 161.
128. Freeman, "Introduction," 1–33; Noble, *Images*, 61–110, 163–80.
129. Noble, *Images*, 165, 169–70 succinctly summarizes the primary focus of the actual Council of Frankfurt, which sought to condemn for a third time the Adoptionist heresy of Felix of Urgel in June 794. The numerous translations in the Noble volume supply a rich repository of useful texts; whenever cited, they are carried over without emendation, as here at 170.
130. Scholz, *Carolingian Chronicles*, 6–8, 73. For more on the royal Frankish annals and their ideological significance, see McKitterick, *History and Memory*, 84–119. McKitterick attributes the revised version of the *Annales regni Francorum* to the author of the entries for 808–20. The revised version E was first classified accordingly by Kurze, who studied their textual transmission in 1895: see Kurze, *Annales regni francorum*.
131. Noble, *Images*, 158–206; an extremely helpful summary of the book's contents can be found on 184–206. Noble expertly argues (at 181–82) that Theodulf's flawed Latin translation of the proceedings of Nicaea II did not impede the exegete's ability to parse correctly the slippery distinction between "*proskynesis* and *latreia*, *veneratio* and *adoratio*." In fact, the clever interpretative turns in Theodulf's assessment of Nicaea II reveal an active engagement not only with this distinction but also with its subtle theological implications. Moreover, tracing the lines of argument Theodulf employed in the development of his thesis ensures that the *Opus Caroli* can be considered more than just a failed courtly record of political significance. The troubled work unveils important Carolingian considerations concerning the nature of Frankish ecclesiastic reform and recondite aspects of Carolingian image theory.
132. Freeman, "Scripture and Images," 163–66. This essay supplies valuable translations of key passages from the *Opus Caroli*. Freeman's translation of the passage addressing writing and imagery follows: "The prophecy that a Virgin shall conceive and bear a son is not to be sought in uncertain and ambiguous things, but to be pondered in the heart. Nor are its hidden mysteries to be explored in pictures, but in holy Scripture." Most published literature references the old edition of the *Opus Caroli: Libri Carolini*, ed. Bastgen, IV.21, 212–13. The reference for the new edition is *Opus Caroli*, ed. Freeman, MGH, Concilia, Tomus II, Supplementum I, IV.21, 539. Also see Chazelle, "Matter, Spirit, and Image," 164–65, and Ommundsen, "Liberal Arts," 175–78.
133. McKitterick, "Text and Image," 297–99.
134. Ganz, "'Pando quod Ignoro,'" 29 provides commentary and the critical translation of the epistle used here. McKitterick, "Text and Image," 297–99 provides commentary and another English translation of the letter. For the original, see Dümmler, *Poetae latini aevi carolini*, 2:196, no. 38. For more historical information about the exchange between Hatto and Hrabanus, see Haarländer, *Rabanus Maurus*, 109.
135. Freeman, "Scripture and Images," 170–71. *Opus Caroli*, ed. Freeman, II.23, 278–79.

136. McKitterick, "Text and Image," 300–301 for the translation and commentary. *Libri Carolini*, ed. Bastgen, III.23, 153; *Opus Caroli*, ed. Freeman, III.23, 446–47.
137. Kandinsky, *Concerning the Spiritual in Art*, 1–18. This demonstrated the philosophical simplicity of the Carolingian image theory, however. Hiberno-Saxon carpet pages were already nascent forms of spiritualized abstraction in the early Middle Ages, intended to facilitate meditative, cenobitic practices.
138. Davis-Weyer, *Early Medieval Art*, 103 provides this translation along with additional passages on the theory of images from the *Opus Caroli. Libri Carolini*, ed. Bastgen, III.23,150–53; *Opus Caroli*, ed. Freeman, III.23, 446.
139. Chazelle, "Matter, Spirit, and Image," 173. All translations from this text are used without emendation unless otherwise indicated.
140. Carruthers, *Craft of Thought*, 24–29.
141. McKitterick, "Text and Image," 298–304.
142. Chazelle, "Matter, Spirit, and Image," 172–75. Incidentally, satyrs offer an interesting case: there are only two disputed biblical passages, that could refer to satyrs. In both cases, however, it is fairly clear that "wild goats" roaming the desert were denoted by the Hebrew: see Isaiah 13:21 and 34:14. Satyrs were listed among the monstrous races in Isidore's *Etymologies* XI.iii.21; see Isidore, *Etymologies*, trans. Barney et al., 245.
143. Chazelle, "Matter, Spirit, and Image," 164–70. Also see the comments in Noble, *Images*, 193–94, concerning *Opus Caroli*, II.26–30.
144. Noble, *Images*, 166, 215 supplies the translation and also underscores the profound approbation this passage received from Charlemagne. Not only was the *Opus Caroli* drafted in the king's voice, but it was also reviewed subsequently and accorded sovereign sanction by Charlemagne and his learned "dream team" of theologically and scientifically minded courtiers. *Opus Caroli*, ed. Freeman, I.9, 254.
145. Chazelle, "Matter, Spirit, and Image," 172–73.
146. Ibid., 166. *Libri Carolini*, ed. Bastgen, II.29, 91.
147. Chazelle, "Matter, Spirit, and Image," 170–71.
148. Noble, *Images*, 187 supplies the translation and context for the passage; also see Chazelle, "Matter, Spirit, and Image," 170–71.
149. Chazelle, "Matter, Spirit, and Image," 170–71. *Libri Carolini*, ed. Bastgen, I.8, 25–26; *Opus Caroli*, ed. Freeman, I.8, 145–48.
150. Eastwood, *Ordering the Heavens*, 32–33, 95–101.
151. Chazelle, "Matter, Spirit, and Image," 171–72.
152. Freeman, "Scripture and Images," 164–65 provides the translation and commentary. *Libri Carolini*, ed. Bastgen, IV.21, 213; *Opus Caroli*, ed. Freeman, IV.21, 540. For more on the heavy descriptive work *tituli* were expected to perform during the Carolingian era, see Ganz, "'Pando quod Ignoro,'" 25–28.
153. For more on the traditional forms of paganism within the Frankish lands, see Salzmann, *On Roman Time*, 142–57; Jones and Pennick, *History of Pagan Europe*, 111–37; Kuhnen, *Religio Romana*, 229–42.
154. The concept of "madness," when applied to those in pursuit of superstitious astrology, actually foreshadows Michel Foucault's description of madness as a bestial display of inhuman desire for alternative access to knowledge; see Foucault, *Madness and Civilization*, 3–37, 279–89, esp. 21. For more on this idea, see Gutting, *Foucault's Archaeology*, 69–73.
155. Davis-Weyer, *Early Medieval Art*, 100 provides the translation, slightly adapted here. See also *Libri Carolini*, ed. Bastgen, III.23, 151; *Opus Caroli*, ed. Freeman, III.23, 441–42.
156. Freeman, "Scripture and Images," 174–75 provides the translation and commentary; *Libri Carolini*, ed. Bastgen, III.23, 152; *Opus Caroli*, ed. Freeman, III.23, 446.
157. Chazelle, "Matter, Spirit, and Image," 172.
158. Migne, *PL* 111, col. 563c; also see Heyse, *Untersuchungen zu den Quellen*, 146, and Reuter, *Text und Bild*, 33–48, 205.
159. Freeman, "Scripture and Images," 176; Ganz, "'Pando quod Ignoro,'" 28.
160. Freeman, "Scripture and Images," 176 provides the translation and commentary. Isidore, *Etymologiarum*, 2:XIX.xvi.1.
161. Ganz, "'Pando quod Ignoro,'" 28–29 provides the translation and commentary. For the text of Hrabanus Maurus, *Commentariorum in Exodum libri quattuor*, see Migne, *PL* 108, col. 95. Also see Diebold, "Carolingian Idol."
162. Kühnel, *End of Time*, 114 similarly argues that during the ninth century, the liberal arts in the Frankish lands "actually lost their liberty and independence.

They were given a purpose and a direction: they were enrolled to enhance faith, to Christianize the kingdom and support the Church."

163. Eastwood and Graßhoff, *Planetary Diagrams*, 27–29.
164. Kühnel, "Carolingian Diagrams," 359–89.
165. Ibid., 375. Also see Kühnel, *End of Time*, 116–59, esp. 139–42.
166. Isidore, *Etymologies*, trans. Barney et al., 99; Isidore, *Etymologiarum*, 1:III.xxiii.2.
167. Fontaine, *Isidore de Séville: Genèse et originalité*, 301–8.
168. Ibid., 301. Isidore, *Etymologies*, trans. Barney et al., 232; Isidore, *Etymologiarum*, 2:XI.i.16.
169. Fontaine, *Isidore de Séville: Genèse et originalité*, 297; Isidore, *De natura rerum*, 24–25.
170. Bierbrauer, *Karolingischen Handschriften*, 1:71–72, cat. 133. Also see Stiegemann and Wemhoff, 799—*Kunst und Kultur*, 714–15.
171. Kühnel, *End of Time*, 133–34.
172. Isidore, *Etymologies*, trans. Barney et al., 90; Isidore, *Etymologiarum*, 1:III.iv.1, 3–4. Also see Wallis, "Images of Order," 45–68.
173. Wallis, "Images of Order," 45–53. For a brief introduction to the Greek roots of these ideas in the philosophies of Pythagoras and Plato, see Matson, *New History of Philosophy*, 1:18–21, 87–97.
174. Wallis, "Images of Order," 50–62.

BIBLIOGRAPHY

Abel, O., J. von Jasmund, and Reinhold Rau, eds. *Quellen zur karolingischen Reichsgeschichte.* Vol. 1, *Die Reichsannalen, Einhard Leben Karls des Grossen, zwei "Leben" Ludwigs, Nithard Geschichten.* 2nd ed. Darmstadt: Wissenschaftliche Buchgesellschaft, 1977.

Alberi, Mary. "The 'Mystery of the Incarnation' and Wisdom's House (Prov. 9:1) in Alcuin's *Disputatio de vera Philosophia.*" *Journal of Theological Studies,* n.s., 48 (1997): 505–16.

Anderson, Benjamin W. "World Image After World Empire: The Ptolemaic Cosmos in the Early Middle Ages, ca. 700–900." Ph.D. diss., Bryn Mawr College, 2012.

Appleby, David. "Instruction and Inspiration Through Images in the Carolingian Period." In *Word, Image, Number: Communication in the Middle Ages,* edited by John J. Contreni and Santa Casciani, 85–111. Micrologus 8. Florence: SISMEL, 2002.

Aratus. *Phaenomena.* Edited and translated by Douglas Kidd. Cambridge Classical Texts and Commentaries 34. Cambridge: Cambridge University Press, 2004.

Augustine. *Concerning the City of God Against the Pagans.* Translated by Henry Bettenson. New York: Penguin, 1984.

———. *Contra academicos, De beata vita, De ordine, De magistro, De libero arbitrio.* Edited by W. M. Green and K. D. Daur. Part II.2, *Aurelii Augustini Opera.* Corpus Christianorum Series Latina 29. Turnhout: Brepols, 1970.

———. *De diversis quaestionibus octoginta tribus, De octo dulcitii quaestionibus.* Part 13.2. Edited by Almut Mutzenbecher. Corpus Christianorum Series Latina 44 A. Turnhout: Brepols, 1975.

———. *Divine Providence and the Problem of Evil: A Translation of St. Augustine's "De Ordine."* Translated and edited by Robert P. Russell. New York: Cosmopolitan Science and Art Service, 1942.

———. *Eighty-Three Different Questions.* Translated by David L. Mosher. The Fathers of the Church 70. Washington, D.C.: Catholic University of America Press, 1982.

———. *Fathers of the Church.* Vol. 2, *Christian Instruction; Admonition and Grace; The Christian Combat; Faith, Hope, and Charity.* Translated by John J. Gavigan, John Courtney Murray, Robert P. Russell, and Bernard M. Peebles. 2nd ed. Washington, D.C.: Catholic University of America Press, 1966.

———. *In Iohannis Evangelium Tractatus CXXIV.* Edited by R. Willems. Corpus Christianorum Series Latina 36. Turnhout: Brepols, 1954.

———. *The Retractions.* Translated by Sister Mary Inez Bogan. The Fathers of the Church 60.

Washington, D.C.: Catholic University of America Press, 1968.

———. *Retractionum Libri II*. Edited by Almut Mutzenbecher. Corpus Christianorum Series Latina 57. Turnhout: Brepols, 1984.

Aurigemma, Luigi. *Le signe zodiacal du Scorpion dans les traditions occidentales de l'Antiquité gréco-latine à la Renaissance*. École des hautes etudes en sciences sociales, Centre de recherches historiques: Civilisations et sociétés 54. Paris: Mouton, 1976.

Auty, Robert, et al., eds. *Lexikon des Mittelalters*. 9 vols. Munich: Artemis Verlag, 1977–98.

Barbero, Alessandro. *Charlemagne: Father of a Continent*. Translated by Allan Cameron. Berkeley: University of California Press, 2004.

Baroffio, Bonifacio Giacomo. "La Biblioteca Capitolare." In *Il Duomo di Monza*, edited by Roberto Conti, vol. 2, *I tesori*, 178–97. Rev. ed. Milan: Electa, 1990.

Barton, Tamsyn. *Ancient Astrology*. New York: Routledge, 1994.

Bastgen, Hubert, ed. *Libri Carolini sive Caroli magni capitulare de imaginibus*. Monumenta Germaniae Historica, Legum Sectio III, Concilia, Tomi II Supplementum. Hannover: Hahn, 1924.

Beaumont, Jacqueline. "The Latin Tradition of the *De Consolatione Philosophiae*." In *Boethius: His Life, Thought, and Influence*, edited by Margaret Gibson, 278–305. Oxford: Blackwell, 1981.

Beck, Roger. *A Brief History of Ancient Astrology*. Malden: Blackwell, 2007.

Beckwith, John. *Early Medieval Art: Carolingian, Ottonian, Romanesque*. New York: Thames and Hudson, 1996.

Bede. *The Reckoning of Time*. Translated by Faith Wallis. Translated Texts for Historians 29. Liverpool: Liverpool University Press, 1999.

Belloni, Annalisa, and Mirella Ferrari. *La Biblioteca Capitolare di Monza*. Medioevo e umanesimo 21. Padua: Editrice Antenore, 1974.

Belting, Hans. "Probleme der Kunstgeschichte Italiens im Frühmittelalter." *Frühmittelalterliche Studien* 1 (1967): 94–143.

Bethe, Erich. *Buch und Bild im Altertum*. Edited by Ernst Kirsten. Amsterdam: Adolf M. Hakkert, 1964.

Bierbrauer, Katharina. *Die vorkarolingischen und karolingischen Handschriften der bayerischen Staatsbibliothek*. 2 vols. Katalog der illuminierten Handschriften der bayerischen Staatsbibliothek in München 1. Wiesbaden: Ludwig Reichert, 1990.

Bischoff, Bernhard. "The Court Library Under Louis the Pious." In Bischoff, *Manuscripts and Libraries in the Age of Charlemagne*, edited and translated by Michael Gorman, 76–92. Cambridge Studies in Palaeography and Codicology 1. Cambridge: Cambridge University Press, 1994.

———. *Latin Palaeography: Antiquity and the Middle Ages*. Translated by Dáibhí Ó Cróinín and David Ganz. Cambridge: Cambridge University Press, 1990.

———. "Paläographische Fragen deutscher Denkmäler der Karolingerzeit." In *Mittelalterliche Studien: Ausgewählte Aufsätze zur Schriftkunde und Literaturgeschichte*, 3:73–111. Stuttgart: Anton Hiersemann, 1981.

Bischoff, Bernhard, Bruce Eastwood, Thomas Klein, Florentine Mütherich, and Pieter F. J. Obbema, eds. *Aratea: Kommentar zum Aratus des Germanicus MS. Voss. Lat. Q. 79 Bibliotheek der Rijksuniversiteit Leiden*. Lucerne: Faksimile Verlag, 1989.

Blume, Dieter. *Regenten des Himmels: Astrologische Bilder in Mittelalter und Renaissance*. Studien aus dem Warburg-Haus 3. Berlin: Akademie Verlag, 2000.

———. "Sichtbares Bild und sichtbare Ordnung: Zur Funktion mittelalterlicher Sternbilderdarstellungen." In *Ästhetik des Unsichtbaren: Bildtheorie und Bildgebrauch in der Vormoderne*, edited by David Ganz, Thomas Lentes, and Georg Henkel, 292–311. KultBild: Visualität und Religion in der Vormoderne 1. Berlin: Reimer, 2004.

Blume, Dieter, Mechthild Haffner, and Wolfgang Metzger. *Sternbilder des Mittelalters: Der gemalte Himmel zwischen Wissenschaft und Phantasie*. 2 vols. Berlin: Akademie Verlag, 2012.

Bober, Harry. "An Illustrated Medieval School-book of Bede's 'De natura rerum.'" *Journal of the Walters Art Gallery* 19/20 (1956–57): 65–97.

Boll, Franz. *Sphaera: Neue griechische Texte und Untersuchungen zur Geschichte der Sternbilder*. 1903. Hildesheim: Georg Olms, 1967.

———. *Sternglaube und Sterndeutung: Die Geschichte und das Wesen der Astrologie.* Edited by W. Gundel. 3rd ed. Leipzig: B. G. Teubner, 1926.

Bonde, Sheila, ed. *Survival of the Gods: Classical Mythology in Medieval Art.* Providence: Department of Art, Brown University, 1987.

Bordona, Jesús Domínguez. *Manuscritos con pinturas.* Vol. 1. Madrid: Centro de Estudios Históricos, 1933.

Boretius, Alfred, ed. *Capitularia regum francorum.* Monumenta Germaniae Historica I, Legum Sectio II. Hannover: Hahn, 1883.

Borst, Arno. "Die Aachener Enzyklopädie von 809 (Lib. comp.)." In Borst, *Schriften zur Komputistik,* 3:1054–334.

———. "Das Aachener Verhör von 809 (Cap. comp.)." In Borst, *Schriften zur Komputistik,* 3:1034–53.

———. "Alkuin und die Enzyklopädie von 809." In Butzer and Lohrmann, *Science in Western and Eastern Civilization,* 53–78.

———. *Das Buch der Naturgeschichte: Plinius und seine Leser im Zeitalter des Pergaments.* Heidelberg: Carl Winter, 1994.

———. "Computus: Zeit und Zahl im Mittelalter." *Deutsches Archiv für Erforschung des Mittelalters* 44 (1988): 1–82.

———. *Die karolingische Kalenderreform.* Monumenta Germaniae Historica Schriften 46. Hannover: Hahnsche Buchhandlung, 1998.

———. "Die Salzburger Enzyklopädie von 818 (Lib. calc.)." In Borst, *Schriften zur Komputistik,* 3:1367–451.

———, ed. *Schriften zur Komputistik im Frankenreich von 721 bis 818.* 3 vols. Monumenta Germaniae Historica, Quellen zur Geistesgeschichte des Mittelalters 21. Hannover: Hahnsche Buchhandlung, 2006.

———. *Der Streit um den karolingischen Kalender.* Monumenta Germaniae Historica Studien und Texte 36. Hannover: Hahnsche Buchhandlung, 2004.

Boschen, Lothar. *Die Annales Prumienses: Ihre nähere und ihre weitere Verwandtschaft.* Düsseldorf: Schwann, 1972.

Brown, Giles. "Introduction: The Carolingian Renaissance." In McKitterick, *Carolingian Culture,* 1–51.

Brown, Michelle P. *The Lindisfarne Gospels: Society, Spirituality, and the Scribe.* Toronto: University of Toronto Press, 2003.

Brown, Peter. *Authority and the Sacred: Aspects of the Christianisation of the Roman World.* Cambridge: Cambridge University Press, 1997.

Brubaker, Leslie, and John Haldon. *Byzantium in the Iconoclast Era, c. 680–850: A History.* Cambridge: Cambridge University Press, 2011.

Bruce-Mitford, R. L. S. "The Art of the Codex Amiatinus." *Journal of the British Archaeological Association,* 3rd ser., 32 (1969): 1–25.

Büchler, J. Lambert, and Carl Georg Dümge, eds. *Archiv der Gesellschaft für altere deutsche Geschichtskunde.* Vol. 3. 1821–22. Hannover: Hahnsche Buchhandlung, 1977.

Bullough, Donald A. *Alcuin: Achievement and Reputation, Being Part of the Ford Lectures Delivered in Oxford in Hilary Term, 1980.* Education and Society in the Middle Ages and Renaissance 16. Leiden: Brill, 2004.

Burckhardt, Max. *Ausgewählte Miniaturen aus mittelalterlichen Handschriften der Universitätsbibliothek Basel.* Basel: Helbing & Lichtenhahn, 1969.

Butzer, Paul Leo, and Dietrich Lohrmann, eds. *Science in Western and Eastern Civilization in Carolingian Times.* Basel: Birkhäuser, 1993.

Bynum, Caroline Walker. *The Resurrection of the Body in Western Christianity, 200–1336.* Lectures on the History of Religions Sponsored by the American Council of Learned Societies, n.s., 15. New York: Columbia University Press, 1995.

Cabaniss, Allen, trans. *Son of Charlemagne: A Contemporary Life of Louis the Pious.* Syracuse: Syracuse University Press, 1961.

Caillet, Jean-Pierre. *L'art carolingien.* Paris: Flammarion, 2005.

Calkins, Robert G. *Illuminated Books of the Middle Ages.* Ithaca: Cornell University Press, 1983.

Camille, Michael. "Word, Text, Image, and the Early Church Fathers in the Egino Codex." In *Testo e immagine nell'Alto Medioevo: 15–21 aprile 1993, Spoleto,* edited by Centro italiano di studi sull'alto medioevo, 1:65–94. Settimane di studio del Centro italiano di studi sull'alto medioevo 41. Spoleto: Presso la Sede del Centro, 1994.

Camps i Sòria, Jordi, ed. *Catalunya a l'època carolíngia: Art i cultura abans del romànic (segles IX i X)*. Barcelona: Museu Nacional d'Art de Catalunya, 1999.

Cancik, Hubert, and Helmuth Schneider, eds. *Der neue Pauly: Enzyklopädie der Antike*. 19 vols. Stuttgart: J. B. Metzler, 1996–2003.

Caramel, L., and M. Mirabella Roberti. *L'arte dall'età romana al Rinascimento*. Vol. 1. Storia di Monza e della Brianza 4. Milan: Edizioni il Polifilo, 1976.

Carruthers, Mary. *The Craft of Thought: Meditation, Rhetoric, and the Making of Images, 400–1200*. Cambridge Studies in Medieval Literature 34. Cambridge: Cambridge University Press, 2008.

Cartelle, Enrique Montero, ed. and trans. *Códice de Metz (Biblioteca Nacional de Madrid 3.307): Una compilación medieval de computo y astronomía*. Scriptorium 5. Madrid: Testimonio Compañia Editorial, 1994.

Cassiodorus. *Institutiones*. Edited by Roger A. B. Mynors. Oxford: Clarendon Press, 1961.

———. *Institutiones divinarum et saecularium litterarum: Einführung in die geistlichen und weltlichen Wissenschaften*. Translated by Wolfgang Bürsgens. 2 vols. Fontes Christiani 39/1–2. Freiburg im Breisgau: Herder, 2003.

———. *"Institutions of Divine and Secular Learning" and "On the Soul."* Translated and edited by James W. Halporn. Translated Texts for Historians 42. Liverpool: Liverpool University Press, 2004.

Catalogue général des *manuscrits des bibliothèques publiques des départements: Publié sous les auspices du Ministre de l'instruction publique*. Vol. 1. Paris: Imprimerie nationale, 1849.

Cavallo, Guglielmo. "La biblioteca longobarda." In *Il futuro dei Longobardi: L'Italia e la costruzione dell'Europa di Carlo Magno*, edited by Carlo Bertelli and Gian Pietro Broglio, 146–61. Milan: Skira, 2000.

Chance, Jane. *Medieval Mythography: From Roman North Africa to the School of Chartres, A.D. 433–1177*. Gainesville: University Press of Florida, 1994.

Charvet, Pascal, and Arnaud Zucker, eds. and trans. *Le ciel: Mythes et histoires des constellations, Les "Catastérismes" d'Ératosthène*. Paris: NiL, 1998.

Chazelle, Celia. "Archbishops Ebo and Hincmar of Reims and the Utrecht Psalter." In *Approaches to Early-Medieval Art*, edited by Lawrence Nees, 97–119. Cambridge, Mass.: Medieval Academy of America, 1998.

———. *The Crucified God in the Carolingian Era: Theology and the Art of Christ's Passion*. Cambridge: Cambridge University Press, 2001.

———. "Matter, Spirit, and Image in the *Libri Carolini*." *Recherches Augustiniennes* 21 (1986): 163–84.

Clemens, Raymond, and Timothy Graham. *Introduction to Manuscript Studies*. Ithaca: Cornell University Press, 2007.

Conant, Kenneth John. *Carolingian and Romanesque Architecture, 800–1200*. New Haven: Yale University Press, 1959.

Contreni, John J. *Carolingian Learning, Masters, and Manuscripts*. Brookfield: Ashgate, 1992.

———. "The Carolingian Renaissance: Education and Literary Culture." In McKitterick, *New Cambridge Medieval History*, vol. 2, *c. 700–c. 900*, 709–57.

———. "The Carolingian School: Letters from the Classroom." In *Carolingian Learning, Masters, and Manuscripts*, 81–111.

———. *The Cathedral School of Laon from 850 to 930: Its Manuscripts and Masters*. Münchener Beiträge zur Mediävistik und Renaissance-Forschung 29. Munich: Arbeo-Gesellschaft, 1978.

———. "John Scottus, Martin Hiberniensis, the Liberal Arts, and Teaching." In *Carolingian Learning, Masters, and Manuscripts*, 1–22.

———. "Three Carolingian Texts Attributed to Laon: Reconsiderations." *Studi Medievali*, 3rd ser., 17, no. 2 (1976): 797–813.

Coon, Lynda L. *Dark Age Bodies: Gender and Monastic Practice in the Early Medieval West*. Philadelphia: University of Pennsylvania Press, 2011.

Cordoliani, A. "Les traités de comput du haut Moyen Âge (526–1003)." *Bulletin du Cange* 17 (1943): 51–72.

Corradini, Richard. "The Rhetoric of Crisis: *Computus* and *Liber Annalis* in Early Ninth-Century Fulda." In *The Construction of Communities in the Early Middle Ages: Texts, Resources, and Artefacts*, edited by Richard Corradini, Max Diesenberger, and Helmut Reimitz, 269–321. The Transformation of the Roman World 12. Leiden: Brill, 2003.

Crivello, Fabrizio. *Le "Omelie sui Vangeli" di Gregorio Magno a Vercelli: La miniature del ms. CXLVIII/8 della*

Biblioteca Capitolare. Archivum Gregorianum 6. Florence: SISMEL, 2005.

Crocker, Richard L. "Carolingian Chant: Roman, Frankish-Roman, Frankish." In *"The Gentle Voices of Teachers": Aspects of Learning in the Carolingian Age*, edited by Richard Sullivan, 142–70. Columbus: Ohio State University Press, 1995.

Crosby, Sumner McKnight. *The Royal Abbey of Saint-Denis from Its Beginnings to the Death of Suger, 475–1151*. Edited by Pamela Z. Blum. New Haven: Yale University Press, 1987.

Davis-Weyer, Caecilia, ed. *Early Medieval Art, 300–1150: Sources and Documents*. Medieval Academy Reprints for Teaching 17. Buffalo: University of Toronto Press, 2003.

Declercq, Georges. *Anno Domini: The Origins of the Christian Era*. Turnhout: Brepols, 2000.

De Jong, Mayke. *The Penitential State: Authority and Atonement in the Age of Louis the Pious, 814–840*. Cambridge: Cambridge University Press, 2009.

Dekker, Elly. "Carolingian Planetary Observations: The Case of the Leiden Planetary Configuration." *Journal for the History of Astronomy* 39, no. 1 (2008): 77–90.

———. *Illustrating the Phaenomena: Celestial Cartography in Antiquity and the Middle Ages*. Oxford: Oxford University Press, 2013.

———. "The Provenance of the Stars in the Leiden *Aratea* Picture Book." *JWCI* 63 (2010–11): 1–37.

Delisle, Léopold. *Inventaire des manuscrits latins conservés à la Bibliothèque nationale sous les numéros 8823–18613, et faisant suite à la série dont le catalogue a été publié en 1744*. Paris: Auguste Durand et Pedone-Lauriel, 1863–71.

Dell'Era, Antonio. "Una rielaborazione dell'Arato latino." *Studi Medievali*, 3rd ser., 20 (1979): 268–301.

Delogu, Paolo. "Lombard and Carolingian Italy." In McKitterick, *New Cambridge Medieval History*, vol. 2, *c. 700–c. 900*, 290–319.

Desportes, Pierre. "La capitale cléricale du haut Moyen Âge (Ve–XIe siècles)." In *Histoire de Reims*, edited by Pierre Desportes, 55–91. Toulouse: Privat, 1983.

Diebold, William. "The Carolingian Idol: Exegetes and Idols." In *Seeing the Invisible in Late Antiquity and the Early Middle Ages: Papers from "Verbal and Pictorial Imaging: Representing and Accessing Experience of the Invisible, 400–1000" (Utrecht, 11–13 December 2003)*, edited by Giselle de Nie, Karl F. Morrison, and Marco Mostert, 445–61. Utrecht Studies in Medieval Literacy 14. Turnhout: Brepols, 2005.

———. *Word and Image: An Introduction to Early Medieval Art*. Boulder: Westview, 2000.

Diem, Albrecht. "The Emergence of Monastic Schools: The Role of Alcuin." In *Alcuin of York: Scholar at the Carolingian Court; Proceedings of the Third Germania Latina Conference Held at the University of Groningen, May 1995*, edited by L. A. J. R. Houwen and A. A. MacDonald, 27–44. Germania Latina 3. Groningen: Egbert Forsten, 1998.

Dodwell, C. R. *Painting in Europe, 800 to 1200*. Baltimore: Penguin, 1971.

Dolan, Marion. "The Role of Illustrated *Aratea* Manuscripts in the Transmission of Astronomical Knowledge in the Middle Ages." Ph.D. diss., University of Pittsburgh, 2007.

Dolezal, Mary-Lyon. "The Elusive Quest for the 'Real Thing': The Chicago Lectionary Project Thirty Years On." *Gesta* 35 (1996): 128–41.

Duckett, Eleanor Shipley. *Carolingian Portraits: A Study in the Ninth Century*. Ann Arbor: University of Michigan Press, 1962.

Dümmler, Ernest, ed. *Epistolae karolini aevi*. Monumenta Germaniae Historica, Epistolarum IV, Karolini Aevi II. Berlin: Weidmann, 1895.

———. *Poetae latini aevi carolini*. Monumenta Germaniae Historica, Poetarum Latinorum Medii Aevi. 2 vols. Berlin: Weidmann, 1881–84.

Dutton, Paul Edward, ed. *Carolingian Civilization: A Reader*. Orchard Park, N.Y.: Broadview, 1999.

———, ed. and trans. *Charlemagne's Courtier: The Complete Einhard*. Readings in Medieval Civilizations and Cultures 3. Orchard Park, N.Y.: Broadview, 1998.

———. "Of Carolingian Kings and Their Stars." In *Charlemagne's Mustache and Other Cultural Clusters of a Dark Age*, edited by Paul Edward Dutton, 93–127. New York: Palgrave, 2004.

Eastwood, Bruce Stansfield. "Astronomy in Christian Latin Europe, c. 500–c. 1150." In Eastwood, *Revival of Planetary Astronomy*, 235–58.

———. "Johannes Scottus Eriugena, Sun-Centered Planets, and Carolingian Astronomy." *Journal for the History of Astronomy* 32 (2001): 281–324.

———. *Ordering the Heavens: Roman Astronomy and Cosmology in the Carolingian Renaissance.* History of Science and Medicine Library 4, Medieval and Early Modern Science 8. Leiden: Brill, 2007.

———. "Origins and Contents of the Leiden Planetary Configuration (MS Voss. Q. 79, f. 93v): An Artistic Astronomical Schema of the Early Middle Ages." In Eastwood, *Revival of Planetary Astronomy,* 1–47.

———. "Plinian Astronomical Diagrams in the Early Middle Ages." In Eastwood, *Revival of Planetary Astronomy,* 141–72.

———. "Plinian Astronomy in the Middle Ages and Renaissance." In Eastwood, *Revival of Planetary Astronomy,* 197–251.

———, ed. *The Revival of Planetary Astronomy in Carolingian and Post-Carolingian Europe.* Burlington: Ashgate, 2002.

Eastwood, Bruce S., and Gerd Graßhoff. *Planetary Diagrams—Descriptions, Models, Theories: From Carolingian Deployments to Copernican Debates.* Max-Planck-Institut für Wissenschaftsgeschichte Preprint 132. Berlin: Max-Planck-Institut für Wissenschaftsgeschichte, 2000.

———. *Planetary Diagrams for Roman Astronomy in Medieval Europe, ca. 800–1500.* Transactions of the American Philosophical Society 94, no. 3. Philadelphia: American Philosophical Society, 2004.

Edgar, Swift, ed. *The Vulgate Bible: Douay-Rheims Translation.* 6 vols. Dumbarton Oaks Medieval Library 1, 4–5, 8, 13, 17, 21. Cambridge, Mass.: Harvard University Press, 2010–13.

Englisch, Brigitte. *Die Artes Liberales im frühen Mittelalter (5.–9. JH.): Das Quadrivium und der Komputus als Indikatoren für Kontinuität und Erneuerung der Exakten Wissenschaften zwischen Antike und Mittelalter.* Sudhoffs Archiv 33. Stuttgart: Franz Steiner, 1994.

———. *Zeiterfassung und Kalenderprogrammatik in der frühen Karolingerzeit: Das Kalendarium der Hs. Köln DB 83–2 und die Synode von Soissons 744.* Instrumenta 8. Stuttgart: Jan Thorbecke, 2002.

Evans, James. *The History and Practice of Ancient Astronomy.* New York: Oxford University Press, 1998.

Ferrari, Mirella. "Centri di trasmissione: Monza, Pavia, Milano, Bobbio." In *La cultura antica nell'occidente latino dal VII all'XI secolo (18–24 aprile 1974),* edited by Centro italiano di studi sull'alto medioevo, 303–20. Settimane di studio del Centro italiano di studi sull'alto medioevo 22. Spoleto: Presso la Sede del Centro, 1975.

Fleckenstein, Josef. "Alcuin im Kreis der Hofgelehrten Karls des Grossen." In Butzer and Lohrmann, *Science in Western and Eastern Civilization,* 3–21.

———. *Die Hofkapelle der deutschen Könige.* 2 vols. Schriften der Monumenta Germaniae Historica 16. Stuttgart: Anton Hiersemann, 1966.

Flint, Valerie I. J. *The Rise of Magic in Early Medieval Europe.* Princeton: Princeton University Press, 1991.

Fontaine, Jacques. *Isidore de Séville et la culture classique dans l'Espagne wisigothique.* 2nd ed. 3 vols. Paris: Études augustiniennes, 1983.

———. "Isidore de Séville et l'astrologie." *Revue des Études Latines* 31 (1953–54): 271–300.

———. *Isidore de Séville: Genèse et originalité de la culture hispanique au temps des Wisigoths.* Turnhout: Brepols, 2000.

Forster, Kurt. "Introduction." In Aby Warburg, *The Renewal of Pagan Antiquity: Contributions to the Cultural History of the European Renaissance,* edited by Kurt W. Forster, translated by David Britt, 1–75. Los Angeles: Getty, 1999.

Foucault, Michel. *Madness and Civilization: A History of Insanity in the Age of Reason.* Translated by Richard Howard. New York: Vintage, 1988.

Freeman, Ann. "*Opus Caroli Regis Contra Synodum:* An Introduction." In Freeman, *Theodulf of Orléans,* 1–123.

———, ed. *Opus Caroli Regis Contra Synodum (Libri Carolini).* Monumenta Germaniae Historica, Concilia, Tomus II, Supplementum I. Hannover: Hahnsche Buchhandlung, 1998.

———. "Scripture and Images in the Libri Carolini." In Freeman, *Theodulf of Orléans,* 163–88.

———. *Theodulf of Orléans: Charlemagne's Spokesman Against the Second Council of Nicaea.* Burlington: Ashgate, 2003.

Fried, Johannes. "The Frankish Kingdoms, 817–911: The East and Middle Kingdoms." In McKitterick, *New Cambridge Medieval History,* vol. 2, *c.* 700–*c.* 900, 142–68.

———. "In den Netzen der Wissensgesellschaft: Das Beispiel des mittelalterlichen Königs- und Fürstenhofes." In *Wissenskulturen: Beiträge zu einem forschungsstrategischen Konzept,* edited by Johannes Fried and Thomas Kailer, 141–93. Berlin: Akademie Verlag, 2003.

———. "Karl der Große, die Artes liberales und die karolingische Renaissance." In *Karl der Grosse und sein Nachwirken: 1200 Jahre Kultur und Wissenschaft in Europa / Charlemagne and His Heritage: 1200 Years of Civilization and Science in Europe,* vol. 1, *Wissen und Weltbild / Scholarship, Worldview, and Understanding,* edited by P. Butzer, M. Kerner, and W. Oberschelp, 25–43. Turnhout: Brepols, 1997.

Gamble, Harry Y. "Bible and Book." In *In the Beginning: Bibles Before the Year 1000,* edited by Michelle Brown, 15–35. Washington, D.C.: Smithsonian Institution, 2006.

Gameson, Richard, ed. *The Early Medieval Bible: Its Production, Decoration, and Use.* Cambridge: Cambridge University Press, 1994.

Ganz, David. *Corbie in the Carolingian Renaissance.* Beihefte der Francia 20. Sigmaringen: Jan Thorbecke, 1990.

———. "'Pando quod Ignoro': In Search of Carolingian Artistic Experience." In *Intellectual Life in the Middle Ages: Essays Presented to Margaret Gibson,* edited by Lesley Smith and Benedicta Ward, 25–32. London: Hambledon Press, 1992.

———. "Theology and the Organisation of Thought." In McKitterick, *New Cambridge Medieval History,* vol. 2, *c.* 700–*c.* 900, 758–85.

Garrison, Mary. "The Emergence of Carolingian Latin Literature and the Court of Charlemagne (780–814)." In McKitterick, *Carolingian Culture,* 111–40.

Gombrich, E. H. *Aby Warburg: An Intellectual Biography.* London: Warburg Institute, University of London, 1970.

Grant, Edward. *The Foundations of Modern Science in the Middle Ages: Their Religious, Institutional, and Intellectual Contexts.* Cambridge: Cambridge University Press, 1996.

Grat, Félix, Jeanne Vielliard, and Suzanne Clémencet, eds. *Annales de Saint-Bertin.* Paris: Klincksieck, 1964.

Gregory of Tours. *Miracula et opera minora.* Edited by Bruno Krusch. Monumenta Germaniae Historica, Scriptorum rerum Merovingicarum I, part II. 1885. Hannover: Hahnsche Buchhandlung, 1969.

Gundel, Hans Georg. *Zodiakos: Tierkreisbilder im Altertum, Kosmische Bezüge und Jenseitsvorstellungen im antiken Alltagsleben.* Kulturgeschichte der antiken Welt 54. Mainz: Philipp von Zabern, 1992.

Gutting, Gary. *Michel Foucault's Archaeology of Scientific Reason.* Cambridge: Cambridge University Press, 1995.

Haarländer, Stephanie, ed. and trans. *Rabanus Maurus zum Kennenlernen: Ein Lesebuch mit einer Einführung in sein Leben und Werk.* Mainz: Bistum Mainz, 2006.

Haffner, Mechthild. *Ein antiker Sternbilderzyklus und seine Tradierung in Handschriften vom frühen Mittelalter bis zum Humanismus: Untersuchungen zu den Illustrationen der "Aratea" des Germanicus.* Studien zur Kunstgeschichte 114. Hildesheim: Georg Olms, 1997.

Hamilton, Edith. *Mythology: Timeless Tales of Gods and Heroes.* New York: Grand Central, 1999.

Hanfmann, George M. A. "The Continuity of Classical Art: Culture, Myth, and Faith." In *Age of Spirituality: A Symposium,* edited by Kurt Weitzmann, 75–99. New York: Metropolitan Museum of Art, 1980.

Hasper, L. W. *Hyginus, philosophus: De imaginibus coeli.* Leipzig, 1861.

Hellmann, S. *Sedulius Scottus.* Quellen und Untersuchungen zur lateinischen Philologie des Mittelalters 1.1. 1906. Frankfurt: Minerva, 1966.

Henderson, George. "Emulation and Invention in Carolingian Art." In McKitterick, *Carolingian Culture,* 248–73.

Henderson, John. *The Medieval World of Isidore of Seville: Truth from Words.* Cambridge: Cambridge University Press, 2007.

Henisch, Bridget Ann. *The Medieval Calendar Year.* University Park: Pennsylvania State University Press, 2002.

Henkel, Nikolaus. *Studien zum Physiologus im Mittelalter.* Hermaea: Germanistische Forschungen, n.s., 38. Tübingen: Max Niemeyer, 1976.

Hermann, Hermann Julius. *Die frühmittelalterlichen Handschriften des Abendlandes.* Die Illuminierten Handschriften und Inkunabeln der Nationalbibliothek in Wien 1. Leipzig: Karl W. Hiersemann, 1923.

Heyse, Elisabeth. *Hrabanus Maurus's Enzyklopädie "De rerum naturis": Untersuchungen zu den Quellen und zur Methode der Kompilation.* Münchener Beiträge zur Mediävistik und Renaissance-Forschung 4. Munich: Arbeo-Gesellschaft, 1969.

Higham, N. J. *The Kingdom of Northumbria, a.d. 350–1100.* Dover: Alan Sutton, 1993.

Himmelmann, Nikolaus. *Antike Götter im Mittelalter.* Trierer Winckelmannsprogramme 7. Mainz: Philipp von Zabern, 1986.

Hoffman, Peter, Joachim Hupe, and Karin Goethert. *Katalog der römischen Mosaike aus Trier und dem Umland.* Trierer Grabungen und Forschungen 16. Trier: Selbstverlag des Rheinischen Landesmuseums, 1999.

Höhl, Claudia. *Ottonische Buchmalerei in Prüm.* Europäische Hochschulschriften ser. 28, vol. 252. Frankfurt am Main: Peter Lang, 1996.

Holter, Kurt. "Der Buchschmuck in Süddeutschland und Oberitalien." In *Karl der Grosse: Lebenswerk und Nachleben,* vol. 3, *Karolingische Kunst,* edited by Wolfgang Braunfels and Hermann Schnitzler, 74–114. Düsseldorf: Schwann, 1966.

Holtz, Louis. "L'œuvre grammaticale d'Alcuin dans le contexte de son temps." In Tremp and Schmuki, *Alkuin von York,* 129–49.

Homburger, Otto. *Die illustrierten Handschriften der Burgerbibliothek Bern: Die vorkarolingischen und karolingischen Handschriften.* Bern: Burgerbibliothek Bern, 1962.

Horden, Peregrine, and Nicholas Purcell. *The Corrupting Sea: A Study of Mediterranean History.* Malden: Blackwell, 2004.

Hourihane, Colum, ed. *Time in the Medieval World: Occupations of the Months and Signs of the Zodiac in the Index of Christian Art.* Princeton: Index of Christian Art, 2007.

Hrabanus Maurus (Rabanus Maurus). "Beati Rabani Mauri Fuldensis Abbatis et Moguntini Archiepiscopi *De Universo* Libri Viginti Duo." *Patrologia Latina,* edited by Jacques-Paul Migne, vol. 111. Patrologia Latina Database, 1996, Chadwyck-Healey Inc., Alexandria, Va. http://pld.chadwyck.com/all/fulltext?action=byid&warn=N&id=Z600171238&div=6&sequence=5&file=../session/1372522678_14079 (accessed June 29, 2013).

———. *De institutione clericorum libri tres.* Edited by Detlev Zimpel. Freiburger Beiträge zur mittelalterlichen Geschichte: Studien und Texte 7. Frankfurt am Main: Peter Lang, 1996.

———. *Expositio in Matthaeum.* Edited by B. Löfstedt. Corpus Christianorum Continuatio Mediaevalis 174 A. Turnhout: Brepols, 2000.

———. *La formazione dei chierici.* Translated and edited by Luigi Samarati. Rome: Città Nuova Editrice, 2002.

———. *Martyrologium.* Edited by John McCulloh. Corpus Christianorum Continuatio Mediaevalis 44. Turnhout: Brepols, 1979.

Hübner, Wolfgang. *Die Begriffe "Astrologie" und "Astronomie" in der Antike: Wortgeschichte und Wissenschaftssystematik, mit einer Hypothese zum Terminus "Quadrivium."* Abhandlungen der geistes- und sozialwissenschaftlichen Klasse 7. Mainz: Akademie der Wissenschaften und der Literatur, 1989.

———. *Zodiacus Christianus: Jüdisch–christliche Adaptionen des Tierkreises von der Antike bis zur Gegenwart.* Beiträge zur klassischen Philologie 144. Königstein im Taunus: Anton Hain Meisenheim, 1983.

Inventario general de manuscritos de la Biblioteca Nacional. Vol. 10. Madrid: Ministerio de Cultura, 1984.

Isidore (of Seville). *De natura rerum liber.* Edited by Gustav Becker. 1857. Amsterdam: Adolf M. Hakkert, 1967.

——— (Isidoro de Sevilla). *Etimologías Libro XIX de Naves, Edificios, y Vestidos.* Translated and edited by Miguel Rodríguez-Pantoja. Paris: Les Belles Lettres, 1995.

———. *Etymologiarum sive Originum Libri XX.* Edited by W. M. Lindsay. 2 vols. 1911. Oxford: Clarendon Press, 1989–90.

———. *The "Etymologies" of Isidore of Seville.* Translated by Stephen A. Barney, W. J. Lewis, J. A. Beach, and Oliver Berghof. Cambridge: Cambridge University Press, 2006.

———. *La natura delle cose.* Translated by Francesco Trisoglio. Rome: Città Nuova Editrice, 2001.

Janini, José, and José Serrano. *Manuscritos litúrgicos de la Biblioteca Nacional.* Madrid: Dirección General de Archivos y Bibliotecas, 1969.

Jansen, Michael, and Frank Pohle, eds. *Die Künste am Hofe Karls des Grossen: Artes liberales et artes mechanicae.* Aachen: Verlag Mainz, 2000.

Jeauneau, Édouard. *Études érigéniennes.* Paris: Études augustiniennes, 1987.

Jones, A. H. M. *The Later Roman Empire, 284–602: A Social, Economic, and Administrative Survey.* 2 vols. Oxford: Blackwell, 1973.

Jones, Charles W. "An Early Medieval Licensing Examination." *History of Education Quarterly* 3 (1963): 19–29.

Jones, Prudence, and Nigel Pennick. *A History of Pagan Europe.* New York: Routledge, 2005.

Kaczynski, Bernice M. *Greek in the Carolingian Age: The St. Gall Manuscripts.* Speculum Anniversary Monographs 13. Cambridge, Mass.: Medieval Academy of America, 1988.

Kandinsky, Wassily. *Concerning the Spiritual in Art.* Edited and translated by M. T. H. Sadler. 1911. New York: Dover, 1977.

Katzenstein, Ranee, and Emilie Savage-Smith. *The Leiden Aratea: Ancient Constellations in a Medieval Manuscript.* Malibu: The J. Paul Getty Museum, 1988.

Kaufmann, Georgius. *De Hygini Memoria Scholiis in Ciceronis Aratum Harleianis Servata.* Breslauer Philologische Abhandlungen 3. Breslau: Wilhelm Koebner, 1888.

Kessler, Herbert L. *Seeing Medieval Art.* Rethinking the Middle Ages 1. Peterborough: Broadview, 2004.

King, P. D., trans. *Charlemagne: Translated Sources.* Lancaster: University of Lancaster Central Print Unit, 1987.

King, Vernon. "An Investigation of Some Astronomical Excerpts From Pliny's Natural History Found in Manuscripts of the Earlier Middle Ages." Undergraduate thesis, St. John's College, 1969.

Kirchner, Joachim. *Beschreibendes Verzeichnis der Miniaturen und des Initialschmuckes in den Phillipps-Handschriften.* Vol. 1. Leipzig: Verlagsbuchhandlung J. J. Weber, 1926.

Koehler, Wilhelm. *Buchmalerei des frühen Mittelalters: Fragmente und Entwürfe aus dem Nachlaß.* Edited by Ernst Kitzinger and Florentine Mütherich. Veröffentlichungen des Zentralinstituts für Kunstgeschichte 5. Munich: Prestel, 1972.

———. *Manuscrit latin 9428, Bibliothèque nationale Paris.* Codices Selecti 49. Graz: Akademische Druck- u. Verlagsanstalt, 1974.

Koehler, Wilhelm, and Florentine Mütherich, eds. *Die karolingischen Miniaturen.* 8 vols. Berlin: Deutscher Verein für Kunstwissenschaft, 1930–2013.

Kühnel, Bianca. "Carolingian Diagrams, Images of the Invisible." In *Seeing the Invisible in Late Antiquity and the Early Middle Ages: Papers from "Verbal and Pictorial Imaging: Representing and Accessing Experience of the Invisible, 400–1000" (Utrecht, 11–13 December 2003),* edited by Giselle de Nie, Karl F. Morrison, and Marco Mostert, 359–89. Utrecht Studies in Medieval Literacy 14. Turnhout: Brepols, 2005.

———. *The End of Time in the Order of Things: Science and Eschatology in Early Medieval Art.* Regensburg: Schnell & Steiner, 2003.

Kuhnen, Hans-Peter, ed. *Religio Romana: Wege zu den Göttern im antiken Trier.* Trier: Rheinisches Landesmuseum, 1996.

Künzl, Ernst. "Der Globus im Römisch-Germanischen Zentralmuseum Mainz: Der bisher einzige komplette Himmelsglobus aus dem griechisch-römischen Altertum / The Globe in the 'Römisch-Germanische Zentralmuseum Mainz': The Only Complete Celestial Globe Found To-Date [*sic*] from Classical Greco-Roman Antiquity." *Der Globusfreund* 45/46 (1998): 7–153.

Kurze, F., ed. *Annales regni francorum.* Monumenta Germaniae Historica, Scriptores rerum Germanicarum in usum scholarum separatim editi 6. Hannover: Hahn, 1895.

Laffitte, Marie-Pierre, and Charlotte Denoël, eds. *Trésors carolingiens: Livres manuscrits de Charlemagne à Charles le Chauve.* Paris: Bibliothèque nationale de France, 2007.

Law, Vivien. "The Study of Grammar." In McKitterick, *Carolingian Culture,* 88–110.

Le Berrurier, Diane O. *The Pictorial Sources of Mythological and Scientific Illustrations in Hrabanus Maurus' "De rerum naturis."* New York: Garland, 1978.

Le Bœuffle, André. *Astronomie / Astrologie: Lexique latin.* Paris: Picard, 1987.

Le Bourdellès, Hubert. *L'Aratus Latinus: Étude sur la culture et la langue latines dans le Nord de la France au VIIIe siècle.* Lille: Presses de l'Université de Lille III, 1985.

———. "De Astronomia more Christiano." *Studi Medievali,* 3rd ser., 32 (1991): 385–444.

Lejbowicz, Max. "Computus: Le nombre et le temps altimédiévaux." In *Le temps, sa mesure et sa perception au Moyen-Âge: Actes du colloque, Orléans 12–13 avril 1991,* edited by Bernard Ribémont, 151–96. Caen: Paradigme, 1992.

———. "Théorie et pratique astronomiques chez Isidore de Séville." In *L'homme et son univers au Moyen Âge: Actes du septième Congrès international de philosophie médiévale (30 août–4 septembre 1982),* edited by Christian Wenin, 2:622–30. Philosophes médiévaux 27. Louvain-la-Neuve: Éditions de l'Institut supérieur de philosophie, 1986.

Lemay, Richard. "The True Place of Astrology in Medieval Science and Philosophy: Towards a Definition." In *Astrology, Science, and Society: Historical Essays,* edited by Patrick Curry, 57–73. Woodbridge: Boydell, 1987.

Le Moigne, François-Yves, ed. *Histoire de Metz.* Toulouse: Privat, 1986.

Lindberg, David C. *The Beginnings of Western Science: The European Scientific Tradition in Philosophical, Religious, and Institutional Context, Prehistory to* A.D. *1450.* 2nd ed. Chicago: University of Chicago Press, 2007.

———. "The Fate of Science in Patristic and Medieval Christendom." In *The Cambridge Companion to Science and Religion,* edited by Peter Harrison, 21–38. Cambridge: Cambridge University Press, 2010.

Lohrmann, Dietrich. "Alcuins Korrespondenz mit Karl dem Grossen über Kalender und Astronomie." In Butzer and Lohrmann, *Science in Western and Eastern Civilization,* 79–114.

Lowden, John. "The Beginnings of Biblical Illustration." In *Imaging the Early Medieval Bible,* edited by John Williams, 9–59. University Park: Pennsylvania State University Press, 2002.

Lowe, E. A., ed. *Codices Latini Antiquiores: A Palaeographical Guide to Latin Manuscripts Prior to the Ninth Century.* 12 vols. Oxford: Clarendon Press, 1959.

Lupus Servatus (de Ferrières.) *Correspondance.* Edited and translated by Léon Levillain. Vol. 1. Paris: Honoré Champion, 1927.

———. *Correspondance.* Edited and translated by Léon Levillain. Vol. 2. Paris: Société d'Édition "Les Belles Lettres," 1935.

Lutz, Cora E. "Remigius' Ideas on the Classification of the Seven Liberal Arts." *Traditio* 12 (1956): 65–86.

Maass, Ernst. *Commentariorum in Aratum Reliquiae.* 1898. Reprint, Berlin: Weidmann, 1958.

Marenbon, John. "Alcuin, the Council of Frankfort, and the Beginnings of Medieval Philosophy." In *Aristotelian Logic, Platonism, and the Context of Early Medieval Philosophy in the West,* edited by John Marenbon, 603–15. Burlington: Ashgate, 2000.

———. "Carolingian Thought." In McKitterick, *Carolingian Culture,* 171–92.

———. *From the Circle of Alcuin to the School of Auxerre: Logic, Theology, and Philosophy in the Early Middle Ages.* Cambridge Studies in Medieval Life and Thought, 3rd ser., 15. Cambridge: Cambridge University Press, 1981.

Mariana, Manuel Sánchez. *Códice de Metz: Tratado de computo y astronomía.* Scriptorium 2. Madrid: Testimonio Compañia Editorial, 1993.

Marrou, Henri Irénée. "Les arts libéraux dans l'antiquité classique." In *Arts libéraux et philosophie au Moyen Âge: Actes du quatrième Congrès international de philosophie médiévale, Université de Montréal, Montréal, Canada 27 août—2 septembre 1967,* edited by Étienne Gilson, 5–33. Montréal: Institut d'études médiévales, 1969.

Martianus Capella. *Martianus Capella and the Seven Liberal Arts.* Translated by William Harris Stahl and Richard Johnson. 2 vols. New York: Columbia University Press, 1971–77.

Martin, Jean. *Histoire du texte des "Phénomènes" d'Aratos.* Études et commentaires 22. Paris: Klincksieck, 1956.

Mathon, Gérard. "Les formes et la signification de la pédagogie des arts libéraux au milieu du IXe siècle: L'enseignement palatin de Jean Scot Érigène." In *Arts libéraux et philosophie au Moyen Âge: Actes du quatrième Congrès international de philosophie médiévale, Université de Montréal, Montréal, Canada 27 août—2 septembre 1967*, edited by Étienne Gilson, 47–70. Montréal: Institut d'études médiévales, 1969.

Matson, Wallace I. *A New History of Philosophy*. Vol. 1, *Ancient and Medieval*. New York: Harcourt Brace Jovanovich, 1987.

Matter, E. Ann. "Alcuin's Theology." In Tremp and Schmuki, *Alkuin von York*, 91–105.

McCluskey, Stephen C. *Astronomies and Cultures in Early Medieval Europe*. Cambridge: Cambridge University Press, 2000.

———. "Astronomies in the Latin West from the Fifth to the Ninth Centuries." In Butzer and Lohrmann, *Science in Western and Eastern Civilization*, 139–60.

———. "Gregory of Tours, Monastic Timekeeping, and Early Christian Attitudes to Astronomy." *Isis* 81 (1990): 8–22.

McCormick, Michael. "The Birth of the Codex and the Apostolic Life-Style." *Scriptorium* 39 (1985): 150–58.

McGurk, Patrick. "Carolingian Astrological Manuscripts." In *Charles the Bald: Court and Kingdom; Papers Based upon a Colloquium Held in London in April 1979*, edited by Margaret T. Gibson, Janet L. Nelson, and David Ganz, 317–32. British Archaeological Reports International 101. Oxford: British Archaeological Reports, 1981.

———. *Catalogue of Astrological and Mythological Illuminated Manuscripts of the Latin Middle Ages: Astrological Manuscripts in Italian Libraries (Other than Rome)*. Vol. 4. London: Warburg Institute, University of London, 1966.

McKitterick, Rosamond, ed. *Carolingian Culture: Emulation and Innovation*. Cambridge: Cambridge University Press, 1994.

———. *The Carolingians and the Written Word*. Cambridge: Cambridge University Press, 1995.

———. *Charlemagne: The Formation of a European Identity*. Cambridge: Cambridge University Press, 2008.

———. *History and Its Audiences: Inaugural Lecture Given by Rosamond McKitterick, Litt.D., Professor of Medieval History in the University of Cambridge, 15 May 2000*. Cambridge: Cambridge University Press, 2000.

———. *History and Memory in the Carolingian World*. Cambridge: Cambridge University Press, 2004.

———, ed. *The New Cambridge Medieval History*. Vol. 2, *c. 700–c. 900*. Cambridge: Cambridge University Press, 1995.

———. "Text and Image in the Carolingian World." In *The Uses of Literacy in Early Medieval Europe*, edited by Rosamond McKitterick, 297–318. Cambridge: Cambridge University Press, 1990.

Melzak, Robert. "Antiquarianism in the Time of Louis the Pious and Its Influence on the Art of Metz." In *Charlemagne's Heir: New Perspectives on the Reign of Louis the Pious (814–40)*, edited by Peter Godman and Roger Collins, 629–40. Oxford: Clarendon Press, 1990.

Milde, Wolfgang. *Der Bibliothekskatalog des Klosters Murbach aus dem 9. Jahrhundert: Ausgabe und Untersuchung von Beziehungen zu Cassiodors "Institutiones."* Beihefte zum Euphorion: Zeitschrift für Literaturgeschichte 4. Heidelberg: Carl Winter, 1968.

Mitchell, John. "The Display of Script and the Uses of Painting in Longobard Italy." In *Testo e immagine nell'alto medioevo: 15–21 aprile 1993*, edited by Centro italiano di studi sull'alto medioevo, 2:887–954. Settimane di studio del Centro italiano di studi sull'alto medioevo 41. Spoleto: Presso la Sede del Centro, 1994.

Mordek, Hubert, Klaus Zechiel-Eckes, and Michael Glatthaar, eds. *Die Admonitio generalis Karls des Großen*. Monumenta Germaniae Historica, Fontes Iuris Germanici Antiqui in usum scholarum separatim editi 16. Hannover: Hahnsche Buchhandlung, 2012.

Moreland, Floyd L., and Rita M. Fleischer. *An Introduction to Latin Literature and Style*. 3rd ed. New York: The Latin/Greek Institute, City University of New York, 1993.

Mostert, Richard, and Marco Mostert. "Using Astronomy as an Aid to Dating Manuscripts: The Example of the Leiden Aratea Planetarium." *Quarendo* 20, no. 4 (1990): 248–61.

Müller, Kathrin. "Irritierende Variabilität: Die mittelalterliche Reproduktion von Wissen in Diagramm." In *Übertragungen: Formen und Konzepte von Reproduktion in Mittelalter und Früher Neuzeit*, edited by Britta Bußmann, Albrecht Hausmann, Annelie Kreft, and Cornelia Logemann, 415–36. Trends in Medieval Philology 5. Berlin: Walter de Gruyter, 2005.

Munro, Dana Carleton, trans. *Laws of Charles the Great.* Translations and Reprints from the Original Sources of European History 6.5. Philadelphia: University of Pennsylvania Press, 1899. New York: AMS Press, 1971.

Murdoch, John E. *Album of Science: Antiquity and the Middle Ages.* New York: Charles Scribner's Sons, 1984.

Mütherich, Florentine. "Die Bilder." In *Aratea: Kommentar zum Aratus des Germanicus MS. Voss. Lat. Q. 79 Bibliotheek der Rijksuniversiteit Leiden*, edited by Bernhard Bischoff, Bruce Eastwood, Thomas Klein, Florentine Mütherich, and Pieter F. J. Obbema, 31–68. Lucerne: Faksimile Verlag, 1989.

———. "Book Illumination at the Court of Louis the Pious." In *Charlemagne's Heir: New Perspectives on the Reign of Louis the Pious (814–40)*, edited by Peter Godman and Roger Collins, 593–604. Oxford: Clarendon Press, 1990.

———. "Carolingian Manuscript Illumination in Rheims." In *The Utrecht Psalter in Medieval Art: Picturing the Psalms of David*, edited by Koert van der Horst, William Noel, and Wilhelmina C. M. Wüstefeld, 104–19. Westrenen: HES, 1996.

———. "Die Erneuerung der Buchmalerei am Hof Karls des Grossen." In *799, Kunst und Kultur der Karolingerzeit: Karl der Große und Papst Leo III. in Paderborn*, edited by Christoph Stiegemann and Matthias Wemhoff, 560–609. Mainz: Philipp von Zabern, 1999.

Mütherich, Florentine, and Joachim E. Gaehde. *Carolingian Painting.* New York: George Braziller, 1976.

Narkiss, Bezalel. "Reconstruction of Some of the Original Quires of the Ashburnham Pentateuch." *Cahiers Archéologiques* 22 (1972): 19–38.

———. "Towards a Further Study of the Ashburnham Pentateuch." *Cahiers Archéologiques* 19 (1969): 45–60.

Nees, Lawrence. "Alcuin and Manuscript Illumination." In Tremp and Schmuki, *Alkuin von York*, 195–228.

———. "Art and Architecture." In McKitterick, *New Cambridge Medieval History*, vol. 2, *c. 700–c. 900*, 809–44.

———. *Early Medieval Art.* Oxford: Oxford University Press, 2002.

———. "Godescalc's Career and the Problems of 'Influence.'" In *Under the Influence: The Concept of Influence and the Study of Illuminated Manuscripts*, edited by John Lowden and Alixe Bovey, 21–43. Turnhout: Brepols, 2007.

———. *The Gundohinus Gospels.* Medieval Academy Books 95. Cambridge, Mass.: Medieval Academy of America, 1989.

———. "On Carolingian Book Painters: The Ottoboni Gospels and Its Transfiguration Master." *Art Bulletin* 83 (2001): 209–39.

———. *A Tainted Mantle: Hercules and the Classical Tradition at the Carolingian Court.* Philadelphia: University of Pennsylvania Press, 1991.

Nelson, Janet L. "Kingship and Royal Government." In McKitterick, *New Cambridge Medieval History*, vol. 2, *c. 700–c. 900*, 383–430.

———. "Kingship, Law, and Liturgy in the Political Thought of Hincmar of Rheims." *The English Historical Review* 92, no. 363 (1977): 241–79.

———. "The Voice of Charlemagne." In *Belief and Culture in the Middle Ages: Studies Presented to Henry Mayr-Harting*, edited by Richard Gameson and Henrietta Leyser, 76–88. Oxford: Oxford University Press, 2001.

Netzer, Nancy. *Cultural Interplay in the Eighth Century: The Trier Gospels and the Making of a Scriptorium at Echternach.* Cambridge: Cambridge University Press, 1994.

Neuß, Wilhelm. "Eine karolingische Kopie antiker Sternzeichenbilder im Codex 3307 der Biblioteca Nacional zu Madrid." *Zeitschrift des deutschen Vereins für Kunstwissenschaft* 8 (1941): 113–40.

———. "Ein Meisterwerk der karolingischen Buchkunst aus der Abtei Prüm in der Biblioteca Nacional zu Madrid." *Spanische Forschungen der Görresgesellschaft,* 1st ser., 8 (1940): 37–64.

Noble, Thomas F. X., trans. and ed. *Charlemagne and Louis the Pious: The Lives by Einhard, Notker, Ermoldus, Thegan, and the Astronomer.* University Park: Pennsylvania State University Press, 2009.

———. *Images, Iconoclasm, and the Carolingians.* Philadelphia: University of Pennsylvania Press, 2009.

Nordenfalk, Carl. "Die Buchmalerei." In *Karl der Grosse: Werk und Wirkung,* edited by Wolfgang Braunfels, 224–308. Düsseldorf: Schwann, 1965.

North, John D. "The Astrolabe." In *Stars, Minds, and Fate: Essays in Ancient and Medieval Cosmology,* edited by John North, 211–20. London: Hambledon Press, 1989.

Obrist, Barbara. *La cosmologie médiévale: Textes et images.* Vol. 1, *Les fondements antiques.* Micrologus 11. Florence: SISMEL, 2004.

———. "Le diagramme Isidorien des Saisons, son contenu physique et les representations figuratives." *Mélanges de l'École Française de Rome: Moyen Âge* 108 (1996): 95–164.

———. "Les manuscrits de 'De cursu stellarum' de Grégoire de Tours et le manuscrit, Laon, Bibliothèque Municipale 422." *Scriptorium* 56 (2002): 335–45.

———. "La représentation carolingienne du zodiaque: À propos du manuscrit de Bâle, Universitätsbibliothek, F III 15a." *Cahiers du Civilisation Médiévale Xe–XIIe Siècles* 44 (2001): 3–33.

O'Connor, Elizabeth Carroll Waldron. "The Star Mantle of Henry II." Ph.D. diss., Columbia University, 1980.

Oexle, O. G. "Die Karolinger und die Stadt des heiligen Arnulf." *Frühmittelalterliche Studien* 1 (1967): 250–364.

Ommundsen, Åslaug. "The Liberal Arts and the Polemic Strategy of the *Opus Caroli Regis Contra Synodum (Libri Carolini)." Symbolae Osloenses* 77, no. 1 (2002): 175–200.

Pace, Valentino. *Arte medievale in Italia meridionale.* Vol. 1, *Campania.* Naples: Liguori Editore, 2007.

———. ed. *L'VIII secolo: Un secolo inquieto: Atti del Convegno internazionale di studi Cividale del Friuli, 4–7 dicembre 2008.* Cividale del Friuli: Comune di Cividale del Friuli, 2010.

Pächt, Otto. *Buchmalerei des Mittelalters.* 4th ed. Edited by Dagmar Thoss and Ulrike Jenni. Munich: Prestel, 2000.

Palmer, James. "Computus After the Paschal Controversy of A.D. 740." In *The Easter Controversy of Late Antiquity and the Early Middle Ages: Its Manuscripts, Texts, and Tables: Proceedings of the 2nd International Conference on the Science of Computus in Ireland and Europe, Galway, 18–20 July, 2008,* edited by Immo Warntjes and Dáibhí Ó Cróinín, 213–41. Studia Traditionis Theologiae: Explorations in Early and Medieval Theology 10. Turnhout: Brepols, 2011.

Panofsky, Erwin. *Renaissance and Renascences in Western Art.* New York: Icon, 1972.

Panofsky, Erwin, and Fritz Saxl. "Classical Mythology in Mediaeval Art." *Metropolitan Museum Studies* 4, no. 2 (1933): 228–80.

Parisse, Michel. "Metz, capitale d'Austrasie (450–925)." In *Histoire de Metz,* edited by François-Yves Le Moigne, 67–88. Toulouse: Privat, 1986.

———. "Remarques sur le destin des assises territoriales de l'évêché de Metz (8e–13e s.)." In *Die alte Diözese Metz / L'ancien diocèse de Metz: Referate eines Kolloquiums in Waldfischbach-Burgalben vom 21. bis 23. März 1990,* edited by Hans-Walter Herrmann, 73–83. Veröffentlichungen der Kommission für Saarländische Landesgeschichte und Volksforschung 19. Saarbrücken: Saarbrücker Druckerei und Verlag, 1993.

Plotzek, Joachim M., Katherina Winnekes, Stefan Kraus, and Ulrike Surmann, eds. *Glaube und Wissen im Mittelalter.* Munich: Hirmer, 1998.

Porcher, Jean. "Book Painting." In *The Carolingian Renaissance,* edited by André Malraux and André Parrot, 70–203. Translated by Stuart Gilbert. New York: George Braziller, 1970.

———. "Book Painting." In *Europe of the Invasions,* edited by André Malraux and André Parrot, 105–206. Translated by Stuart Gilbert and James Emmons. New York: George Braziller, 1969.

———. "La peinture provinciale." In *Karl der Grosse: Lebenswerk und Nachleben,* vol. 3, *Karolingische Kunst,* edited by Wolfgang Braunfels

and Hermann Schnitzler, 54–73. Düsseldorf: Schwann, 1966.
Ptolemy. *Ptolemy's Almagest.* Translated by G. J. Toomer. Princeton: Princeton University Press, 1998.
Rabanus Maurus. *See* Hrabanus Maurus.
Ramírez-Weaver, Eric. "Classical Constellations in Carolingian Codices: Investigating the Celestial Imagery of Madrid, Biblioteca Nacional, MS 3307." In *Negotiating Secular and Sacred in Medieval Art: Christian, Islamic, and Buddhist,* edited by Alicia Walker and Amanda Luyster, 103–28. Burlington: Ashgate, 2009.
Rampley, Matthew. "From Symbol to Allegory: Aby Warburg's Theory of Art." *Art Bulletin* 79, no. 1 (1997): 41–55.
Rau, Reinhold, ed. *Quellen zur karolingischen Reichsgeschichte.* 3 vols. Darmstadt: Wissenschaftliche Buchgesellschaft, 1955–60.
Remigius. *Remigii Autissiodorensis Commentum in Martianum Capellam.* Edited by Cora E. Lutz. 2 vols. Leiden: Brill, 1962–65.
Reudenbach, Bruno. *Das Godescalc-Evangelistar: Ein Buch für die Reformpolitik Karls des Großen.* Frankfurt am Main: Fischer Taschenbuch, 1998.
Reuter, Marianne. *Text und Bild im Codex 132 der Bibliothek von Montecassino "Liber Rabani de originibus rerum": Untersuchungen zur mittelalterlichen Illustrationspraxis.* Münchener Beiträge zur Mediävistik und Renaissance-Forschung 34. Munich: Arbeo-Gesellschaft, 1984.
Ribémont, Bernard. *Littérature et encyclopédies du Moyen Âge.* Medievalia 42. Orléans: Paradigme, 2002.
———. *Les origines des encyclopédies médiévales: D'Isidore de Séville aux Carolingiens.* Nouvelle bibliothèque du Moyen Âge 61. Paris: Honoré Champion, 2001.
Riché, Pierre. Écoles et enseignement dans le Haut Moyen Âge: Fin du Ve siècle–milieu du XIe siècle. 3rd ed. Paris: Picard Éditeur, 1999.
Ridpath, Ian. *Stars and Planets.* 2nd ed. London: Dorling Kindersley, 2002.
Rissel, Maria. *Rezeption antiker und patristischer Wissenschaft bei Hrabanus Maurus: Studien zur karolingischen Geistesgeschichte.* Lateinische Sprache und Literatur des Mittelalters 7. Frankfurt am Main: Peter Lang, 1976.
Roberts, Colin H., and T. C. Skeat. *The Birth of the Codex.* Oxford: Oxford University Press, 1983.
Rose, Valentin. *Verzeichnis der von der königlichen Bibliothek zu Berlin erworbenen Meerman-Handschriften des Sir Thomas Phillipps.* Berlin: A. W. Schade's Buchdruckerei, 1892.
Rosenthal, Erwin. *The Illuminations of the Vergilius Romanus (Cod. Vat. Lat. 3867): A Stylistic and Iconographical Analysis.* Zurich: Urs Graf-Verlag Dietikon, 1972.
Rück, Karl. *Auszüge aus der Naturgeschichte des C. Plinius Secundus in einem astronomisch-komputistischen Sammelwerke des achten Jahrhunderts.* Programm des Königlichen Ludwigs-Gymnasiums 1887–88. Munich: Akademische Buchdruckerei von F. Staub, 1888.
Salazar, Christine, ed. *Brill's New Pauly: Encyclopedia of the Ancient World.* 16 vols. Leiden: Brill, 2002–10.
Salzmann, Michele Renee. *On Roman Time: The Codex-Calendar of 354 and the Rhythms of Urban Life in Late Antiquity.* Berkeley: University of California Press, 1990.
Sauerländer, Willibald. "'Barbari ad portas': Panofsky in den fünfziger Jahren." In *Erwin Panofsky: Beiträge des Symposions Hamburg 1992,* edited by Bruno Reudenbach, 123–37. Schriften des Warburg-Archivs im Kunstgeschichtlichen Seminar der Universität Hamburg 3. Berlin: Akademie Verlag, 1994.
Savage-Smith, Emilie. *Islamicate Celestial Globes: Their History, Construction, and Use.* Smithsonian Studies in History and Technology 46. Washington, D.C.: Smithsonian Institution, 1985.
Saxl, Fritz. *Lectures.* Vol. 1. London: Warburg Institute, University of London, 1957.
———. *Verzeichnis astrologischer und mythologischer illustrierter Handschriften des lateinischen Mittelalters.* Vol. 1, *In römischen Bibliotheken.* Sitzungsberichte der Heidelberger Akademie der Wissenschaften Stiftung Heinrich Lanz Philosophisch-historische Klasse 1915. 1915. Nendeln, Liechtenstein: Kraus Reprint, 1978.
———. *Verzeichnis astrologischer und mythologischer illustrierter Handschriften des lateinischen Mittelalters.* Vol. 2, *Die Handschriften der National-Bibliothek in Wien.* Sitzungsberichte der Heidelberger

Akademie der Wissenschaften Philosophisch-historische Klasse 1925–26. Heidelberg: Carl Winters Universitätsbuchhandlung, 1927.

Saxl, Fritz, and Hans Meier. *Verzeichnis astrologischer und mythologischer illustrierter Handschriften des lateinischen Mittelalters.* Vol. 3, *Handschriften in englischen Bibliotheken.* Edited by Harry Bober. London: Warburg Institute, University of London, 1953.

Scherrer, Gustav. *Verzeichniss der Handschriften der Stiftsbibliothek von St. Gallen.* 1875. Hildesheim: Georg Olms, 1975.

Schieffer, Rudolf. *Die Karolinger.* 3rd ed. Stuttgart: W. Kohlhammer, 2000.

Schmitz, Gerhard. "Bonifatius und Alkuin: Ein Beitrag zur Glaubensverkündigung in der Karolingerzeit." In Tremp and Schmuki, *Alkuin von York,* 73–89.

Schmuki, Karl, Peter Ochsenbein, and Cornel Dora. *Cimelia Sangallensia: Hundert Kostbarkeiten aus der Stiftsbibliothek St. Gallen.* St. Gallen: Verlag am Klosterhof, 2000.

Scholz, Bernhard Walter, trans. *Carolingian Chronicles: Royal Frankish Annals and Nithard's Histories.* Ann Arbor: University of Michigan Press, 1970.

Schramm, Percy Ernst. *Die deutschen Kaiser und Könige in Bildern ihrer Zeit. 751–1190.* Edited by Florentine Mütherich. Rev. ed. Munich: Prestel, 1983.

———. *Kaiser, Könige und Päpste: Gesammelte Aufsätze zur Geschichte des Mittelalters.* Vol. 1. Stuttgart: Anton Hiersemann, 1968.

———. *Kaiser, Rom und Renovatio.* 2 vols. Leipzig: B. G. Teubner, 1929.

———. *Sphaira, Globus, Reichsapfel: Wanderung und Wandlung eines Herrscherschaftszeichens von Caesar bis zu Elisabeth II.* Stuttgart: Anton Hiersemann, 1958.

Schramm, Percy Ernst, and Florentine Mütherich. *Denkmale der deutschen Könige und Kaiser: Ein Beitrag zur Herrschergeschichte von Karl dem Großen bis Friedrich II. 768–1250.* 2nd ed. Veröffentlichungen des Zentralinstituts für Kunstgeschichte in München 2. Munich: Prestel, 1981.

Sclafer, Jacqueline, and Marie-Pierre Laffitte, eds. *Catalogue général des manuscrits latins nos 8823–8921.* Paris: Bibliothèque nationale de France, 1997.

Sedulius Scottus. *On Christian Rulers and the Poems.* Translated and edited by Edward Gerard Doyle. Medieval and Renaissance Texts and Studies 17. Binghamton: SUNY Binghamton, 1983.

Shanzer, Danuta. *A Philosophical and Literary Commentary on Martianus Capella's "De Nuptiis Philologiae et Mercurii," Book 1.* Berkeley: University of California Press, 1986.

Sharpe, William D. "Isidore of Seville: The Medical Writings." *Transactions of the American Philosophical Society,* n.s., 54, no. 2 (1964): 1–75.

Springsfeld, Kerstin. *Alkuins Einfluß auf die Komputistik zur Zeit Karls des Großen.* Sudhoffs Archiv 48. Stuttgart: Franz Steiner, 2002.

———. "Karl der Große, Alkuin und die Zeitrechnung." *Berichte zur Wissenschaftsgeschichte* 27 (2004): 53–66.

Stalley, Roger. *Early Medieval Architecture.* Oxford: Oxford University Press, 1999.

Stern, Henri. *Le calendrier de 354: Étude sur son texte et ses illustrations.* Institut français d'archéologie de Beyrouth: Bibliothèque archéologique et historique 55. Paris: Librairie Orientaliste Paul Geuthner, 1953.

Stiegemann, Christoph, and Matthias Wemhoff, eds. *799—Kunst und Kultur der Karolingerzeit: Karl der Große und Papst Leo III. in Paderborn.* Mainz: Philipp von Zabern, 1999.

Story, Joanna, ed. *Charlemagne: Empire and Society.* Manchester: Manchester University Press, 2005.

Stückelberger, Alfred. *Bild und Wort: Das illustrierte Fachbuch in der antiken Naturwissenschaft, Medizin und Technik.* Kulturgeschichte der antiken Welt 62. Mainz: Philipp von Zabern, 1994.

Teeuwen, Mariken. *Harmony and the Music of the Spheres: The "Ars Musica" in Ninth-Century Commentaries on Martianus Capella.* Mittellateinische Studien und Texte 30. Leiden: Brill, 2002.

———. "The Study of Martianus Capella's *De Nuptiis* in the Ninth Century." In *Learned Antiquity: Scholarship and Society in the Near East, the Greco-Roman World, and the Early Medieval West,* edited by Alasdair A. MacDonald, Michael W. Twomey, and Gerrit J. Reinink, 185–94. Groningen Studies in Cultural Change 5. Leuven: Peeters, 2003.

Theodulf of Orléans. *Theodulf of Orléans: The Verse.* Translated by Theodore M. Andersson. Medieval and Renaissance Texts and Studies 450. Tempe: Arizona Center for Medieval and Renaissance Studies, 2014.

Thiele, Georg. *Antike Himmelsbilder: Mit Forschungen zu Hipparchos, Aratos und seinen Fortsetzern und Beiträgen zur Kunstgeschichte des Sternhimmels.* Berlin: Weidmannsche Buchhandlung, 1898.

Thorndike, Lynn. *A History of Magic and Experimental Science During the First Thirteen Centuries of Our Era.* Vol. 1. New York: Columbia University Press, 1943.

Tremp, Ernst, and Karl Schmuki, eds. *Alkuin von York und die geistige Grundlegung Europas: Akten der Tagung vom 30. September bis zum 2. Oktober 2004 in der Stiftsbibliothek St. Gallen.* Monasterium Sancti Galli 5. St. Gallen: Verlag am Klosterhof, 2010.

Tremp, Ernst, Karl Schmuki, and Theres Flury, eds. *Benediktinisches Mönchtum: Ausstellung zum Gedenken an die Aufhebung der Fürstabtei St. Gallen vor 200 Jahren.* St. Gallen: Verlag am Klosterhof, 2005.

Ullmann, Walter. *The Carolingian Renaissance and the Idea of Kingship: The Birkbeck Lectures, 1968–69.* London: Methuen, 1969.

Underwood, Paul A. "The Fountain of Life in Manuscripts of the Gospels." *Dumbarton Oaks Papers* 5 (1950): 41–138.

Unruh, Frank. "Zwischen Esus und Jesus: Der Wandel der antiken Religion im Spiegel schriftlicher Zeugnisse." In *Religio Romana: Wege zu den Göttern im antiken Trier,* edited by Hans-Peter Kuhnen, 38–71. Trier: Rheinisches Landesmuseum, 1996.

Unterkircher, Franz. *Inventar der illuminierten Handschriften, Inkunabeln und Frühdrucke der Österreichischen Nationalbibliothek: Die Abendländischen Handschriften.* Vol. 2. Vienna: Georg Prachner, 1957.

———. *Zur Ikonographie und Liturgie des Drogo-Sakramentars (Paris, Bibliothèque nationale, Ms. lat. 9428).* Interpretationes ad codices 1. Graz: Akademische Druck- u. Verlagsantalt, 1977.

Van der Horst, Koert, William Noel, and Wilhelmina C. M. Wüstefeld, eds. *The Utrecht Psalter in Medieval Art: Picturing the Psalms of David.* Westrenen: HES, 1996.

Van de Vyver, André. "Hucbald de Saint-Amand, écolâtre, et l'invention du Nombre d'or." In *Mélanges Auguste Pelzer: Études d'histoire littéraire et doctrinale de la Scholastique médiévale offertes à Monseigneur Auguste Pelzer à l'occasion de son soixante-dixième anniversaire,* 61–79. Recueil de travaux d'histoire et de philologie, 3rd ser., 26. Louvain: Bibliothèque de l'université, 1947.

Vattasso, Marcus, and Pius Franchi de' Cavalieri. *Bibliothecae Apostolicae Vaticanae: Codices Manu Scripti Recensiti, Codices Vaticani Latini: Codices 1–678.* Vol. 1. Rome: Typis Vaticanis, 1902.

Verkerk, C. L. "*Aratea:* A Review of the Literature Concerning MS. Vossianus lat. q. 79 in Leiden University Library." *Journal of Medieval History* 6 (1980): 245–87.

Verkerk, Dorothy Hoogland. "Biblical Manuscripts in Rome 400–700 and the Ashburnham Pentateuch." In *Imaging the Early Medieval Bible,* edited by John Williams, 97–120. University Park: Pennsylvania State University Press, 1999.

Vieillard-Troiekouroff, May. "Art carolingien et art roman parisiens: Les illustrations astrologiques jointes aux Chroniques de Saint-Denis et de Saint-Germain-des- Prés (IXe–XIe siècle)." *Cahiers Archéologiques* 16 (1966): 77–105.

Vitruvius. *The Ten Books on Architecture.* Translated by Morris Hicky Morgan. 1914. New York: Dover, 1960.

Von Euw, Anton. "Alkuin als Lehrer der Komputistik und Rhetorik Karls des Grossen im Spiegel der St. Galler Handschriften." In Tremp and Schmuki, *Alkuin von York,* 251–62.

———. "Beda Venerabilis: Naturlehre, historiographische und zeitrechnerische Werke." In *Glaube und Wissen im Mittelalter,* edited by Joachim M. Plotzek, Katharina Winnekes, Stefan Kraus, and Ulrike Surmann, 129–35. Munich: Hirmer, 1998.

———. *Der Leidener Aratus: Antike Sternbilder in einer karolingischen Handschrift.* Munich: Bayerische Staatsbibliothek, 1988.

———. *Die St. Galler Buchkunst von 8. bis zum Ende des 11. Jahrhunderts.* Vol. 1. Monasterium Sancti Galli 3. St. Gallen: Verlag am Klosterhof, 2008.

———. "Zeitrechnung, Naturlehre und Himmelskunde zur Zeit Hildebalds von Köln (vor 787–818)."

Jahrbuch des kölnischen Geschichtsvereins 69 (1998): 1–30.
von Steiger, Christoph, and Otto Homburger, eds. *Physiologus bernensis: Voll-Faksimile-Ausgabe des Codex Bongarsianus 318 der Burgerbibliothek Bern.* Basel: Alkuin Verlag, 1964.
Wallis, Faith. "Images of Order in the Medieval *Computus.*" In *Ideas of Order in the Middle Ages,* edited by Warren Ginsberg, 45–68. Acta 15. Binghamton: Center for Medieval and Early Renaissance Studies, SUNY Binghamton, 1990.
Warntjes, Immo. *The Munich Computus: Text and Translation; Irish Computistics Between Isidore of Seville and the Venerable Bede and Its Reception in Carolingian Times.* Sudhoffs Archiv 59. Stuttgart: Franz Steiner, 2010.
Weber, Winfried. *Constantinische Deckengemälde aus dem römischen Palast unter dem Trierer Dom.* 4th ed. Trier: Bischöfliches Dom- und Diözesanmuseum, 2000.
Weitzmann, Kurt. *Illustrations in Roll and Codex: A Study of the Origin and Method of Text Illustration.* Studies in Manuscript Illumination 2. Princeton: Princeton University Press, 1970.
———. *Late Antique and Early Christian Book Illumination.* New York: George Braziller, 1977.
Willjung, Harald, ed. *Das Konzil von Aachen 809.* Monumenta Germaniae Historica, Concilia, vol. II, supp. II. Hannover: Hahnsche Buchhandlung, 1998.
Wilmart, Andreas. *Bibliothecae Apostolicae Vaticanae: Codices Manu Scripti Recensiti, Codices Reginenses Latini.* Vol. 2. Vatican City: Vatican Library, 1945.
Woodruff, H. "The Physiologus of Bern: A Survival of Alexandrian Style in a 9th c. MS." *Art Bulletin* 12 (1930): 226–53.
Wormald, Francis. *The Miniatures in the Gospels of St. Augustine.* Cambridge: Cambridge University Press, 1954.
Zinner, Ernst. *Verzeichnis der astronomischen Handschriften in des deutschen Kulturgebietes.* Munich: C. H. Beck'sche Verlagsbuchhandlung, 1925.

INDEX

Page numbers in italics indicate illustrations.

TYPESET BY

REGINA STARACE

PRINTED AND BOUND BY

REGENT PUBLISHING SERVICES LTD

COMPOSED IN

BRIOSO PRO AND WHITNEY

PRINTED ON

140 GSM FSC MIX MATT ART

BOUND IN

JHT